AF780412

Pollination Biology

Pollination Biology

Guest Editor

Yu Gao

MDPI

Basel • Beijing • Wuhan • Barcelona • Belgrade • Novi Sad • Cluj • Manchester

Guest Editor
Yu Gao
College of Plant Protection
Jilin Agricultural University
Changchun
China

Editorial Office
MDPI AG
Grosspeteranlage 5
4052 Basel, Switzerland

This is a reprint of the Special Issue, published open access by the journal *Biology* (ISSN 2079-7737), freely accessible at: https://www.mdpi.com/journal/biology/special_issues/92W016861L.

For citation purposes, cite each article independently as indicated on the article page online and as indicated below:

Lastname, A.A.; Lastname, B.B. Article Title. *Journal Name* **Year**, *Volume Number*, Page Range.

ISBN 978-3-7258-5797-5 (Hbk)
ISBN 978-3-7258-5798-2 (PDF)
https://doi.org/10.3390/books978-3-7258-5798-2

Contents

About the Editor

Yu Gao

Yu Gao is an Associate Professor in the College of Plant Protection at Jilin Agricultural University and holds a Ph.D. from the Chinese Academy of Agricultural Sciences. His primary research focuses on plant–insect interactions, integrated pest management theory and technology, and the development and utilization of functional components from edible insects.

He serves as the Position Scientist of the China Agriculture Research System of MOF and MARA; Deputy Director of the Key Laboratory of Soybean Pest and Disease Control at the Ministry of Agriculture and Rural Affairs; Member of the Expert Advisory Group for Plant Protection at the Ministry of Agriculture and Rural Affairs; Council Member of the Soybean Specialized Committee of the Crop Science Society of China; Deputy Secretary-General and Executive Director of the Jilin Provincial Society of Plant Protection; Member of the Soybean Specialized Committee of the Ninth Jilin Provincial Crop Variety Approval Committee; Science and Technology Correspondent of Jilin Province; external expert of the national tea quality safety Engineering Technology Research Center of the Ministry of Science and Technology; Domestic Visiting Scholar at Nanjing Agricultural University; and a core member of the first batch of National University Huang Danian-Style Teaching Teams and Jilin Provincial University Innovation Teams. He also serves as an Editorial Board Member for journals such as *Biology* and *Agrochemicals*, and as a Young Editorial Board Member for *Oil Crop Science*, the *Journal of Jilin Agricultural University*, and *Soybean Science and Technology*. To date, he has published more than 90 papers as first or corresponding author, including one ESI 'Highly Cited Paper', and has co-authored 8 college teaching materials and books.

biology

MDPI

Editorial

Pollination Biology: From Pollinators and Floral Traits to Landscape Management

Yu Gao [1,2]

[1] College of Plant Protection, Jilin Agricultural University, Changchun 130118, China; gaothrips@jlau.edu.cn
[2] Key Laboratory of Soybean Disease and Pest Control, Ministry of Agriculture and Rural Affairs, Changchun 130118, China

Pollination biology examines the mechanisms, agents, and ecological significance of pollination [1,2]. As key ecosystem actors, pollinators provide vital ecological services and help maintain the dynamic balance and stability of natural systems. They are essential for the reproduction of many plant species and the production of fruits and seeds [3,4]. Their health is tightly linked to green agriculture, food security, and human nutrition. Insect pollinators alone are responsible for 80–85% of all animal-mediated pollination, a service widely recognized by both scientists and the public [5,6]. Against this backdrop, the nine papers (eight articles and one editorial) assembled in this Special Issue on *Pollination Biology* weave compelling research findings that spans scales and disciplines. Together, these studies deepen our understanding of pollination processes and offer insights into how this knowledge can inform practical, pollinator-friendly management strategies.

Contribution 1: Huancas et al. deliver the first comprehensive review of mathematical models for pollinator dynamics [7]. Their bibliometric map highlights four dominant toolkits—ODEs, PDEs, network-patch frameworks, and emerging stochastic or fractional-order models—while at the same time identifying gaps in optimal control and stochastic approaches. The findings highlight modeling trends, ecological insights, and policy applications in agriculture and conservation. This study provides a comprehensive typology of mathematical models for pollinator dynamics, bridging theoretical ecology and real-world decision-making.

Contribution 2: Layek et al. investigate seasonal shifts in floral traits, pollinator assemblages, and fruit set of the wild *Solanum sisymbriifolium* across the Rarh plains of West Bengal [8]. Their year-round monitoring reveals that although hot-season flowers are smaller and produce less nectar, large-bodied bees dominate during this period and drive the highest reproductive success. This research uniquely links microclimate-driven floral plasticity to real-time pollinator efficiency.

Contribution 3: Ejaz et al. test whether adding managed honey bee hives can improve seed yields in Punjab alfalfa fields [9]. Across three commercial farms, they trial 0–4 hives per hectare, finding a linear increase in seed yield, peaking at 38% boost with four hives. Notably, wild pollinators contributed less than 12% of total visits. The study highlights the practical value of hive supplementation as a scalable strategy to enhance forage crop profitability in Pakistan.

Contribution 4: Laboisse et al. chart how honey bees exploit floral resources across space and time in the Banja Luka region of Bosnia [10]. Using RFID-tagged workers and hive scales for two years, they show that 90% of foraging stays within 1.2 km of the hive. Bees flexibly shift their daily activity to match the phenology of sequentially blooming crops like apple, acacia, sunflower, and heather. This high-resolution calendar provides valuable guidance for synchronizing hive placement with peak nectar flows.

Contribution 5: Bukhari et al. examine the impact of nitrophos fertilizer on floral traits and pollinator visitation in Pakistan's onion seed fields [11]. While higher fertilizer doses increase flower size, they simultaneously dilute nectar sugar concentration, reducing visits from wild bees and hoverflies by 22%, and ultimately lowering seed yield by 14% compared to moderate fertilization. This study uniquely reveals that excessive nitrogen input can disrupt plant–pollinator synergy.

Contribution 6: Layek et al. explore how seasonal changes in temperature, humidity, and photoperiod influence the floral biology and pollination of *Turnera ulmifolia* in India [12]. Long, humid days enhance pollen viability and bee visits, boosting fruit set. However, nectar thickens above 35 °C, deterring hoverflies and leading to a summer dip in yields. This work illustrates how climate-driven shifts can desynchronize plant–pollinator interaction.

Contribution 7: Zhang et al. investigate the pollination biology of *Iris setosa*, a cold-hardy ornamental native to Jilin, China [13]. Combining field observations, hand-pollination experiments, and floral scent analysis, they show that large violet perianths attract bumblebees via visual contrast, while eugenol-rich fragrances reinforce this attraction. The species is largely self-incompatible; hive supplementation raises seed set from 18% to 62%, pointing to managed bumblebees as a practical tool for commercial seed production of this northern iris.

Contribution 8: Eggers et al. identify a recessive nuclear gene (al) on the short arm of chromosome 6 in potato that causes antherless cytoplasmic male sterility (CMS) when interacting with P-type cytoplasm, but remains fertile with A- or T-type cytoplasm [14]. Exploiting this CMS system eliminates the need for manual emasculation of female parents, improves hybrid seed production, and significantly reduces berry set in F_1 hybrids under field conditions. This work presents a valuable tool for diploid hybrid potato breeding.

Contribution 9: Leung & Reid examine Canada's regulatory landscape for the import and movement of managed bees, focusing on risks to native pollinators in Yukon [15]. They highlight the lack of ecological risk assessments and permitting requirements in the territory. Their proposed Yukon-specific framework—combining seasonal transport limits, pathogen screening, and incentives for domestic bumblebee rearing—offers a replicable model for northern regions aiming to balance agricultural needs with pollinator conservation.

This Special Issue offers, at best, a modest snapshot of an extraordinarily complex and rapidly evolving field. While the breadth of topics covered and the depth of individual studies reflect the momentum in pollination biology, we are acutely aware that any Special Issue can only scratch the surface. These contributions are intended not as final answers, but as conversation starters. They remind us that integrative, interdisciplinary collaboration is not a luxury but a necessity if we are to deepen our still-fragmentary understanding of these remarkable organisms and craft management strategies humble enough to accommodate nature's surprises. Pollination research remains stubbornly place-based; a shift in bloom phenology, a turnover in pollinator guilds, or even a subtle change in slope aspect can overturn yesterday's 'universal' finding tomorrow. These geographic fingerprints limit how far any one dataset can travel, but they also provide the calibration marks that global models so desperately need. By treating regional idiosyncrasies as adjustable parameters rather than immovable barriers—and by openly sharing both our successes and our missteps—we aim to advance the discipline toward more robust and generalizable insights. Pollination research must move beyond simply mapping 'who pollinates whom'. At the molecular, individual, and population levels, we need to understand how environmental factors, chemical signals, morphological fit, and behavioral decisions collectively influence pollen transfer. These mechanisms must then be translated into parameters for larger-scale models. Finally, through collaboration with agronomists, plant breeders, land managers, and policymakers, we must translate these forecasts into actionable spatial plans, cropping systems, seed technologies, and transnational policy frameworks.

Funding: This work was supported by the Earmarked Fund for China Agriculture Research System of MOF and MARA (Grant number CARS–04).

Institutional Review Board Statement: Not applicable.

Informed Consent Statement: Not applicable.

Data Availability Statement: The original contributions presented in this study are included in the list of contributions. Further inquiries can be directed to the corresponding author.

Acknowledgments: Special thanks are due to the authors, the reviewers, and the editorial staff of Biology for their great contributions to this Special Issue.

Conflicts of Interest: The author declares no conflicts of interest.

References

1. Huang, S.; Guo, Y. New advances in pollination biology and the studies in China. *Chin. Sci. Bull.* **2000**, *45*, 1441–1447. [CrossRef]
2. Fattorini, R.; Glover, B.J. Molecular mechanisms of pollination biology. *Annu. Rev. Plant Biol.* **2020**, *71*, 487–515. [CrossRef] [PubMed]
3. Klein, A.-M.; Vaissière, B.E.; Cane, J.H.; Steffan-Dewenter, I.; Cunningham, S.A.; Kremen, C.; Tscharntke, T. Importance of pollinators in changing landscapes for world crops. *Proc. R. Soc. B Biol. Sci.* **2007**, *274*, 303–313. [CrossRef] [PubMed]
4. Kennedy, C.M.; Lonsdorf, E.; Neel, M.C.; Williams, N.M.; Ricketts, T.H.; Winfree, R.; Bommarco, R.; Brittain, C.; Burley, A.L.; Cariveau, D.; et al. A global quantitative synthesis of local and landscape effects on wild bee pollinators in agroecosystems. *Ecol. Lett.* **2013**, *16*, 584–599. [CrossRef] [PubMed]
5. Garibaldi, L.A.; Gemmill-Herren, B.; D'Annolfo, R.; Graeub, B.E.; Cunningham, S.A.; Breeze, T.D. Farming approaches for greater biodiversity, livelihoods, and food security. *Trends Ecol. Evol.* **2017**, *32*, 68–80. [CrossRef] [PubMed]
6. Eilers, E.J.; Kremen, C.; Smith Greenleaf, S.; Garber, A.K.; Klein, A.-M. Contribution of pollinator-mediated crops to nutrients in the human food supply. *PLoS ONE* **2011**, *6*, e21363. [CrossRef] [PubMed]
7. Huancas, F.; Coronel, A.; Lozada, E.; Torres, J. Mathematical Modeling of Population Dynamics of Pollinators: A Survey. *Biology* **2025**, *14*, 1308. [CrossRef] [PubMed]
8. Layek, U.; Majhi, P.; Das, A.; Karmakar, P.; Kundu, A. Seasonal Variation in Flower Traits, Visitor Traits, and Reproductive Success of *Solanum sisymbriifolium* Lamarck (Solanaceae) in the Rarh Region of West Bengal, India. *Biology* **2025**, *14*, 865. [CrossRef] [PubMed]
9. Ejaz, K.; Ali, M.; Khan, F.Z.A.; Mozūratis, R. Enhancing Alfalfa (*Medicago sativa*) Seed Yield: The Effect of Honey Bee (*Apis mellifera*) Supplementation and Efficiency of Other Pollinators. *Biology* **2025**, *14*, 599. [CrossRef] [PubMed]
10. Laboisse, S.; Vaillant, M.; Cazenave, C.; Kelečević, B.; Chevalier, I.; Andres, L. Space and Time Dynamics of Honeybee (*Apis mellifera* L.)-Melliferous Resource Interactions Within a Foraging Area: A Case Study in the Banja Luka Region (Bosnia & Herzegovina). *Biology* **2025**, *14*, 422. [CrossRef] [PubMed]
11. Bukhari, S.F.; Ali, M.; Khan, F.Z.A.; Mozūratis, R. Effect of Nitrophos Fertilizer on Pollinator Dynamics and Onion Seed Yield. *Biology* **2025**, *14*, 119. [CrossRef] [PubMed]
12. Layek, U.; Das, N.; Samanta, A.; Karmakar, P. Impact of Seasonal Atmospheric Factors and Photoperiod on Floral Biology, Plant–Pollinator Interactions, and Plant Reproduction on *Turnera ulmifolia* L. (Passifloraceae). *Biology* **2025**, *14*, 100. [CrossRef] [PubMed]
13. Zhang, X.; Liu, R.; Chen, L.; Pei, T.; Gao, Y.; Lu, X.; Zhou, Y. Effects of Floral Characters on the Pollination Biology and Breeding System of *Iris setosa* (Iridaceae): A Cold-Tolerant Ornamental Species from Jilin Province. *Biology* **2025**, *14*, 2. [CrossRef] [PubMed]
14. Eggers, E.-J.; Su, Y.; van der Poel, E.; Flipsen, M.; de Vries, M.E.; Bachem, C.W.B.; Visser, R.G.F.; Lindhout, P. Identification, Elucidation and Deployment of a Cytoplasmic Male Sterility System for Hybrid Potato. *Biology* **2024**, *13*, 447. [CrossRef] [PubMed]
15. Leung, M.; Reid, D. Reducing Risks to Native Pollinators by Introduced Bees: A Review of Canada's Legislation with Recommendations for Yukon Territory. *Biology* **2025**, *14*, 282. [CrossRef] [PubMed]

biology

MDPI

Article

Identification, Elucidation and Deployment of a Cytoplasmic Male Sterility System for Hybrid Potato

Ernst-Jan Eggers [1,2,3,*], Ying Su [1], Esmee van der Poel [2], Martijn Flipsen [4], Michiel E. de Vries [1], Christian W. B. Bachem [1,2], Richard G. F. Visser [2] and Pim Lindhout [1,*]

1 Solynta, Dreijenlaan 2, 6703 HA Wageningen, The Netherlands; christian.bachem@wur.nl (C.W.B.B.)
2 Plant Breeding, Wageningen University & Research, P.O. Box 386, 6700 AJ Wageningen, The Netherlands; richard.visser@wur.nl (R.G.F.V.)
3 Graduate School Experimental Plant Sciences, Wageningen University & Research, 6708 PB Wageningen, The Netherlands
4 Hogeschool Arnhem Nijmegen, Laan van Scheut 2, 6525 EM Nijmegen, The Netherlands
* Correspondence: ernstjan.eggers@solynta.com (E.-J.E.); pim.lindhout@solynta.com (P.L.)

Simple Summary: Conventional potato breeding has produced only limited genetic gain due to the polyploid nature of the crop. In recent years, hybrid potato breeding at the diploid level has been developed to overcome this limited genetic gain. In diploid potato breeding, homozygous inbred lines are developed by self-fertilization, enabling incremental improvements of the material in each generation. This type of breeding requires self-fertility, which makes hybridization of inbred lines labor-cumbersome and results in hybrids that produce many undesirable berries in the field. In many crop species, cytoplasmic male sterility is used to produce maternal inbred lines that are male sterile. In this study, we explore the antherless cytoplasmic male sterility system in potato. We identify a recessive locus that is required for sterility and we show that this trait is expressed in *Phureja* cytoplasm but not in *Andigena* or *Tuberosum* cytoplasm. We implemented this system in hybrid seed production and show that the resulting hybrids set far fewer berries in the field than male fertile controls.

Abstract: Recent advances in diploid F_1 hybrid potato breeding rely on the production of inbred lines using the *S-locus inhibitor* (*Sli*) gene. As a result of this method, female parent lines are self-fertile and require emasculation before hybrid seed production. The resulting F_1 hybrids are self-fertile as well and produce many undesirable berries in the field. Utilization of cytoplasmic male sterility would eliminate the need for emasculation, resulting in more efficient hybrid seed production and male sterile F_1 hybrids. We observed plants that completely lacked anthers in an F_2 population derived from an interspecific cross between diploid *S. tuberosum* and *S. microdontum*. We studied the *antherless* trait to determine its suitability for use in hybrid potato breeding. We mapped the causal locus to the short arm of Chromosome 6, developed KASP markers for the *antherless* (*al*) locus and introduced it into lines with T and A cytoplasm. We found that *antherless* type male sterility is not expressed in T and A cytoplasm, proving that it is a form of CMS. We hybridized male sterile *al/al* plants with P cytoplasm with pollen from *al/al* plants with T and A cytoplasm and we show that the resulting hybrids set significantly fewer berries in the field. Here, we show that the antherless CMS system can be readily deployed in diploid F_1 hybrid potato breeding to improve hybridization efficiency and reduce berry set in the field.

Keywords: diploid potato breeding; cytoplasmic male sterility; QTL analysis; berry and seed production

1. Introduction

In recent years, potato (*Solanum tuberosum*) breeding has seen a shift towards breeding on the diploid level [1–6]. Conventional potato breeding is usually performed at the tetraploid level, where two heterozygous clones are crossed and selections take place among large numbers of F_1 offspring. It requires many locations and years (10–15 years) of testing to select a single commercial cultivar and genetic gain is relatively low [6–8]. The push towards inbred line-based diploid F_1 hybrid breeding is expected to speed up the breeding process and improve genetic gain by allowing continuous improvement of inbred lines. In addition, bulking up a new cultivar is fast as millions of hybrid seeds can be produced per year. Elite inbred lines can be further improved by introgression of favorable genes, such as resistance genes, by means of a backcrossing program [9].

To allow inbreeding of self-incompatible diploid potato clones, these clones can be crossed with a line containing the dominant allele of the *Sli* (*S-locus inhibitor*) gene, which enables self-fertilization [10–12]. The *Sli* gene, which encodes an F-box PP2-B10 protein, is expressed in pollen and interacts with the maternal components of the gametophytic self-incompatibility (GSI) system, the S-RNases, during pollination, leading to a breakdown of the GSI system [13,14]. From a breeding point of view, this system works well and has resulted in the production of inbred lines in multiple programs [4–6,15–18]. However, as a consequence of the presence of *Sli*, maternal inbred lines are self-fertile, which necessitates emasculation for hybrid seed production. The manual emasculation process takes time and may lead to reduced hybrid seed set due to the mechanical stress imposed on the flower buds. Another, more detrimental, effect is that the resulting F_1 hybrid offspring are self-fertile as well. While F_1 hybrid fertility is essential in crops where seeds are the end product, for potato, where the tubers are the end product, this is not the case. Primarily, this is because the seeds in the berries can remain viable in the field and produce a volunteer crop, which may act as a reservoir for diseases and disrupt crop rotation schemes [19]. Additionally, the growth of these berries and seeds may sequester photo-assimilates that might otherwise have been directed to tuber bulking [20]. A straightforward route to resolve these problems would be to introduce cytoplasmic male sterility (CMS) to the maternal lines as is commonly performed in many F_1 hybrid crops [21]. CMS systems exploit incongruities between organellar genomes and nuclear genomes, which lead to a failure to produce fertile pollen.

Sterilizing cytoplasms have been identified in many crop species, like rice, wheat, soybean, sunflower, rapeseed and several more [21]. CMS has been successfully implemented in breeding programs [22–26]. However, deployment of CMS in crops has not always gone smoothly. For instance, CMS has been associated with deleterious effects such as susceptibility to southern corn blight in maize and the inefficiencies of pollen donors or insect vectors in soybean [27,28]. In crop species where the end product is seed, implementations of CMS rely on the *three-line system* in which a *CMS* line is maintained by an isogenic *maintainer* line and male fertility in the hybrid is restored by crossing the CMS line to the *restorer* [29–32]. In potato, the end product is the belowground tubers, and restorer lines are not required. In fact, a major goal of implementing CMS in hybrid potato breeding would be to avoid fertility in the F_1 hybrid to ensure minimal berry and seed production in the field.

Among potato and its wild relatives, many cytoplasm types are present. In a study of cultivated potato and closely related wild species, Hosaka and Sanetomo distinguished 129 chloroplast DNA (ctDNA) types and 63 mitochondrial DNA (mtDNA) types using RFLP, CAPS, SSR and SCAR markers, resulting in 164 unique cytoplasm types [33]. Hosaka and Sanetomo proposed to group these 164 cytoplasm types into 6 types (T, D, P, A, M and W). Cytoplasm types T, D and W have been associated with male sterility in multiple breeding populations, whereas P and A cytoplasm are usually associated with male fertility [34–39]. Recently, Santayana et al. observed segregation for male fertility among CIP potato breeding populations with T and D cytoplasm and identified parental lines that potentially contain nuclear *Restorer of Fertility* (*Rf*) genes for D and T cytoplasm [40]. On the cytoplasmic side, Sanetomo et al. identified a recombinant mitochondrial DNA molecule, RC-I, the presence of which is completely associated with tetrad sterility (T-CMS) among interspecific hybrids derived from *S. stoloniferum* Schltdl. & Bouché [41]. Interestingly, T-CMS must be the result of an interaction between a dominant *S. tuberosum* nuclear gene and the mitochondrial RC-I molecule, since hexaploid interspecific hybrids containing the complete *S. stoloniferum* genome still show T-CMS and thus should contain all dominant *Rf* genes present in the *S. stoloniferum* genome [42]. For the present study, only three cytoplasm types are relevant: A, P and T cytoplasm, originating from *S. tuberosum* ssp *andigenum* Hawkes, *S. phureja* Juz. et Buk. and *S. tuberosum*, respectively.

While most genotypes with P cytoplasm are male fertile, Endelman and Jansky observed segregation for anther length in an F_2 derived from the cross DM × M6, whereby DM carries the P cytoplasm. In this population, 23% of the progeny had short anthers (SA) that did not shed any pollen. QTL analysis revealed a single recessive allele (*sa*) on the short arm of Chromosome 6 to be responsible for this SA phenotype. Interestingly, the recessive allele was inherited from the male fertile M6 genotype whose self-fertilized progeny do not show the SA phenotype, leading the authors to suggest that the SA phenotype is the result of an interaction between the recessive *sa* allele from M6 and the P cytoplasm from DM [43]. At Solynta, we observed complete absence of anthers (which we named antherless) in F_2 genotypes derived from an interspecific hybrid *S. tuberosum* × *S. microdontum* subs. *gigantophyllum*. The antherless phenotype is similar to the short anther phenotype, except that the level of malformation varies, ranging from complete absence of anthers to somewhat malformed anthers. Given the importance of CMS for diploid hybrid potato breeding, we set out to elucidate the genetics and applicability of the antherless trait. We aimed to localize the causal locus in the nuclear genome via QTL analysis and to determine whether it is a form of CMS by introducing the trait to lines with non-P cytoplasm types. Here, we report on the mapping of the causal gene and the characterization of this CMS system for fertility traits. Proof of principle was obtained by exploiting this CMS system to produce male sterile maternal lines and hybrids.

2. Materials and Methods

2.1. Plant Materials

An overview of all plant materials used in this study is available in Table S1. All *S. tuberosum* genotypes used in this study are derived from the founders of the Solynta breeding program as described in Lindhout et al. (2016) [1]. The donor of the antherless trait was derived from a CGN accession of *S. microdontum* subs. *gigantophyllum*, which is available from CGN (Wageningen, The Netherlands) under accession number CGN18200. *S. microdontum* subs. *gigantophyllum* is a potato wild relative with resistance against late blight and wart disease [44,45]. It is diploid and has an endosperm balance number of 2 and is therefore crossable with diploid *S. tuberosum*.

2.2. Crossing Conditions

To avoid any unwanted out-crossing and control plant growth as best as possible, experiments involving crossing were conducted in a greenhouse. Greenhouse conditions were the same as described in Eggers et al. (2021) [13].

2.3. Phenotypic Analysis of Male Fertility

Anther malformation was assessed on a scale from 0 to 3, where a score of 0 means complete absence of anthers and a score of 3 means normal anthers (Figure 1a). Flowers with anthers (scores 1–3) were vibrated using an electronic toothbrush while collecting the pollen in a micro-centrifuge tube. The amount of pollen shed was scored on a scale from 0–3 (where a score of 0 means no pollen shed, and a score of 3 means abundant pollen shed). Self-pollinations were made using the collected pollen, and the micro-centrifuge tubes with leftover pollen were placed in sealed containers with abundant silica gel beads to dry. To assess the viability of the pollen, one drop of acetocarmine was added to the pollen and incubated for one minute. The samples were then vortexed and 10 µL of the mix were pipetted into a counting chamber and observed with normal light microscopy.

Stained and unstained pollen were counted and viability was scored using the formula: $Viable\ pollen\ (\%) = \frac{Stainable\ pollen}{Total\ pollen} \times 100$.

2.4. SeqSNP Genotyping, Linkage Analysis and QTL Mapping of Population BC2(P)-1

Leaf discs from 249 individuals of population BC2(P)-1 were sampled in 96-well plates and were genotyped via SeqSNP™ (LGC Genomics GmbH, Berlin, Germany) [46]. SNPs were extracted from the SeqSNP™ reads using the method described in Adams et al. (2023) [47], resulting in 2116 SNPs. SNPs that were homozygous alternate between F2-1 and Solyntus were selected. SNPs with more than 10% missing data and duplicate SNPs originating from the same SeqSNP™ probes were removed, resulting in a genotypic dataset of 244 SNPs (Table S2). This dataset was converted to Joinmap coding and genetic maps for all 12 chromosomes were created using Joinmap 4.1 [48] with population type BC1 and default settings (Table S3). QTL mapping was performed using interval mapping in MapQTL6 [49].

2.5. KASP Marker Development and Genotyping

Between 30 and 50 mg of leaf material was collected from genotype F2-1 and submitted for DNA extraction, library preparation and Illumina PE150 sequencing with 30× coverage by Novogene UK Company LTD (Cambridge, UK). The reads were mapped to DM4.03 with BWA and variants were called using BCFtools. Several thousand high-quality variants in F2-1 and Solyntus were inspected in jBrowse to determine suitability for KASP genotyping. For KASP genotyping, leaf discs from population BC2(P)-1 were sampled in 96-well plates and submitted to VHLgenetics (Wageningen, The Netherlands) for DNA extraction and KASP analysis as described in Eggers et al. (2021) [13]. The quality of the resulting KASP marker data was assessed using SNPviewer (lgcgroup.com/products/genotyping-software/snpviewer); markers that did not segregate or showed unexpected segregation were discarded from further analysis. For the remaining markers, segregation ratios were tested using $\chi 2$ tests with the null hypothesis that the antherless locus segregates in mendelian 1:2:1 ratio. *p*-values were calculated from the $\chi 2$ to determine the likelihood that the null hypothesis of mendelian 1:2:1 segregation is true.

2.6. Field Experiment

The field trial was designed with plots of four ridges of 32 plants per plot with 8 plants per ridge and with plots replicated in two randomized blocks. The plants were spaced at 25 cm on each ridge, and the distance between ridges was 75 cm. Plots were separated by an empty row, one row of cultivar Bergerac, and then another empty row. The design included the 13 antherless proof-of-concept hybrids and three male fertile hybrids as controls. The seeds were sown and raised in a greenhouse and transplanted six weeks after sowing to a field in Heelsum, The Netherlands on the 17th of May 2023. Crop handling was performed according to standard agricultural practices, as described by Kacheyo et al. (2023) [50]. At 108 days after transplanting, we harvested all berries from the plants on the middle two ridges of each plot, including the berries that were already detached from the plants but were present between the two middle ridges. All berries were bulked per plot in a mesh bag and fresh weight was determined and used for analysis. The significance of the difference in berry weight per plant between the *antherless* plants and male fertile controls was determined using a *t*-test. Two datapoints with more than two times the standard deviation from the mean among the antherless plants were removed for the *t*-test.

3. Results

3.1. Identification of the Antherless Phenotype and Development of a Mapping Population

In a breeding program at Solynta, we identified plants lacking anthers in an F2 population derived from a cross between *S. tuberosum* × *S. microdontum* subs. *gigantophyllum* (F2-1). We designated this phenotype "*antherless*". In this small population of 50 plants, we identified three individuals that completely lacked anthers (Figure 1a, score = 0), suggesting that the phenotype is caused by two recessive loci or one locus with severely skewed inheritance. We set out to characterize the loci that are responsible for the antherless phenotype, to elucidate the genetics and to determine the contribution of cytoplasm types. To generate a dedicated mapping population, we pollinated one antherless plant, F2-1, with pollen from the *S. tuberosum* genotype Solyntus [51]. In the resulting BC_1 population BC1(P), all plants had normal anthers, confirming the recessive nature of the antherless gene(s). To study this locus in more detail, we backcrossed BC1(P)-1 to F2-1 to generate population BC2(P)-1.

3.2. Segregation of the Antherless Phenotype in Population BC2(P)-1

In the BC2(P)-1 population, we distinguished four anther phenotypes, complete absence of anthers (score 0), severely reduced anthers (score 1), moderately reduced anthers (score 2) and normal anthers (score 3) (Figure 1a). Plants that lacked anthers or showed reduced anthers (scores 0–2) never released any pollen upon vibration with an electric toothbrush. We grew 252 individuals from this BC2 population in the greenhouse and obtained 225 flowering plants (Table S4). We identified 84 genotypes which lacked or had reduced anthers, whereas the remaining 141 plants all had normal anthers. Among the 84 plants with malformed anthers, 32 showed complete absence of anthers, 35 had severely reduced anthers, and 17 plants had moderately reduced anthers. We genotyped the population using SeqSNP, resulting in 656 segregating markers. After removing non-informative markers, we generated a map with 153 SNPs covering 12 chromosomes which we used for QTL mapping of the anther phenotype. We found one highly significant QTL on the top of Chromosome 6 (LOD = 80.92 Figure 1c). Interestingly, other flower-related traits, such as number of floral buds, bud abortion and corolla shape, mapped to this locus as well (Table S5). Further inspection of individual recombinants showed that the causal locus is located in a 25.12 Mb interval between 5.59 and 30.71 Mb on the reference genome DM6.1 [52]. All plants that are heterozygous in this interval have normal anthers (score = 3), whereas those individuals that are homozygous for the *S. microdontum* haplotype all have reduced or absent anthers (score = 0, 1 or 2), indicating that in this population, antherless is effectively a monogenic recessive trait (Figure 1c). This locus is designated *Al*. We observed significant segregation distortion at the top of Chromosome 6, where the *Al* locus is located ($\chi2 = 14.33$, $p < 0.001$) (Figure 1d).

To reduce the size of the interval, we screened 2011 new seedlings from a closely related BC2 (BC2(P)-2). We screened this population for recombination in the interval using four KASP markers, two of which were located at the proximal side and the other two located at the distal side of the 25.12 Mb interval. We identified only 16 putative recombinants in this region. As the number of these putative recombinants compared to the size of the population (n = 2011) was low, these results could also be explained by some rare errors in the marker analyses. Therefore, we genotyped these 16 putative recombinants with 25 additional KASP markers. Indeed, the genotypic data of these additional markers showed that 2 of the 16 putative recombinants had a recombination just outside the interval at the centromeric side, whereas the other 14 likely resulted from genotyping errors in the initial genotyping with the flanking markers (Table S6).

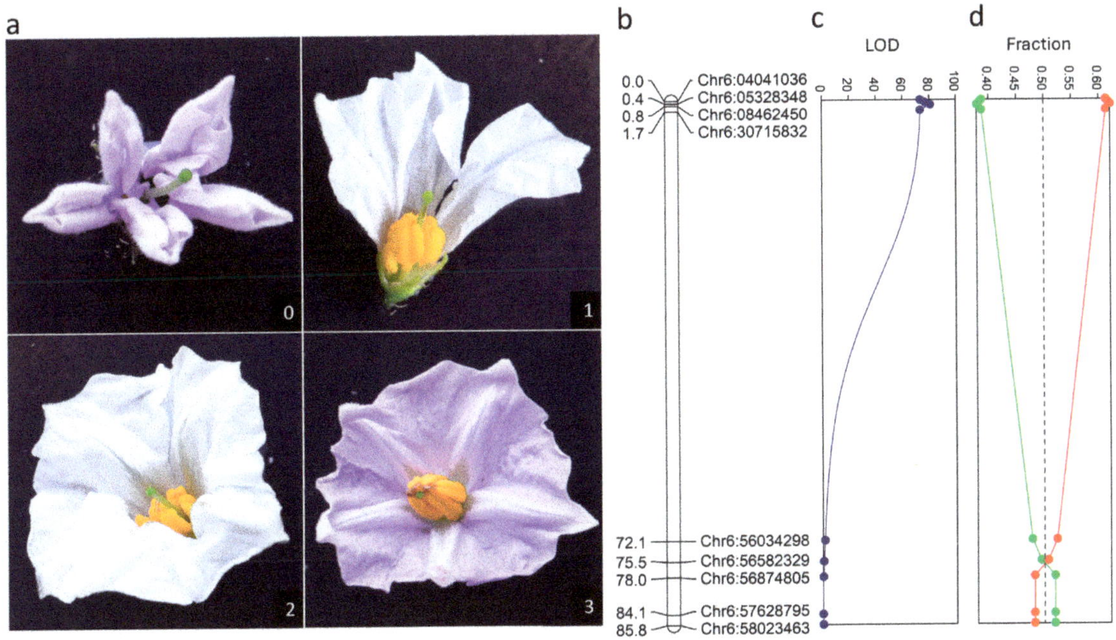

Figure 1. Population BC2(P)-1 segregates for anther length. (**a**) Anther phenotypes in BC_1 population BC2(P)-1 scored from 0 (complete absence of anthers) to 3 (normal anthers). (**b**) Genetic map of Chromosome 6. (**c**) QTL analysis for anther score reveals a significant QTL on the top of Chromosome 6. (**d**) Segregation distortion on the top of Chromosome 6 reduces the number of antherless plants. In red, the fraction of plants heterozygous per locus is shown and in green the fraction of plants homozygous for the antherless donor allele is shown.

3.3. Expression of the Antherless Phenotype in A and T Cytoplasm Types

The results so far show that homozygosity for the recessive *al* allele leads to the antherless phenotype in material with P cytoplasm. However, to apply the antherless system in breeding for male sterile hybrids, it is important to know whether it is a form of CMS, and we set out to introduce the *al* allele in lines with different cytoplasm types. We selected one clone from the proprietary Solynta genebank, D02, that has the A-type cytoplasm, and D03, D12 and two dihaploids from cultivar VR808 that have the T-type cytoplasm. These cytoplasm types were identified using pedigree analyses and were confirmed by using the multiplex PCR markers as described by Hosaka and Sanetomo [53] (Figure S1). We crossed these five cytoplasm type genitors as females with BC1-1 (*Alal*) and used six *al*-specific KASP markers to identify BC_2 plants that received the recessive *al* allele from BC1-1. We selected eight self-fertile *Alal* heterozygous BC_2 plants and self-fertilized these to proceed to the BC_2F_2 generation. The BC_2F_2 populations were genotyped with four KASP markers in the interval to determine the genotype at the *Al* locus. Interestingly, we observed distorted segregation in both directions; two populations, BC2F2(A)-1 and BC2F2(A)-2, showed significant distortion towards the *Al* allele, five populations showed no significant distortion, and one population showed significant distortion in favor of the *al* allele (Table 1). From all eight populations, we selected all plants that were homozygous for the *antherless* haplotype (*alal*) and transplanted these to pots. For the populations for which we had fewer than 25 *alal* plants, we added *Alal* and *AlAl* plants to fill up to at least 25 genotypes per population, allowing us to compare the effects of the genotype of the *Al* locus on the anther phenotype (Table S7).

Table 1. Segregation distortion in F_2 populations with T and A cytoplasm.

Population	*Antherless* Genotype			$\chi 2$	*p*-Value
	AlAl	*Alal*	*alal*		
BC2F2(A)-1	33	38	6	18.95	0.00008
BC2F2(A)-2	23	34	7	8.25	0.01616
BC2F2(T)-6	16	21	8	3.04	0.21823
BC2F2(T)-1	12	23	10	0.2	0.90484
BC2F2(T)-2	10	19	9	0.05	0.97404
BC2F2(T)-3	11	22	11	0	1
BC2F2(T)-4	8	31	20	5.03	0.08071
BC2F2(T)-5	1	45	39	34.27	0.00001

Generally, we observed poor fertility in the F_2 populations, with 80 of the 200 BC_2F_2 plants not producing any open flowers. Among the flowering F_2 plants, anther morphology segregated, ranging from severely reduced anthers to normal anthers, which we scored using the same 0–3 scale. Pollen release segregated as well, with many plants not releasing any pollen at all. Interestingly, the level of anther malformation and pollen release was independent of the genotype of the *Al* locus, suggesting that factors other than the antherless locus cause reduced fertility in these populations. Most importantly, from the 120 flowering F_2 plants, 8 produced seed after self-pollination, of which three plants were homozygous *alal* while the other four were heterozygous *Alal* (Figure 2, Table 2). Two of the *alal* homozygous self-fertile F_2 plants have the A-type cytoplasm, and the third has the T-type cytoplasm. These results are in line with the hypothesis that the homozygous *alal* genotype results in male sterility in the P cytoplasm type but not in T and A cytoplasm.

Figure 2. Flowers of alal homozygous genotypes with A and T cytoplasm have anthers that produce pollen and set self-seed.

To provide further evidence for the male fertility of *alal* plants with A and T cytoplasm, we planted five tubers each from four *alal* F_2 plants with A cytoplasm and two *alal* F_2 plants with the T cytoplasm. In addition, we sowed F3 seed from seven self-fertile F_2 plants (two with A cytoplasm and fixed for the *al* allele and five with T cytoplasm of which four are fixed for the *al* allele, and two of which segregate for *Al* and *al* (Table 2)). We genotyped the tuber-grown BC_2F_2 plants, as well the TPS-grown F_3 seedlings with the same KASP markers that we used in the TPS-raised BC_2F_2 populations. All tuber-grown F_2 plants and F_3 populations derived from *alal* F_2 plants were confirmed to be *alal*, whereas the F_3 populations derived from *Alal* heterozygous F_2 plants segregated for the *Al* locus. Interestingly, the BC_2F_2 plants raised from tubers from genotypes BC2F2(A)-1-29, BC2F2(A)-2-44 and BC2F2(T)-3-16 showed improved self-fertility compared to the same genotypes raised from TPS, suggesting improved vigor of the tuber-raised plants compared to seedling-raised plants (Table 3). The F_3 populations suffered from inbreeding depression showing overall poor fertility. Nevertheless, seven *alal* homozygous F3 plants with A cytoplasm and six *alal* homozygous F_3 plants with T cytoplasm produced berries and seed upon self-pollination, ranging from 15 to 297 seeds per plant (Table 4).

Table 2. Fertility characteristics of eight self-fertile F2 genotypes and their genotype at the *Al* locus.

Genotype	Cytoplasm Type	# Observed Flowers	Anther Phenotype (0–3)	Pollen Shed (0–3)	# Self-Berries	# Spontaneous Berries	# Seeds	*Antherless* Genotype	F3 Population
BC2F2(A)-1-29	A	20	3	3	5	2	23	*alal*	BC2F3(A)-1
BC2F2(A)-2-44	A	11	2	2	3	0	305	*alal*	BC2F3(A)-2
BC2F2(T)-2-07	T	23	3	1	2	1	106	*Alal*	BC2F3(T)-1
BC2F2(T)-2-20	T	25	3	3	5	2	220	*Alal*	BC2F3(T)-2
BC2F2(T)-3-06	T	6	3	1	3	0	97	*Alal*	BC2F3(T)-3
BC2F2(T)-4-16	T	30	3	3	10	16	140	*alal*	BC2F3(T)-4
BC2F2(T)-4-17	T	14	2	3	2	0	6	*Alal*	
BC2F2(T)-6-02	T	27	2	1	2	1	9	*Alal*	BC2F3(T)-5

Table 3. Flowering, berry and seed set of tuber-grown BC_2F_2 genotypes.

Genotype	Cytoplasm Type	Antherless Genotype	# Flowers	Anther Phenotype	Pollen Shed (0–3)	# Selfings	# Berries	# Seeds	Pollen Viability
BC2F2(A)-1-29-C1	A	*alal*	11	3	1-2	6	4	1296	0.9
BC2F2(A)-1-29-C2	A	*alal*	7	3	1	4	2	135	0.9
BC2F2(A)-1-29-C3	A	*alal*	17	3	1	4	1	22	0.9
BC2F2(A)-1-29-C4	A	*alal*	9	3	1-2	3	3	4	0.9
BC2F2(A)-1-29-C5	A	*alal*	11	3	1-2	1	1	49	0.9
BC2F2(A)-2-09-C1	A	*alal*	17	2-3	1	6	26	12	0.8
BC2F2(A)-2-09-C2	A	*alal*	0	N.D	N.D	0	N.D	N.D	N.D
BC2F2(A)-2-09-C3	A	*alal*	3	3	1	1	13	1	0.8
BC2F2(A)-2-09-C4	A	*alal*	7	2-3	1-2	2	25	0	0.8
BC2F2(A)-2-09-C5	A	*alal*	11	2-3	1	1	12	3	0.8
BC2F2(A)-2-39-C1	A	*alal*	4	2	0	0	N.D	N.D	N.D.
BC2F2(A)-2-39-C2	A	*alal*	3	2	0	0	N.D	N.D	N.D.
BC2F2(A)-2-39-C3	A	*alal*	0	N.D	N.D	0	N.D	N.D	N.D.
BC2F2(A)-2-39-C4	A	*alal*	1	2	0	0	N.D	N.D	N.D.
BC2F2(A)-2-39-C5	A	*alal*	0	N.D	N.D	0	N.D	N.D	N.D.
BC2F2(A)-2-44-C1	A	*alal*	32	2-3	3	17	19	740	0.9
BC2F2(A)-2-44-C2	A	*alal*	32	2-3	3	24	28	2457	0.9
BC2F2(A)-2-44-C3	A	*alal*	11	2-3	3	10	23	2320	0.9
BC2F2(A)-2-44-C4	A	*alal*	15	2-3	3	12	19	719	0.9
BC2F2(A)-2-44-C5	A	*alal*	24	2-3	3	13	34	1246	0.9
BC2F2(T)-3-07-C1	T	*alal*	17	3	1	4	2	18	N.D.
BC2F2(T)-3-07-C2	T	*alal*	20	3	1	6	3	40	N.D.
BC2F2(T)-3-07-C3	T	*alal*	24	3	0-1	3	2	3	N.D.
BC2F2(T)-3-07-C4	T	*alal*	33	3	0-1	8	4	82	N.D.
BC2F2(T)-3-07-C5	T	*alal*	38	3	0-1	8	7	75	N.D.
BC2F2(T)-4-16-C1	T	*alal*	52	3	3	16	21	550	0.9
BC2F2(T)-4-16-C2	T	*alal*	47	3	3	17	21	906	0.9
BC2F2(T)-4-16-C3	T	*alal*	40	3	3	10	7	177	0.9
BC2F2(T)-4-16-C4	T	*alal*	44	3	3	14	10	199	0.9
BC2F2(T)-4-16-C5	T	*alal*	43	3	3	20	37	843	0.9

Table 4. Flowering, berry and seed set of TPS-grown BC_2F_3 genotypes.

Genotype	Parent	Cytoplasm Type	Antherless Genotype	# Flowers	Anther Phenotype (0–3)	Pollen Shed (0–3)	# Selfings	# Berries	# Seeds	Pollen Viability
BC2F3(A)-1-11	BC2F2(A)-1-29	A	*alal*	21	3	1	5	5	15	N.D.
BC2F3(A)-1-12	BC2F2(A)-1-29	A	*alal*	12	3	3	9	3	48	N.D.
BC2F3(A)-2-04	BC2F2(A)-2-44	A	*alal*	9	2	2	7	14	98	N.D.
BC2F3(A)-2-05	BC2F2(A)-2-44	A	*alal*	2	2	3	2	1	125	N.D.
BC2F3(A)-2-14	BC2F2(A)-2-44	A	*alal*	5	2	3	5	3	242	85%
BC2F3(A)-2-17	BC2F2(A)-2-44	A	*alal*	7	2	3	7	10	278	90%
BC2F3(T)-2-06	BC2F2(T)-2-20	T	*AlAl*	5	3	3	5	4	83	60%
BC2F3(T)-2-09	BC2F2(T)-2-20	T	*Alal*	9	3	2-3	8	5	30	70%

Table 4. *Cont.*

Genotype	Parent	Cytoplasm Type	Antherless Genotype	# Flowers	Anther Phenotype (0–3)	Pollen Shed (0–3)	# Selfings	# Berries	# Seeds	Pollen Viability
BC2F3(T)-2-24	BC2F2(T)-2-20	T	*Alal*	13	3	2-3	18	11	336	70%
BC2F3(T)-3-27	BC2F2(T)-3-06	T	*Alal*	22	2	3	18	12	1514	90%
BC2F3(T)-3-29	BC2F2(T)-3-06	T	*Alal*	45	3	1-2	22	21	707	90%
BC2F3(T)-4-01	BC2F2(T)-4-16	T	*alal*	35	2-3	2-3	24	10	55	85%
BC2F3(T)-4-07	BC2F2(T)-4-16	T	*alal*	32	3	2-3	19	1	10	60%
BC2F3(T)-4-13	BC2F2(T)-4-16	T	*alal*	15	3	3	24	21	122	70%
BC2F3(T)-4-25	BC2F2(T)-4-16	T	*alal*	17	3	1-2	7	10	59	60%
BC2F3(T)-4-31	BC2F2(T)-4-16	T	*alal*	49	3	2	23	11	69	N.D.
BC2F3(T)-5-02	BC2F2(T)-4-16	T	*alal*	22	3	3	19	13	297	90%

Taken together, the apparent male fertility of the tuber-raised *alal* homozygous F_2 plants and the seedling-raised *alal* homozygous F_3 plants provide clear evidence that homozygosity for the *al* allele does not result in male sterility in T and A cytoplasm, providing convincing evidence that the antherless trait is a form of cytoplasmic male sterility.

3.4. Application and Deployment of the Antherless Gene

With the antherless CMS system, an improved inbred line-based hybrid breeding system becomes feasible, whereby male sterile inbred lines and hybrids are generated. After an initial cross between an elite inbred line with P cytoplasm and an *al* donor line, markers can be used to select heterozygous *Alal* progeny after the initial cross and after each round of self-fertilization or backcrossing. When a sufficient level of homozygosity has been achieved, a final round of self-fertilization can be performed and homozygous *alal* progeny can be selected using the same markers, resulting in a male sterile maternal inbred line. Similarly, after an initial cross between an elite line with T or A cytoplasm and an *al* donor, the markers can be used to select for heterozygosity during backcrossing or homozygosity during inbreeding, resulting in male fertile *alal* inbred lines. With this system, emasculation of the maternal lines is no longer necessary and hybridization will result in male sterile F_1 hybrids. To determine the feasibility of this system, we fertilized five male sterile *alal* homozygous lines in P cytoplasm with pollen from four male fertile *alal* homozygous lines with A or T cytoplasm and obtained seeds from 13 crosses (Table S8). Next, we planted these 13 male sterile hybrids together with male fertile control hybrids in a field trial and harvested all berries at the end of the growing season. While the male sterile antherless hybrids did set some berries, the total be yield was significantly lower than those of male fertile control hybrids (Figure 3).

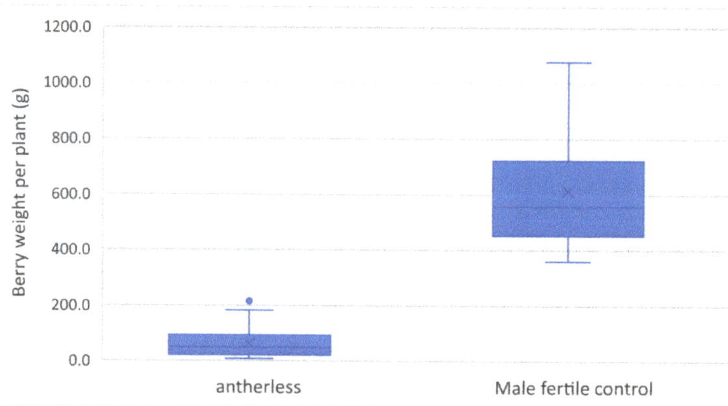

Figure 3. Berry yield of antherless POC hybrids and male fertile controls in a field trial. Antherless hybrids (n = 26) produce significantly fewer berries than male fertile controls (n = 8) ($p < 0.001$).

The boxes represent the 1st and 3rd quartiles, the horizontal lines in the box represent the medians, and the × represent the means.

4. Discussion

Diploid hybrid potato breeding is based on the generation of homozygous inbred parent lines by continuous self-fertilization and hybrid cultivars by crossing these parent lines. For inbreeding, self-fertilized berry set is crucial, while this is an undesired character for the cultivation of potato in the field. In this study, we investigated the antherless CMS system which can be used to generate potato hybrids that set few berries in the field. We identified a recessive allele (*al*) on the short arm of Chromosome 6 of *S. microdontum* subsp. *gigantophyllum* which results in complete male sterility due to malformed or absent anthers in the presence of the P cytoplasm (Figure 1a).

Previously, Endelman and Jansky identified the *Sa* locus, involved in male sterility, which was mapped to the short arm of Chromosome 6 in an F_2 population from the cross between DM and M6. The recessive allele *sa* originates from the M6 parent, and the short anther phenotype is likely a result of an interaction between the *Sa* locus and the P cytoplasm from DM [43]. Given the similarity between the antherless and short anther traits of the *Al* and the *Sa* locus, it is possible that these loci are identical. However, as both loci originate from distant *Solanum* species, these loci might also be different. Therefore, here we use the designation "*Al* locus" to refer to the *antherless* locus from *S. microdontum* subsp. *gigantophyllum*

In this study, we have found that the antherless trait is expressed in P, but not in T and A cytoplasm, proving that antherless is a form of CMS and that it is the result of an interaction between the nuclear genome and the cytoplasm. It is likely that the dominant allele of the *Al* locus encodes a fertility restorer that is required for fertility in genotypes carrying the P cytoplasm. Unfortunately, our efforts to fine-map the *Al* locus were hindered by a lack of recombination in the region of the *Al* locus. Among 2011 BC_2 individuals, we did not find any true recombinants in the 25.12 Mb interval on the short arm of Chromosome 6 (Table S3). The suppression of recombination in this interval on Chromosome 6 is observed in many potato genetic maps, suggesting that this region is pericentromeric and that fine-mapping approaches will not reduce the size of the interval [16,43,54–57]. Thus, identification of the causal gene requires an alternative strategy. One strategy would be to perform untargeted mutagenesis on seed from a population derived from a cross between an *alal* and an *AlAl* individual wherein the maternal parent has the P cytoplasm. All progeny from such a cross are heterozygous *Alal* and should have normal male fertile anthers. However, when the dominant *Al* allele of such a genotype is knocked out, it results in an *alal* genotype, which is antherless and can easily be identified among the fertile plants of this backcross population. The causal mutation can then be identified by comparing sequence data from the mutant with the parental genomes. Such a strategy has been successful in the cloning of several genes from genomic regions with suppressed recombination and could be used to identify the *al* allele [58–60]. Alternatively, a targeted mutagenesis approach could reveal the causal gene for the antherless phenotype. This approach would rely on knowledge about nuclear Restorer of Fertility (Rf) genes involved in CMS in other crops and targeted knock-out of candidate genes using CRISPR-Cas. Multiple *Rf* genes have been identified, and the majority of those encode pentatricopeptide repeat (PPR) proteins [61,62]. PPR encoding genes have been implicated in restoration of fertility in crop species such as pepper [63], rice [64], radish [65,66], sorghum [67], soybean [68], cotton [69,70], and others. These *Rf-PPR* exert their function in mitochondria, where they bind CMS inducing mitochondrial mRNAs, leading to reduced levels of sterilizing protein via mRNA destabilization, modification or translation inhibition [64,71–73]. In potato, Anisimova et al. identified 38 sequence fragments with homology to *petunia Rf-PPR592* and *Capsicum annuum CaPPR6*. These fragments map to five genomic loci of the potato reference genome DM [74]. Interestingly two of the identified *Rf-PPR* genes are located within the interval of the *Al* locus and are good candidates as causal genes underlying the antherless and short anther traits.

The *Al* locus is located on the short arm of Chromosome 6 in a pericentromeric region where other floral traits like bud abortion and corolla shape mapped to as well. It is unclear whether the same causal genes are involved in anther malformation, bud abortion and corolla shape. It is not possible to remove any linked deleterious alleles from the interval due to the lack of recombination. However, in this study we did not observe obvious deleterious linkage drag associated with the antherless locus, but it is possible that linkage drag may reveal itself in more advanced materials and more detailed agronomical studies. We observed variation in the severity of anther malformation among the *alal* homozygous plants from population BC_2(P)-1, but the genetic background of this variation remains unknown. The severity of anther malformation is relatively stable within individual plants and their clonal offspring, but genetic analysis within the *alal* homozygous group did not reveal any significant QTL (Table S5). It is possible that the combined effect of multiple segregating small-effect loci are responsible for the observed variation in anther malformation, although epigenetic control of such loci cannot be excluded. Further studies with large populations which are fixed for the *al* allele could resolve this question.

The implementation of CMS in hybrid potato breeding will provide two major advantages: (1) seed production becomes more efficient by eliminating the need to emasculate maternal flowers and (2) ware crop production will benefit from significantly reduced berry set on the field. In contrast to crops like maize and rice, where the seed are the commercial end product, implementation of CMS in potato does not require fertility to be restored in the hybrid, and hence a *three-line system* is redundant. Here, we show that the antherless CMS system can be used to develop *alal* male sterile maternal lines with P cytoplasm and male fertile *alal* lines with T or A cytoplasm and that the hybrids derived from crosses between these parental lines hardly set any berries in the field. Male sterile maternal *alal* lines can be generated by introduction of the *al* allele in the breeding germplasm and subsequent marker-assisted selection for *Alal* heterozygosity upon inbreeding. When such an inbred line has met the criteria for becoming a female parent, breeders can easily select *alal* homozygotes based on the clear anther phenotype and proceed with test crosses or commercial seed production. We observed a reduction in fertility upon inbreeding in the BC_2F_2 and BC_2F_3 populations; it is not clear to what extent this is caused by the cytoplasm type. While it is possible that T and A cytoplasm affect fertility upon inbreeding, the BC_2F_2 and BC_2F_3 populations were derived from crosses with non-inbred cytoplasm donors which likely contained significant recessive genetic load. This genetic load may have resulted in inbreeding depression and the associated reduction in fertility, which are commonly observed during inbreeding of potato, especially in early inbred generations [75–78]. So, we successfully inbred *alal* lines with T and A cytoplasm up to the BC_2F_4 generation, used the *alal* BC_2F_2 as male parents in hybrid crosses with male sterile *alal* lines and showed that the resulting hybrids produce far fewer berries than male fertile controls. However, the antherless hybrids did still set some berries, likely caused by insect or wind-driven cross-pollination. Many male fertile diploid potato genotypes were present in and around our field experiment, providing ample fertile pollen which could have fertilized the antherless hybrids. In future commercial application of the antherless CMS system, the produced CMS hybrids will likely be grown in fields without male fertile diploid pollen donors where the CMS hybrids would likely set fewer or no berries at all. We consider this an effective proof of principle of the antherless CMS system. Further research is needed to determine how inbreeding depression in T and A cytoplasm affects fertility and plant vigor and whether linkage drag is an issue for the *Al* locus. Future research may also focus on identification of the causal gene and several good leads are available.

5. Conclusions

In conclusion, in this study we identified the recessive *al* locus on chr06 that causes male sterility in diploid potato with P cytoplasm but not in A and T cytoplasm. Efforts to more precisely localize the locus were not successful due to a lack of recombination in this region of chr06. We implemented the antherless CMS system in a hybrid breeding scheme and show that the resulting F_1 hybrid set significantly fewer berries under field conditions. Further exploration of this trait could focus on the identification of the causal nuclear gene via a candidate gene approach and CRISPR-Cas-induced knock-out, as well as the identification of the corresponding mitochondrial or chloroplast gene.

Supplementary Materials: The following supporting information can be downloaded at: https://www.mdpi.com/article/10.3390/biology13060447/s1, Figure S1: Cytoplasm type determination using multiplex PCR markers as described by Hosaka and Sanetomo; Table S1: Plant materials used in this study; Table S2: Genotypic data of population BC2(P)-1; Table S3: Genetic map of population BC2(P)-1; Table S4: Phenotypic data from population BC2(P)-1; Table S5: QTL identified in population BC2(P)-1; Table S6: Genotypic data from putative recombinants from population BC2(P)-2; Table S7: Phenotypic data from BC2F2 populations with A and T cytoplasm; Table S8: Berry weights per plant of antherless and male fertile control plants.

Author Contributions: Conceptualization, E.-J.E., M.E.d.V., Y.S., P.L. and R.G.F.V.; methodology, E.-J.E., M.E.d.V. and P.L.; formal analysis, E.-J.E.; investigation, E.-J.E., E.v.d.P. and M.F.; resources, Y.S.; data curation, E.-J.E., E.v.d.P. and M.F.; writing—original draft preparation, E.-J.E.; writing—review and editing, E.-J.E., M.E.d.V., C.W.B.B., R.G.F.V. and P.L.; visualization, E.-J.E.; supervision, M.E.d.V., P.L., C.W.B.B. and R.G.F.V.; project administration, M.E.d.V., R.G.F.V. and E.-J.E.; funding acquisition, E.-J.E., P.L. and R.G.F.V. All authors have read and agreed to the published version of the manuscript.

Funding: This research was funded by the Dutch Research Council, grant number NWA.17.023.

Institutional Review Board Statement: Not applicable.

Informed Consent Statement: Not applicable.

Data Availability Statement: All data are contained within the article or Supplementary Materials.

Acknowledgments: We acknowledge the help from Shadee Martis and Alycia Klerkx for their efforts in the phenotypic analysis of the recombinant screening population and the BC2F3 populations. We also acknowledge the Solynta bioinformatics, greenhouse and agronomy teams for their help with WGS data analysis, marker development, plant raising and maintenance and field trial execution.

Conflicts of Interest: Authors Ernst-Jan Eggers, Ying Su, Michiel Erik de Vries, Christian Bachem and Pim Lindhout were employed by the company Solynta. Authors Esmee van der Poel, Martijn Flipsen and Richard Visser declare that the research was conducted in the absence of any commercial or financial relationships that could be construed as a potential conflict of interest. The funders had no role in the design of the study; in the collection, analyses, or interpretation of data; in the writing of the manuscript; or in the decision to publish the results.

References

1. Lindhout, P.; de Vries, M.; ter Maat, M.; Ying, S.; Viquez-Zamora, M.; van Heusden, S. Hybrid Potato Breeding for Improved Varieties. In *Achieving Sustainable Cultivation of Potatoes*; Burleigh Dodds Science Publishing: London, UK, 2016; Volume 1, pp. 1–24. [CrossRef]
2. Zhang, C.; Yang, Z.; Tang, D.; Zhu, Y.; Wang, P.; Li, D.; Zhu, G.; Xiong, X.; Shang, Y.; Li, C.; et al. Genome Design of Hybrid Potato. *Cell* **2021**, *184*, 3873–3883.e12. [CrossRef] [PubMed]
3. Jansky, S.H.; Charkowski, A.O.; Douches, D.S.; Gusmini, G.; Richael, C.; Bethke, P.C.; Spooner, D.M.; Novy, R.G.; De Jong, H.; De Jong, W.S.; et al. Reinventing Potato as a Diploid Inbred Line-Based Crop. *Crop Sci.* **2016**, *56*, 1412–1422. [CrossRef]
4. Alsahlany, M.; Enciso-Rodriguez, F.; Lopez-Cruz, M.; Coombs, J.; Douches, D.S. Developing Self-Compatible Diploid Potato Germplasm through Recurrent Selection. *Euphytica* **2021**, *217*, 47. [CrossRef]
5. Hosaka, K.; Sanetomo, R. Creation of a Highly Homozygous Diploid Potato Using the S Locus Inhibitor (Sli) Gene. *Euphytica* **2020**, *216*, 169. [CrossRef]
6. Lindhout, P.; Meijer, D.; Schotte, T.; Hutten, R.C.B.; Visser, R.G.F.; van Eck, H.J. Towards F1 Hybrid Seed Potato Breeding. *Potato Res.* **2011**, *54*, 301–312. [CrossRef]
7. Douches, D.S.; Maas, D.; Jastrzebski, K.; Chase, R.W. Assessment of Potato Breeding Progress in the USA over the Last Century. *Crop Sci.* **1996**, *36*, 1544–1552. [CrossRef]
8. Jansky, S. Breeding, Genetics, and Cultivar Development. In *Advances in Potato Chemistry and Technology*; Singh, J., Kaur, L., Eds.; Academic Press: New York, NY, USA, 2009; pp. 26–27.

9. Su, Y.; Viquez-Zamora, M.; den Uil, D.; Sinnige, J.; Kruyt, H.; Vossen, J.; Lindhout, P.; van Heusden, S. Introgression of Genes for Resistance against Phytophthora Infestans in Diploid Potato. *Am. J. Potato Res.* **2020**, *97*, 33–42. [CrossRef]
10. Hosaka, K.; Hanneman, R.E. Genetics of Self-Compatibility in a Self-Incompatible Wild Diploid Potato Species Solanum Chacoense. 1. Detection of an S Locus Inhibitor (Sli) Gene. *Euphytica* **1998**, *99*, 191–197. [CrossRef]
11. Clot, C.R.; Polzer, C.; Prodhomme, C.; Schuit, C.; Engelen, C.J.M.; Hutten, R.C.B.; van Eck, H.J. The Origin and Widespread Occurrence of Sli-Based Self-Compatibility in Potato. *Theor. Appl. Genet.* **2020**, *133*, 2713–2728. [CrossRef]
12. Birhman, R.K.; Hosaka, K. Production of Inbred Progenies of Diploid Potatoes Using an S -Locus Inhibitor (Sli) Gene, and Their Characterization. *Genome* **2000**, *502*, 495–502. [CrossRef]
13. Eggers, E.J.; van der Burgt, A.; van Heusden, S.A.W.; de Vries, M.E.; Visser, R.G.F.; Bachem, C.W.B.; Lindhout, P. Neofunctionalisation of the Sli Gene Leads to Self-Compatibility and Facilitates Precision Breeding in Potato. *Nat. Commun.* **2021**, *12*, 4141. [CrossRef] [PubMed]
14. Ma, L.; Zhang, C.; Zhang, B.; Tang, F.; Li, F.; Liao, Q.; Tang, D.; Peng, Z.; Jia, Y.; Gao, M.; et al. A NonS-Locus F-Box Gene Breaks Self-Incompatibility in Diploid Potatoes. *Nat. Commun.* **2021**, *12*, 4142. [CrossRef] [PubMed]
15. Phumichai, C.; Mori, M.; Kobayashi, A.; Kamijima, O.; Hosaka, K. Toward the Development of Highly Homozygous Diploid Potato Lines Using the Self-Compatibility Controlling Sli Gene. *Genome* **2005**, *48*, 977–984. [CrossRef] [PubMed]
16. Kaiser, N.R.; Billings, G.; Coombs, J.; Buell, C.R.; Enciso-Rodríguez, F.; Douches, D.S. Self-Fertility and Resistance to the Colorado Potato Beetle (Leptinotarsa Decemlineata) in a Diploid Solanum Chacoense Recombinant Inbred Line Population. *Crop Sci.* **2021**, *61*, 3392–3414. [CrossRef]
17. Jansky, S.H.; Chung, Y.S.; Kittipadukal, P. M6: A Diploid Potato Inbred Line for Use in Breeding and Genetics Research. *J. Plant Regist.* **2014**, *8*, 195–199. [CrossRef]
18. Zhang, C.; Wang, P.; Tang, D.; Yang, Z.; Lu, F.; Qi, J.; Tawari, N.R.; Shang, Y.; Li, C.; Huang, S. The Genetic Basis of Inbreeding Depression in Potato. *Nat. Genet.* **2019**, *51*, 374–378. [CrossRef] [PubMed]
19. Veerman, A.; Van Loon, C.D. Prevention of Berry Formation in Potato Plants (*Solanum tuberosum* L.) by Single Foliar Applications of Herbicides or Growth Regulators. *Potato Res.* **1993**, *36*, 135–142. [CrossRef]
20. Bartholdi, W.L. *Influence of Flowering and Fruiting Upon Vegetative Growth and Tuber Yield in the Potato*; Minn. Tech. Bull 150; University of Minnesota: Minneapolis, MN, USA, 1942.
21. Ter Steeg, E.M.S.; Struik, P.C.; Visser, R.G.F.; Lindhout, P. Crucial Factors for the Feasibility of Commercial Hybrid Breeding in Food Crops. *Nat. Plants* **2022**, *8*, 463–473. [CrossRef] [PubMed]
22. Gramaje, L.V.; Caguiat, J.D.; Enriquez, J.O.S.; dela Cruz, Q.D.; Millas, R.A.; Carampatana, J.E.; Tabanao, D.A.A. Heterosis and Combining Ability Analysis in CMS Hybrid Rice. *Euphytica* **2020**, *216*, 1–22. [CrossRef]
23. Martín, A.C.; Atienza, S.G.; Ramírez, M.C.; Barro, F.; Martín, A. Male Fertility Restoration of Wheat in Hordeum Chilense Cytoplasm Is Associated with 6HchS Chromosome Addition. *Aust. J. Agric. Res.* **2008**, *59*, 206–213. [CrossRef]
24. Martín, A.C.; Castillo, A.; Atienza, S.G.; Rodríguez-Suárez, C. A Cytoplasmic Male Sterility (CMS) System in Durum Wheat. *Mol. Breed.* **2018**, *38*, 90. [CrossRef]
25. Tyagi, V.; Dhillon, S.K.; Kaushik, P.; Kaur, G. Characterization for Drought Tolerance and Physiological Efficiency in Novel Cytoplasmic Male Sterile Sources of Sunflower (*Helianthus annuus* L.). *Agronomy* **2018**, *8*, 232. [CrossRef]
26. Li, J.; Nadeem, M.; Sun, G.; Wang, X.; Qiu, L. Male Sterility in Soybean: Occurrence, Molecular Basis and Utilization. *Plant Breed.* **2019**, *138*, 659–676. [CrossRef]
27. Bruns, H.A. Southern Corn Leaf Blight: A Story Worth Retelling. *Agron. J.* **2017**, *109*, 1218–1224. [CrossRef]
28. Graybosch, R.A.; Palmer, R.G. Male Sterility in Soybean (*Glycine max*). I. Phenotypic Expression of the Ms2 Mutant. *Am. J. Bot.* **1985**, *72*, 1738–1750. [CrossRef]
29. Balaji Suresh, P.; Srikanth, B.; Hemanth Kishore, V.; Subhakara Rao, I.; Vemireddy, L.R.; Dharika, N.; Sundaram, R.M.; Ramesha, M.S.; Sambasiva Rao, K.R.S.; Viraktamath, B.C.; et al. Fine Mapping of Rf3 and Rf4 Fertility Restorer Loci of WA-CMS of Rice (*Oryza sativa* L.) and Validation of the Developed Marker System for Identification of Restorer Lines. *Euphytica* **2012**, *187*, 421–435. [CrossRef]
30. Wu, J.; Zhang, M.; Zhang, B.; Zhang, X.; Guo, L.; Qi, T.; Wang, H.; Zhang, J.; Xing, C. Genome-Wide Comparative Transcriptome Analysis of CMS-D2 and Its Maintainer and Restorer Lines in Upland Cotton. *BMC Genom.* **2017**, *18*, 454. [CrossRef]
31. El-Rady, A.; Ayman, G.; Soliman, G. Identification of Maintainer and Restorer Lines Based on CMS System for Developing Hybrid Wheat in Egypt. *Egypt. J. Agric. Res.* **2023**, *101*, 322–330. [CrossRef]
32. Kim, Y.J.; Zhang, D. Molecular Control of Male Fertility for Crop Hybrid Breeding. *Trends Plant Sci.* **2018**, *23*, 53–65. [CrossRef] [PubMed]
33. Hosaka, K.; Sanetomo, R. Comparative Differentiation in Mitochondrial and Chloroplast DNA among Cultivated Potatoes and Closely Related Wild Species. *Genes Genet. Syst.* **2009**, *84*, 371–378. [CrossRef]
34. Dionne, L.A. Cytoplasmic Sterility in Derivatives of Solanum Demissum. *Am. Potato J.* **1961**, *38*, 117–120. [CrossRef]
35. Ortiz, R.; Iwanaga, M.; Peloquin, S.J. Male Sterility and 2n Pollen in 4x Progenies Derived from 4x × 2x and 4x × 4x Crosses in Potatoes. *Potato Res.* **1993**, *36*, 227–236. [CrossRef]
36. Lössl, A.; Götz, M.; Braun, A.; Wenzel, G. Molecular Markers for Cytoplasm in Potato: Male Sterility and Contribution of Different Plastid-Mitochondrial Configurations to Starch Production. *Euphytica* **2000**, *116*, 221–230. [CrossRef]

37. Hoopes, R.W.; Plaisted, R.L.; Cubillos, A.G. Yield and Fertility of Reciprocal-Cross Tuberosum-Andigena Hybrids. *Am. Potato J.* **1980**, *57*, 275–284. [CrossRef]
38. Staub, J.E.; Grun, P.; Amoah, V. Cytoplasmic Evaluations during Substitution Backcrossing in Solanum. *Potato Res.* **1982**, *25*, 299–319. [CrossRef]
39. Grun, P. Cytoplasmic Sterilities That Separate the Group Tuberosum Cultivated Potato from Its Putative Tetraploid Ancestor. *Evolution* **1973**, *27*, 633. [CrossRef]
40. Santayana, M.; Aponte, M.; Kante, M.; Gastelo, M. Cytoplasmic Male Sterility Incidence in Potato Breeding Populations with Late Blight Resistance and Identification of Breeding Lines with a Potential Fertility Restorer Mechanism. *Plants* **2022**, *11*, 3093. [CrossRef] [PubMed]
41. Sanetomo, R.; Akai, K.; Nashiki, A. Discovery of a Novel Mitochondrial DNA Molecule Associated with Tetrad Pollen Sterility in Potato. *BMC Plant Biol.* **2022**, *22*, 302. [CrossRef]
42. Sanetomo, R.; Nashiki, A. Identification of the Tetrad-Sterility-Causing Solanum Stoloniferum Schltdl. & Bouché Cytoplasm in Interspecific Hybrids with *S. tuberosum* L. *Genet Resour. Crop Evol.* **2021**, *68*, 3383–3397. [CrossRef]
43. Endelman, J.B.; Jansky, S.H. Genetic Mapping with an Inbred Line-Derived F2 Population in Potato. *Theor. Appl. Genet.* **2016**, *129*, 935–943. [CrossRef]
44. Spooner, D.M.; Ghislain, M.; Simon, R.; Jansky, S.H.; Gavrilenko, T. Systematics, Diversity, Genetics, and Evolution of Wild and Cultivated Potatoes. *Bot. Rev.* **2014**, *80*, 283–383. [CrossRef]
45. Lin, X.; Torres Ascurra, Y.C.; Fillianti, H.; Dethier, L.; de Rond, L.; Domazakis, E.; Aguilera-Galvez, C.; Kiros, A.Y.; Jacobsen, E.; Visser, R.G.F.; et al. Recognition of Pep-13/25 MAMPs of *Phytophthora* Localizes to an RLK Locus in *Solanum microdontum*. *Front. Plant Sci.* **2023**, *13*, 1037030. [CrossRef] [PubMed]
46. Parmentier, J.; Fartmann, B.; Winkler, S.; Arvidsson, S.; Osterkamp, S.; Zimmermann, W.; Heath, J.D. SeqSNP, a Massively Parallel Marker Screening Approach–High-Speed and High-Throughput Genomic Selection for Plants and Livestock. In Proceedings of the Plant and Animal Genome XXVI Conference, San Diego, CA, USA, 13–17 January 2018.
47. Adams, J.; de Vries, M.; van Eeuwijk, F. Efficient Genomic Prediction of Yield and Dry Matter in Hybrid Potato. *Plants* **2023**, *12*, 2617. [CrossRef] [PubMed]
48. Van Ooijen, J.W. *JoinMap®4, Software for the Calculation of Genetic Linkage Maps in Experimental Populations*; Kyazma BV: Wagening, The Netherlands, 2006.
49. van Ooijen, J.W. Accuracy of Mapping Quantitative Trait Loci in Autogamous Species. *Theor. Appl. Genet.* **1992**, *84*, 803–811. [CrossRef] [PubMed]
50. Kacheyo, O.C.; de Vries, M.E.; van Dijk, L.C.M.; Schneider, H.M.; Struik, P.C. Agronomic Consequences of Growing Field-Transplanted Hybrid Potato Seedlings. *Crop Sci* **2024**, *64*, 1093–1111. [CrossRef]
51. van Lieshout, N.; van der Burgt, A.; de Vries, M.E.; ter Maat, M.; Eickholt, D.; Esselink, D.; van Kaauwen, M.P.W.; Kodde, L.P.; Visser, R.G.F.; Lindhout, P.; et al. Solyntus, the New Highly Contiguous Reference Genome for Potato (*Solanum tuberosum*). *G3 Genes Genomes Genet.* **2020**, *10*, 3489–3495. [CrossRef]
52. Pham, G.M.; Hamilton, J.P.; Wood, J.C.; Burke, J.T.; Zhao, H.; Vaillancourt, B.; Ou, S.; Jiang, J.; Robin Buell, C. Construction of a Chromosome-Scale Long-Read Reference Genome Assembly for Potato. *Gigascience* **2020**, *9*, 1–11. [CrossRef]
53. Hosaka, K.; Sanetomo, R. Development of a Rapid Identification Method for Potato Cytoplasm and Its Use for Evaluating Japanese Collections. *Theor. Appl. Genet.* **2012**, *125*, 1237–1251. [CrossRef] [PubMed]
54. Massa, A.N.; Manrique-Carpintero, N.C.; Coombs, J.J.; Zarka, D.G.; Boone, A.E.; Kirk, W.W.; Hackett, C.A.; Bryan, G.J.; Douches, D.S. Genetic Linkage Mapping of Economically Important Traits in Cultivated Tetraploid Potato (*Solanum tuberosum* L.). *G3 Genes Genomes Genet.* **2015**, *5*, 2357–2364. [CrossRef]
55. da Silva Pereira, G.; Mollinari, M.; Schumann, M.J.; Clough, M.E.; Zeng, Z.B.; Yencho, G.C. The Recombination Landscape and Multiple QTL Mapping in a Solanum Tuberosum Cv. 'Atlantic'-Derived F1 Population. *Heredity* **2021**, *126*, 817–830. [CrossRef]
56. Bourke, P.M.; Voorrips, R.E.; Kranenburg, T.; Jansen, J.; Visser, R.G.F.; Maliepaard, C. Integrating Haplotype-Specific Linkage Maps in Tetraploid Species Using SNP Markers. *Theor. Appl. Genet.* **2016**, *129*, 2211–2226. [CrossRef] [PubMed]
57. Sharma, S.K.; Bolser, D.; de Boer, J.; Sønderkær, M.; Amoros, W.; Carboni, M.F.; D'Ambrosio, J.M.; de la Cruz, G.; Di Genova, A.; Douches, D.S.; et al. Construction of Reference Chromosome-Scale Pseudomolecules for Potato: Integrating the Potato Genome with Genetic and Physical Maps. *G3 Genes Genomes Genet.* **2013**, *3*, 2031–2047. [CrossRef]
58. Harkess, A.; Zhou, J.; Xu, C.; Bowers, J.E.; Van Der Hulst, R.; Ayyampalayam, S.; Mercati, F.; Riccardi, P.; McKain, M.R.; Kakrana, A.; et al. The Asparagus Genome Sheds Light on the Origin and Evolution of a Young Y Chromosome. *Nat. Commun.* **2017**, *8*, 1279. [CrossRef] [PubMed]
59. Harkess, A.; Huang, K.; van der Hulst, R.; Tissen, B.; Caplan, J.L.; Koppula, A.; Batish, M.; Meyers, B.C.; Leebens-Mack, J. Sex Determination by Two Y-Linked Genes in Garden Asparagus. *Plant Cell* **2020**, *32*, 1790–1796. [CrossRef] [PubMed]
60. Underwood, C.J.; Vijverberg, K.; Rigola, D.; Okamoto, S.; Oplaat, C.; Camp, R.H.M.O.d.; Radoeva, T.; Schauer, S.E.; Fierens, J.; Jansen, K.; et al. A PARTHENOGENESIS Allele from Apomictic Dandelion Can Induce Egg Cell Division without Fertilization in Lettuce. *Nat. Genet.* **2022**, *54*, 84–93. [CrossRef]
61. Dahan, J.; Mireau, H. The Rf and Rf-like PPR in Higher Plants, a Fast-Evolving Subclass of PPR Genes. *RNA Biol.* **2013**, *10*, 1469–1476. [CrossRef] [PubMed]

62. Gaborieau, L.; Brown, G.G.; Mireau, H. The Propensity of Pentatricopeptide Repeat Genes to Evolve into Restorers of Cytoplasmic Male Sterility. *Front. Plant Sci.* **2016**, *7*, 225677. [CrossRef]
63. Jo, Y.D.; Ha, Y.; Lee, J.H.; Park, M.; Bergsma, A.C.; Choi, H.I.; Goritschnig, S.; Kloosterman, B.; van Dijk, P.J.; Choi, D.; et al. Fine Mapping of Restorer-of-Fertility in Pepper (*Capsicum annuum* L.) Identified a Candidate Gene Encoding a Pentatricopeptide Repeat (PPR)-Containing Protein. *Theor. Appl. Genet.* **2016**, *129*, 2003–2017. [CrossRef]
64. Wang, Z.; Zou, Y.; Li, X.; Zhang, Q.; Chen, L.; Wu, H.; Su, D.; Chen, Y.; Guo, J.; Luo, D.; et al. Cytoplasmic Male Sterility of Rice with Boro II Cytoplasm Is Caused by a Cytotoxic Peptide and Is Restored by Two Related PPR Motif Genes via Distinct Modes of MRNA Silencing. *Plant Cell* **2006**, *18*, 676–687. [CrossRef]
65. Brown, G.G.; Formanová, N.; Jin, H.; Wargachuk, R.; Dendy, C.; Patil, P.; Laforest, M.; Zhang, J.; Cheung, W.Y.; Landry, B.S. The Radish Rfo Restorer Gene of Ogura Cytoplasmic Male Sterility Encodes a Protein with Multiple Pentatricopeptide Repeats. *Plant J.* **2003**, *35*, 262–272. [CrossRef]
66. Koizuka, N.; Imai, R.; Fujimoto, H.; Hayakawa, T.; Kimura, Y.; Kohno-Murase, J.; Sakai, T.; Kawasaki, S.; Imamura, J. Genetic Characterization of a Pentatricopeptide Repeat Protein Gene, Orf687, That Restores Fertility in the Cytoplasmic Male-Sterile Kosena Radish. *Plant J.* **2003**, *34*, 407–415. [CrossRef] [PubMed]
67. Madugula, P.; Uttam, A.G.; Tonapi, V.A.; Ragimasalawada, M. Fine Mapping of Rf2, a Major Locus Controlling Pollen Fertility Restoration in Sorghum A1 Cytoplasm, Encodes a PPR Gene and Its Validation through Expression Analysis. *Plant Breed.* **2018**, *137*, 148–161. [CrossRef]
68. Wang, D.; Zhang, L.; Li, J.; Hu, G.; Wu, Q.; Jiang, H.; Huang, Z. The Restorer Gene for Soybean M-Type Cytoplasmic Male Sterility, Rf-m, Is Located in a PPR Gene-Rich Region on Chromosome 16. *Plant Breed.* **2016**, *135*, 342–348. [CrossRef]
69. Gao, B.; Ren, G.; Wen, T.; Li, H.; Zhang, X.; Lin, Z. A Super PPR Cluster for Restoring Fertility Revealed by Genetic Mapping, Homocap-Seq and de Novo Assembly in Cotton. *Theor. Appl. Genet.* **2022**, *135*, 637–652. [CrossRef] [PubMed]
70. Zhang, B.; Zhang, X.; Guo, L.; Qi, T.; Wang, H.; Tang, H.; Qiao, X.; Shahzad, K.; Xing, C.; Wu, J. Genome-Wide Analysis of Rf-PPR-like (RFL) Genes and a New InDel Marker Development for Rf1 Gene in Cytoplasmic Male Sterile CMS-D2 Upland Cotton. *J. Cotton Res.* **2018**, *1*, 1–11. [CrossRef]
71. Wang, C.; Lezhneva, L.; Arnal, N.; Quadrado, M.; Mireau, H. The Radish Ogura Fertility Restorer Impedes Translation Elongation along Its Cognate CMS-Causing MRNA. *Proc. Natl. Acad. Sci. USA* **2021**, *118*, e2105274118. [CrossRef] [PubMed]
72. Liu, Z.; Yang, Z.; Wang, X.; Li, K.; An, H.; Liu, J.; Yang, G.; Fu, T.; Yi, B.; Hong, D. A Mitochondria-Targeted PPR Protein Restores Pol Cytoplasmic Male Sterility by Reducing Orf224 Transcript Levels in Oilseed Rape. *Mol. Plant* **2016**, *9*, 1082–1084. [CrossRef]
73. Tang, H.; Luo, D.; Zhou, D.; Zhang, Q.; Tian, D.; Zheng, X.; Chen, L.; Liu, Y.-G. The Rice Restorer Rf4 for Wild-Abortive Cytoplasmic Male Sterility Encodes a Mitochondrial-Localized PPR Protein That Functions in Reduction of WA352 Transcripts. *Mol. Plant* **2014**, *9*, 1497–1500. [CrossRef]
74. Anisimova, I.N.; Alpatieva, N.V.; Karabitsina, Y.I.; Gavrilenko, T.A. Nucleotide Sequence Polymorphism in the RFL-PPR Genes of Potato. *J. Genet.* **2019**, *98*, 87. [CrossRef]
75. Phumichai, C.; Hosaka, K. Cryptic Improvement for Fertility by Continuous Selfing of Diploid Potatoes Using Sli Gene. *Euphytica* **2006**, *149*, 251–258. [CrossRef]
76. Peterson, B.A.; Holt, S.H.; Laimbeer, F.P.E.; Doulis, A.G.; Coombs, J.; Douches, D.S.; Hardigan, M.A.; Buell, C.R.; Veilleux, R.E. Self-Fertility in a Cultivated Diploid Potato Population Examined with the Infinium 8303 Potato Single-Nucleotide Polymorphism Array. *Plant Genome* **2016**, *9*, 1–13. [CrossRef] [PubMed]
77. De Jong, H.; Rowe, P.R. Inbreeding in Cultivated Diploid Potatoes. *Potato Res.* **1971**, *14*, 74–83. [CrossRef]
78. Hutten, R.C.B.; Soppe, W.J.J.; Hermsen, J.G.T.; Jacobsen, E. Evaluation of Dihaploid Populations from Potato Varieties and Breeding Lines. *Potato Res.* **1995**, *38*, 77–86. [CrossRef]

biology

MDPI

Article

Effects of Floral Characters on the Pollination Biology and Breeding System of *Iris setosa* (Iridaceae): A Cold-Tolerant Ornamental Species from Jilin Province

Xiyue Zhang [1], Ruoqi Liu [1], Lifei Chen [2], Tianhao Pei [3], Yu Gao [3], Xi Lu [2,*] and Yunwei Zhou [2,*]

1 College of Forestry and Grassland Science, Jilin Agricultural University, 2888 Xincheng Street, Changchun 130118, China; zxyiris123@163.com (X.Z.); 18843262894@163.com (R.L.)
2 College of Horticulture, Jilin Agricultural University, 2888 Xincheng Street, Changchun 130118, China; zexichen@163.com
3 College of Plant Protection, Jilin Agricultural University, Changchun 130118, China; 20231486@mails.jlau.edu.cn (T.P.); gaothrips@jlau.edu.cn (Y.G.)
* Correspondence: luxi@jlau.edu.cn (X.L.); yunweizh@jlau.edu.cn (Y.Z.)

Simple Summary: *Iris setosa* is a cold-tolerant ornamental species, whose pollination system remains vague. Therefore, the authors aimed to determine the impact of flower characteristics on pollination biology as well as on the breeding system. In this paper, the authors conducted an in-depth study on the phenological period and flowering dynamics, floral organ characteristics, pollen viability and stigma receptivity, breeding habits, and flower-visiting insects of *I. setosa*. Furthermore, our study project shows a high adaptive evolutionary trend between floral morphological characters and pollination biology of *I. setosa* species.

Abstract: Floral phenology and features are intricately linked to pollinator behavior and pollination systems. *Iris setosa* is one of the ornamental irises of the family Iridaceae with beautiful flowers and leaves, and little research has been reported on its pollination biology. This study analyzed how phenology, floral features, breeding systems, and pollinator visits affect reproductive success of *I. setosa* populations in Jilin Province. Field observations and pollination studies demonstrated that *I. setosa* reached the bud stage in late May, with an average flowering time of 30 days. The anthers were outwardly dehiscent toward the outer edge of the style branches. In herkogamy, the relative locations of the anthers and stigma remained unchanged during flower opening. The stamens matured first. The pollen was most viable and the stigmas were most receptive on the first day of flowering. The nectar had the maximum sugar content. The sexual reproduction system was mainly outcrossing, with some self-compatibility and a need for pollinators. After artificial self-pollination, fluorescent microscopy revealed the winding of pollen tubes. The predominant flower-visiting insects were *Apis mellifera*, *Megachile* sp., *Syrphus corollae*, *Episyrphus balteatus*, and *Lasioglossum* sp., among which *A. mellifera*, *Megachile* sp., and *Lasioglossum* sp. were effective pollinators. Understanding the pollination mechanisms and strategies of *I. setosa* provides basic reference data on the potential for reproduction, and conservation efforts.

Keywords: flowering characteristics; breeding system; flower-visiting insects; pollination; flower development

1. Introduction

Plants produce flowers until visitation yields the amount of pollen necessary to produce the maximum number of seeds that resources can support [1]; thus, plants are simultaneously limited by resources and pollen. The frequency and potential consequences of pollen limitation for plant populations and communities have been intensely explored during the last few decades [2,3]. The pollinator limitation of female reproduction occurs when an inadequate supply of pollen limits the fruit set below the level possible given the plant's available resources. A decrease in pollinator frequency (i.e., number of visits) likely decreases the quantity of pollen deposited onto the stigma [4], reducing the fruit and seed sets [5–7]. In addition, the plant's ability to attract pollinators via flower morphological features can be crucial to fitness [8]. The limitation of fruit and seed production due to insufficient pollinator visitation is common and ubiquitous across plants [9].

China is one of the distribution centers of *Iris*. Despite being the largest genus in the family Iridaceae, with more than 250 species [10], there is little information on the role that pollinators play in the reproductive success of *Iris*. *Iris setosa* is a rhizomatous geophyte that belongs to the genus *Iris* and is widely distributed in northern temperate zones, such as North America and East Asia [11]. The inner perianth of *I. setosa* is small and almost invisible compared with other species of the family. The derived features of Iridaceae include two morphological characteristics: a unifacial isobilateral leaf and the presence of only three stamens [12]. Over time, the inner and outer whorls of the tepals have become progressively differentiated in *Iris*. Stripes, dots, or color gradients on the outer tepals produce nectar guides, a distinct feature intended to attract pollination insects to the flowers. In addition, *Iris* is a typical nectar plant. The nectaries are typically found between the base of the corolla tube and the base of the style, or they are spread across the inner wall of the corolla tube [13]. Most known *Iris* species have nectaries, such as *Iris sibirica*, which uses nectar guides to attract insects [14,15]. Many species are primarily attracted by floral resources and perianth pigments, with various flower fragrances serving as an attractant. For some pollinators, however, flower physical characteristics, particularly functional floral symmetry, can be just as significant.

The intricate flowers of the Iridaceae family are finely adapted for pollination by various animals, including hummingbirds, sunbirds, beetles, butterflies, moths, wasps, and bees. This intimate connection between flower form and pollination biology reveals how the wide range of flower colors, shapes, and scents are vital to the lives of the species [16]. Hymenoptera (mainly bees) pollinate the majority of Iridaceae plants. For example, bumblebees are one of the main pollinators of *Iris lutescens* [17,18]. Plants depend on one insect species or a small number of ecologically related species for pollination [19]. The diversity of floral characteristics of different *Iris* species leads to different attraction behaviors of pollinators. For example, the loose corolla of *Iris tectorum* is compatible with the pollination behavior of the large insects *Xylocopa appendiculata* and *X. tranquebarorum* [20]; *Gladiolus* (Iridaceae) are primarily or exclusively pollinated by flies with elongated mouthparts [21]; the corolla of salty *Iris halophila* is more tightly packed, and, therefore, the only effective pollinators are *Anthophora* sp. [22]; and *Iris pseudacorus* has narrower pollination pathways and attracts pollinators such as *Episyrphus balteatus*, which are flatter in size [23]. Pollinator limitation has a substantial impact on the reproductive success of cross-pollinated *Iris* species [24]. Cross-pollination of *I. sibirica* is carried out by insects, mainly bumblebees, honeybees, and male solitary bees, and, less often, by wasps and ants [25]. This ensures the success of reproduction. Some Iridaceae plants are known to be pollinated by bees and wasps, apparently seeking nectar rewards, while others are pollinated by beetles and flies, apparently seeking carrion for brood sites or food [26].

In this paper, we measured observations of local floral visitors to better understand *I. setosa*'s pollination techniques. The impacts of floral morphology and reproductive systems on pollination were studied to better understand how these elements influence reproductive success. This research aimed to (i) characterize floral resources foraged by visitors and pinpoint the most important group of pollinators based on their foraging behavior and frequency of visits; (ii) perform pollination experiments simulating the high frequency of visits of pollinators across flowers, assess the reproductive success between treatments by comparing fruit sets, and follow the plant population throughout the entire season to describe its phenology; and (iii) define the breeding system of *I. setosa*.

2. Materials and Methods

2.1. Study Sites and Species

Iris setosa, is in the Iridaceae. The experiment was conducted from April to July 2023–2024 in Jilin Agricultural University, China.

2.2. Observations on Floral Phenology and Morphology

We observed flowering phenology for individual flowers and entire plants in April–July 2023–2024. The flowering of 87 *I. setosa* populations was defined as three stages (10% of plants flowering at the same time was the first flowering period, 50% of plants flowering at the same time was the full flowering period, and less than 10% still flowering was the last flowering period). On the first day of flowering, three out of 10 randomly selected plants with similar growth were used to measure the length and width of 30 flowers in terms of flower diameter, perianth segments, filaments, anthers, pistils and ovaries of 30 flowers were measured with vernier calipers.

2.3. Determination of Pollen Viability and Stigma Receptivity

Selected unopened buds were marked, and anthers and stigmas from five flowers were taken at day before bloom, first day of bloom, second day of bloom, third day of bloom. TTC (2, 3, 5-Triphenyltetramlium chloride, Beijing, China) staining was used to test pollen viability on various flowering days, and pollen viability was computed using pollen coloration rate [27]. Stigma receptivity of *I. setosa* was detected using the benzidine-hydrogen peroxide (Beijing, China) method and examined under a microscope. The results showed that the stigma was strongly receptive if a significant number of air bubbles developed surrounding it [28].

2.4. Determination of Pollen Viability at Different Storage Temperatures

In mid to early June 2023, the soon-to-be-split *I. setosa* anthers were collected in four parts. We loaded them into 5 mL centrifuge tubes with caps and sealed with a small amount of silica gel desiccant. They were stored at room temperature 25 °C, 4 °C, −20 °C and −80 °C. After 1, 3, 5, 7, 9, 11, 13 and 365 days from the date of storage, five anthers were extracted from each of the four different temperature treatments. TTC staining was performed to determine pollen viability and calculate pollen storage life. In the laboratory, *I. setosa* anthers were placed on sulfate paper to disperse pollen naturally. We gently flicked the anthers with forceps to collect the scattered pollen and placed on a glass slide. We added 1–2 drops of TTC stain with a rubber-tipped dropper and mixed well with a dissecting needle. We covered the slide and stain for 20 min, protected from light. Then, we observed in the low magnification of a light microscope. Pollen grains with high viability were dark red, pollen grains with low viability were light red, and pollen without viability was colorless.

2.5. Estimation of Outcrossing Index (OCI) and the Pollen-Ovule Ratio

The Outcrossing index (OCI) was estimated using the criteria suggested by Dafni. A [29], as followers: (1) Flower diameters ≤ 1 mm were recorded as 0, 1–2 mm as 1, 2.1–6 mm as 2, and >6 mm as 3. (2) Simultaneous maturation of both pistil and stamen, or pistil first, was recorded as 0, and stamen first, as 1. (3) Pistil and stamen are recorded as 0 for the same height and 1 for herkogamy. When OCI is 0, the breeding habit is cleistogamy. When OCI is 1, the breeding habit is self-pollination. When OCI is 2, the breeding habit is facultative selfing. When OCI is 3, the breeding habit is self-compatibility and pollinators are sometimes required. When OCI is greater than or equal to 4, the breeding habit is mainly outcrossing, with some self-compatibility and need pollinators.

In *I. setosa* flower buds, randomly selected stamens were immersed in 1 mol·L^{-1} HCl (Guangzhou, China). The anther walls were removed by hydrolysis in a water bath at 60 °C for 1 h. Distilled water was added to the stamen suspension and the volume was fixed to 10 mL. 10 μL of the suspension was pipetted onto a slide, and the amount of pollen per stamen was observed and measured microscopically. The pollen count per stamen was 1000 by multiplying the mean value X. The ovaries of randomly selected *I. setosa* flower buds were dissected through the carpels with a dissecting knife under a stereomicroscope (Leica DM2500, Wetzlar, Germany) 20 times, the number of ovules was counted and recorded, and the average value was computed. The type of *I. setosa* breeding system was evaluated using Cruden RW's criterion [30].

2.6. Sampling and Calculation of Nectar Characteristics

Flowers that opened on the same day were chosen to assess the nectar content and nectar sugar concentration in their natural state Wang [31]. Ten plants were chosen, and two neighboring flowers were chosen at random from each plant, yielding a total of 20 blooms for measurement. Flowers were chosen based on the following criteria: they had to be freshly opened and show no evident signs of insect feeding. Measurement of 24 h storage capacity: the bud-bearing meristems of *I. setosa* were bagged with different sizes of mesh bags, and nectar traits were measured on the first day after the flowers opened, and the nectar volume and sugar content of entire flowers were measured and estimated.

Nectar volume was measured by aspirating all nectar secreted by the nectar glands using a disposable capillary tube (5 μL/32 mm, ringcaps®, Hirschmann, Germany)and measuring the length of the column with a vernier caliper.

Sugar concentration was measured by aspirating nectar using a 5 μL capillary tube and measuring it with a hand-held glucometer (0–32% Brix, LICHEN, Shanghai, China). The temperature was 23.9–28.5 °C. The nectar volume is calculated by converting the length of the measured nectar column (L) to the nectar volume (μL, the length of a 5 μL capillary tube is 32 mm) using the following formula:

$$\text{Nectar volume} = L \times 5/32 \quad (1)$$

$$Y = 0.00226 + 0.00937X + 0.0000585X^2 \quad (2)$$

The sugar content of nectar (Z, mg) was calculated from the measured values of nectar volume, where X is the nectar sugar concentration (%) and Y is the sugar content of nectar per microliter (mg/μL), using the following equation [32]:

$$Z = V \times Y \text{ (V is the nectar volume, μL)} \quad (3)$$

2.7. Sampling and Preservation of Pollen Tubes

We used fluorescence monitoring to observe pollen tube germination and growth in natural conditions, as well as artificial self-pollination and artificial cross-pollination of *I. setosa*. *Iris setosa* styles and ovaries were collected at 1, 2, 4, 6, and 8 h after pollination. Four samples were taken at each time point and kept in FAA (Fuzhou, China) fixative at 4 °C.

The fixative was poured out and rinsed with distilled water 2–3 times before observation. 1 mol·L^{-1} NaOH (Tianjin, China) was added and placed in a 60 °C water bath to soften for about 1 h. 1% Aniline Blue (Water Soluble) was used for staining, and after staining for 24 h at room temperature under dark conditions, the samples were slowly placed on slides and the coverslips were pressed. The samples were observed under an Olympus microscope (Leica DM2500, Wetzlar, Germany) and photographed.

2.8. Observations on Flower Visiting Insect Species and Behavior

In order to understand the types of *I. setosa* flower visitors and their collection behaviors of pollen and nectar, the collection behaviors of different flower visitors were observed on sunny days in June 2023–2024 at 08:00–11:00 a.m. and 12:00–17:00 p.m. Two plots (1 × 1 m) were randomly established with approximately 30 flowers per plot. These plots were observed for 60 min per day for a total of 28 observations. A camera (Canon EOS 70D, Tokyo, Japan) was used to record the foraging behavior of several floral visitors. Furthermore, the number of flower visits and the number of flowers visited by different flower visitors per 30 min were recorded, and each flower visitor's visit rate (the number of times a flower was visited every 30 min) was calculated by dividing the total number of flowers observed by the number of flowers visited in each 30-min period. The primary pollinators were identified based on the frequency of flower visits and foraging behavior. Bees bearing pollen masses were caught and identified as prospective pollinators based on whether the insect touched the stigma's pollinable surface during flower visits, resulting in cross-pollination.

2.9. Artificial Pollination Test

Pollinator limitation studies were conducted on *I. setosa* populations in 2023 and 2024 by comparing hand-pollinated and naturally pollinated flowers. Potential animal pollinators or seed dispersers were identified through in-person observation and video camera use. The treatment is as follows:

T1 is control (CK) natural pollination. T2 treatment is direct bagging of flower buds to detect the need for pollinators in the natural state. T3 treatment is removal of males with forceps before flowering and bagging to detect the presence of apomixis. T4 treatment is removal of males with forceps before flowering without bagging to detect the extent to which pollinators contribute to seed set in the natural state. T5 treatment was to remove the male with forceps before flowering, set a bag, and artificial self-pollination on the next day to detect the presence of self-compatibility. T6 treatment was to remove the male with forceps before flowering, set a bag, and artificial geitonogamy on the next day. T7 treatment was to remove the male with forceps before flowering, set a bag, and artificial xenogamy on the next day. Each treatment group needed 20 flowers, bagged immediately after treatment, and the paper bag could be removed after 15–20 d. Fruit set rate, seed number and seed set rate of each group were counted.

2.10. Statistical Analysis

Data from this experiment were statistically analyzed using SPSS 26.0 (IBM Corp., Chicago, IL, USA) statistical analysis software. One-way ANOVA was used to analyze whether there were significant differences in the number of seeds and seed set between the different treatments of the pollination experiment.

3. Results

3.1. Flowering Characteristics and Flowering Period Characteristics

As shown in Figure 1, the average diameter of *Iris setosa* flowers was approximately 67.9 mm (Table 1). The outer tepals were three, approximately 45.1 mm long and 23.1 mm wide, with a pattern of blue-purple pinstripes at their bases as nectar guides (Figure 1A). The inner tepals were three, approximately 10.1 mm long and 4.5 mm wide (Figure 1B). The stamens were three, approximately 20.3 mm high, growing close to the stylar branches. The anthers were purple, outwardly dehiscent, and situated on the outer sides of the styles, just underneath the stigmas. The filaments were as long as the anthers and white with a purple halo. The stylar branches were flattened and arching with a pronounced bend (Figure 2). The style color was lighter than the tepals, forming a semi-enclosed passage with an outer perianth for pollinating insects to enter. The height was ca. 29.9 mm, the apical lobes were nearly square, and the margins were sparsely dentate. Three stigmas were located at the base of the apical lobes of the style, with a height of ca. 2.2 mm. The stigmas folded upward at maturity (Figure 2). The nectaries were located in the region between the base of the corolla tube and that of the style. The ovary was inferior, ca. 11.7 mm high, long-cylindrical, yellowish-green, and three-loculed, with median placentation and numerous anatropous ovules (Figure 1C).

Iris setosa began to sprout into the nutritive growth stage in late April in 2023–2024 (Figure 3A,B), entered the stem extraction stage in mid-May (Figure 3C), and began to show buds in late May (Figure 3D). In 2024, the initial flowering period was from 28 May to 1 June (Figure 3E), the blooming period was from 2 June to 13 June (Figure 3F), and the end flowering period was from 14 June to 23 June (Figure 3G). In 2023, the initial flowering period was from 25 May to 30 May, the blooming period was from 31 May to 11 June, and the end flowering period was from 12 June to 20 June (Figure 3G). Since the average temperature in May–June in 2023 was 3–4 °C higher than that in 2024, the overall flowering period was 2–3 days earlier than that in 2024. During single-flower opening, the buds gradually expanded (Figure 4A,B), and the outer perianth dispersed one by one, with the apical part of the perianth drooping until all three outer perianths drooped (Figure 4C–E). After 2–4 d of flowering, the whole perianth curled up, the stamens dried up (Figure 4) and gradually withered (Figure 4G–J), and the ovary of the pollinated flower began to expand (Figure 4K,L). The flowering periods of single flowers and single plants were 2–4 d and 10–14 d, respectively.

Figure 1. The characteristics of the flower of *I. setosa*. (**A**) The overall structure of the corolla of *I. setosa*, (**B**) top view of corolla, (**C**) floral unit. sti. Stigma, st. Stamen, pi. Pistil, hg. nectar guide, ct. Corolla tube, ov. Ovary, op. Outer perianth, ip. Inner perianth, an. Anther, fi. Filament.

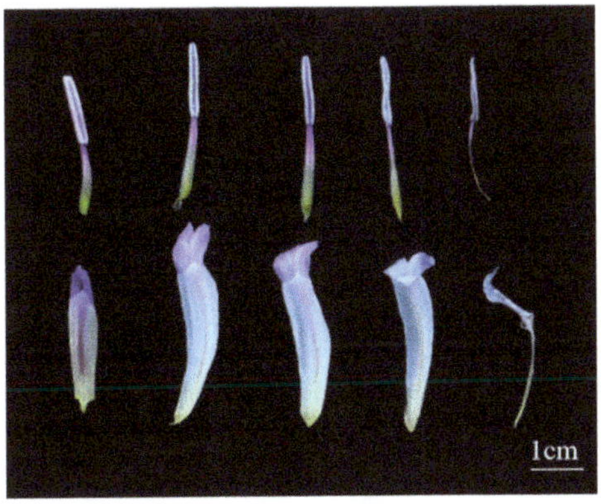

Figure 2. Changes of stamens and pistils in *I. setosa* during flowering. From left to right, 3 p.m. the day before bloom, 9 a.m. on the day of bloom, 3 p.m. on the day of bloom, 9 a.m. on the first day of bloom, and 9 a.m. on the second day of bloom.

Table 1. Floral parameters of *I. setosa*.

Floral Organ	mm Size
Outer perianth length	45.1 ± 3.8
Outer perianth breadth	23.1 ± 1.9
Inner perianth length	10.1 ± 0.6
Inner perianth breadth	4.5 ± 0.6
Ovary height	11.7 ± 1.8
Style height	29.9 ± 2.2
Stigma height	2.2 ± 0.4
Stamen height	20.3 ± 1.7
Flower diameter	67.9 ± 9.1

Data are mean ± SD.

Figure 3. Annual phenology of *I. setosa* group. (**A**,**B**) nutritive growth stage, (**C**) stem extraction stage, (**D**) squaring period, (**E**) initial flowering period, (**F**) blooming period, (**G**) end flowering period, (**H**,**I**) fruit period.

Figure 4. The flowering process of *I. setosa*. (**A**) bracts, buds, (**B**) buds about to open, (**C–E**) perianths are scattered one by one, (**F**) blooming flowers, (**G–J**) flowers gradually withered, (**K**,**L**) ovary enlarged.

3.2. Pollen Viability and Stigma Receptivity

Pollen viability is an important parameter for evaluating pollen quality. As shown in Table 2, TTC (2, 3, 5-triphenyltetramlium chloride) staining experiments were conducted on flowers collected on the day before bloom, the first day of bloom, the second day of bloom, and the third day of bloom. The results show that *I. setosa* pollen was already highly viable on the day before flowering (Figure 5). The first day of flowering had the highest viability of (96 ± 2)%, which was the best time for pollination. Overall, the viability showed an increasing and then decreasing trend (Table 2).

Table 2. Pollen viability and stigma receptive for *I. setosa*.

Time/d	Pollen Viability/%	Stigma Receptive
−1	91 ± 3	–
1	96 ± 2	+++
2	66 ± 9	++
3	29 ± 7	+

"–" indicates that the stigma is not receptive, "+" indicates that the stigma is receptive, "++" indicates that the stigma is highly receptive, "+++" indicates that the stigma is extremely receptive.

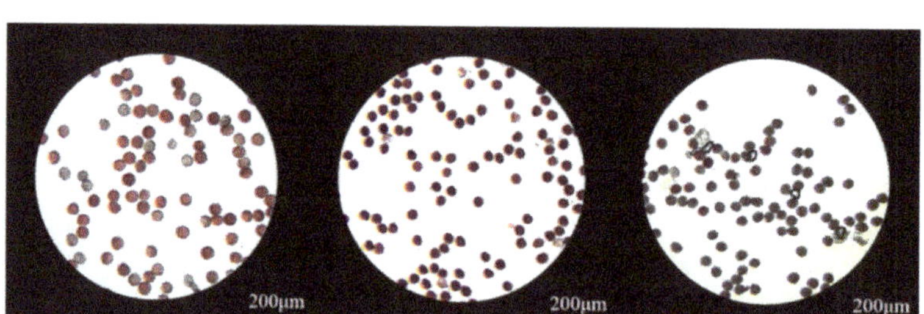

Figure 5. Pollen vitality of *I. setosa*.

Iris setosa stigmas were not receptive on the day before flowering (Figure 6 and Table 2). The stigmas were most receptive on the first day of flowering. Thereafter, stigmatic receptivity declined over time and could last until the third day of flowering, suggesting that *I. setosa* exhibits dichogamy.

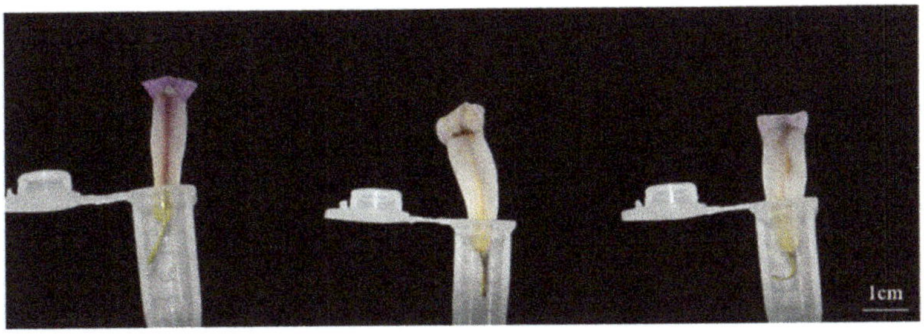

Figure 6. Stigma receptivity of *I. setosa*.

3.3. Effect of Different Storage Temperatures on the Determination of Pollen Viability

As shown in Figure 7, the pollen viability rate of *I. setosa* pollen was 62% on the first day at room temperature (25 °C), and pollen viability was lost after 24 h. The lifespan of *I. setosa* pollen could be prolonged under the three preservation conditions of low-temperature storage (4 °C), low-temperature freezing storage (−20 °C), and ultra-low-temperature freezing storage (−80 °C), which could last approximately 24 h, at most, under room-temperature storage conditions. The figure shows that the pollen viability rate of *I. setosa* under cryopreservation, cryogenic freezing, and ultra-low-temperature freezing did not significantly differ at the start of the preservation period but steadily decreased thereafter. The viability rate of *I. setosa* pollen did not significantly differ from that of cryopreserved and ultra-low-temperature-frozen pollen at the beginning of the storage period. The viability rate of cryopreserved and cryofrozen pollen gradually decreased with the increase in the storage time. Moreover, the viability rate of the pollen in cryopreserved and cryofrozen conditions decreased more slowly than that under other storage conditions. The longest storage time was 365 days, and the average pollen viability rate reached 74% in 365 d.

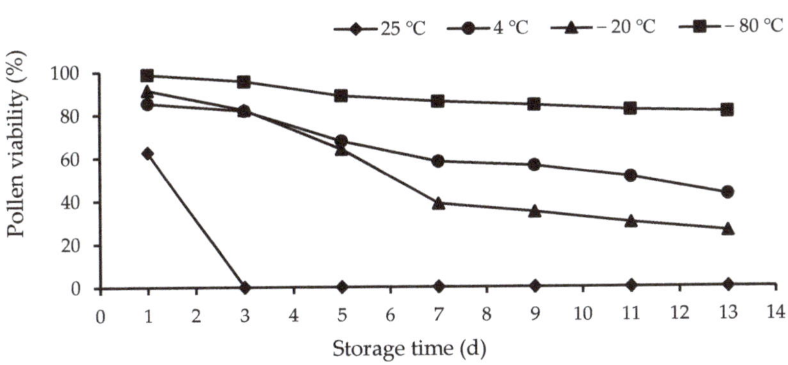

Figure 7. Changes in pollen viability of *I. setosa* under different storage temperatures.

3.4. Outcrossing Index and Pollen-Ovule Ratio

The pendant petals of the corolla of *I. setosa* are the most enlarged part of the flower and can provide a basis for measuring the size of the floral organ. The length and width of the pendant petals were about 45 mm and 23 mm, respectively (Table 1), and were larger than 6 mm. Additionally, the outcrossing index (OCI) was recorded as three, and the stamen-first was recorded as one. The anthers were spatially separated from the stigma, and the OCI was recorded as one. In summary, the OCI of *I. setosa* was five, indicating that the sexual reproduction system of *I. setosa* was mainly outcrossing, with some self-compatibility and need for pollinators. The average pollen number of single flowers of *I. setosa* was 34,200, the average ovule number was 88, and the pollen-ovule ratio (P-O) was 388.6, indicating that the sexual breeding system of *I. setosa* is facultative selfing or facultative xenogamy.

3.5. Nectar Characteristics

The calculation of the nectar volume and sugar content of *I. setosa* on the day of flowering in the natural state showed that the nectar volume was 19.6 ± 4 μL. The sugar concentration and content of the nectar were 17% and 3.5 ± 0.7 mg, respectively.

3.6. Fluorescence Microscopy of Pollen Tube Growth

The fluorescence observation of pollen tube growth was carried out on *I. setosa* after natural flowering and *I. setosa* after artificial self-crossing and cross-pollination, respectively. The results are shown in Figures 8–10.

The growth of pollen tubes after the natural flowering of *I. setosa* is shown in Figure 8. When *I. setosa* naturally flowered for 0 d in the morning, pollen was observed on the stigma of the pistil. At this time, the anthers dispersed the pollen completely, probably due to the insects feeding on the nectar or pollen, transferring the pollen they carried to the stigma (Figure 8A). By the afternoon, we observed pollen sprouting out of the pollen tubes and growing in bundles (Figure 8B). Based on the previous section, *I. setosa* pollen was most active on the day of flowering, prompting pollen tubes that continued to extend downward. By 1 d of natural flowering in the morning and 1 d in the afternoon, the pollen tubes reached the vicinity of the ovule. Subsequently, the pollen tubes entered the embryo sac to complete fertilization (Figure 8C,D).

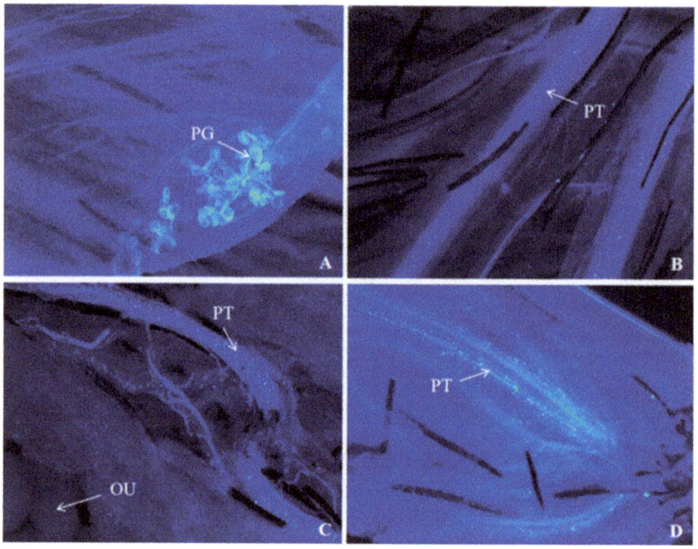

Figure 8. Fluorescence observation of pollen tube growth after natural flowering of *I. setosa*. (**A**) Natural flowering 0 d in the morning, (**B**) Natural flowering 0 d in the afternoon, (**C**) Natural flowering 1 d in the morning, (**D**) Natural flowering 1 d in the afternoon. PG: Pollen grain, PT: Pollen tube, OU: Ovule.

The results of the fluorescence observation of pollen tube growth after self-pollination demonstrate the following: After self-pollination for 1 h, *I. setosa* pollen sprouted pollen tubes on the stigma (Figure 9B). At 2 h of self-pollination, the pollen tubes grew in bundles (Figure 9C), and the pollen tube growth trajectory was irregular. A small number of pollen tubes were observed sprouting and entangled with each other in clusters after 4 h of pollination (Figure 9D), and pollen tube anomalies appeared. Pollen tubes continued to grow downward in the style (Figure 9E,F). A small number of pollen tubes appeared to be bent and broken within the channel (Figure 9G,H). After 6 h of pollination, the pollen tubes reached the ovule (Figure 9I). Hence, the pollen tubes took 4–6 h to cross the style, and no substances that inhibit pollen tube growth were present in the style.

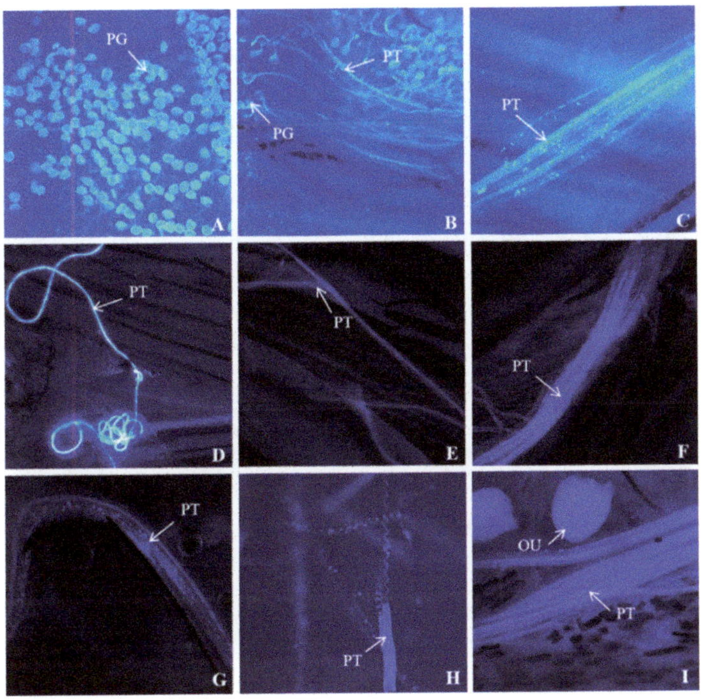

Figure 9. Fluorescence observation of pollen tube growth of *I. setosa* after artificial self-pollination. (**A**) Pollen morphology under fluorescence, (**B**) Pollen germinated on the stigma at 1 h after pollination, (**C**) Pollen tube bundle growth at 2 h after pollination, (**D**) Pollen tube folds intertwined at 4 h after pollination, (**E**) Pollen tubes entered the style, (**F**) Pollen tube continues to grow downward, (**G**) Pollen tube bending, (**H**) Pollen tube fracture, (**I**) Pollen tubes reaches the ovule at 6 h after pollination. PG: Pollen grain, PT: Pollen tube, OU: Ovule.

The results of the fluorescence observation of pollen tube growth after cross-pollination demonstrate the following: After cross-pollination for 1 h, *I. setosa* pollen sprouted pollen tubes on the stigma (Figure 10B). At 2 h of cross-pollination, the pollen tubes grew in bundles (Figure 10C). Numerous pollen tubes were observed to sprout and twist back and forth with each other 4 h after pollination (Figure 10D), and the pollen tubes continued to grow downward and appeared to bend (Figure 10E,F). The pollen tubes entered the bottom of the style and reached the ovule 6 h after pollination (Figure 10G,H). After 8 h of pollination, many pollen tubes wrapped around the ovule (Figure 10I).

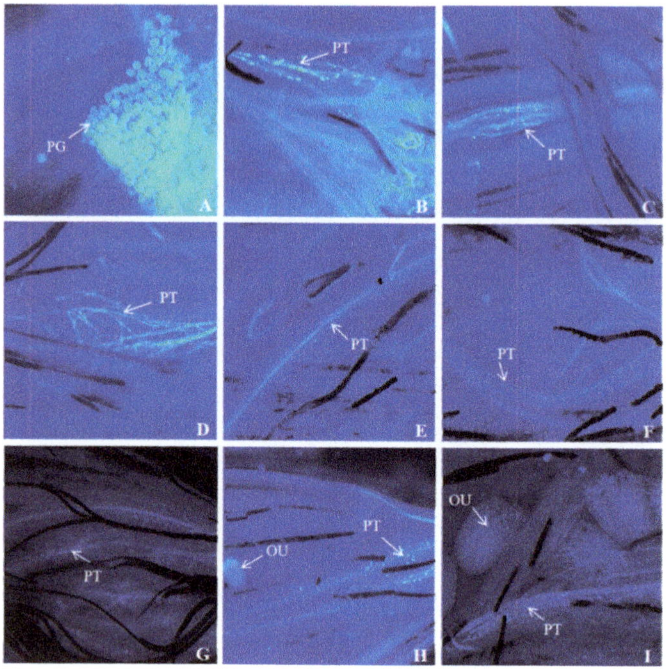

Figure 10. Fluorescence observation of pollen tube growth of *I. setosa* after artificial cross pollination. (**A**) Pollen morphology under fluorescence, (**B**) Pollen germinated on the stigma at 1 h after pollination, (**C**) Pollen tube bundle growth at 2 h after pollination, (**D**) Pollen tube folds intertwined at 4 h after pollination, (**E**) Pollen tube continues to grow downward, (**F**) Pollen tube bending, (**G**) Pollen tube reaches the bottom of the style, (**H**) Pollen tubes reaches the ovule at 6 h after pollination, (**I**) Pollen tube wrapped around the ovule at 8 h after pollination. PG: Pollen grain, PT: Pollen tube, OU: Ovule.

In summary, the insect vector can pass *I. setosa* pollen to the pistil stigma under natural conditions. However, pollen viability is an important prerequisite for pollen tube growth. The pollen sprouts pollen tubes only when viable. These pollen tubes gradually extend downward to the ovary ovule to complete fertilization. Then, the *I. setosa* plant can bear fruits after natural flowering. In the fluorescence observation of *I. setosa* after self-pollination, abnormal phenomena, such as pollen tube folding and twisting, breaking and bending, and dried pollen grains not sprouting, were found, which made it difficult to complete the fertilization of the pollen tubes and reduced the fruiting rate of *I. setosa* after self-pollination.

3.7. Observations of Flower-Visiting Insect Species and Behavior

Photographs of flower-visiting insects were taken during the experimental observations. As shown in Figure 11, the species of flower-visiting insects were mainly *Apis mellifera* (Figure 11A), *Megachile* sp. (Figure 11B), *Syrphus corollae* (Figure 11C), *Episyrphus balteatus* (Figure 11D), *Lasioglossum* sp. (Figure 11E), and others. Small numbers of flower-visiting insects such as Mordellidae (Figure 11F), ants, and beetles were also observed, alongside a small number of spiders, which may be predators of flower-visiting insects.

Figure 11. Flower visiting insect species. (**A**) *Apis mellifera*, (**B**) *Megachile* sp., (**C**) *Syrphus corollae*, (**D**) *Episyrphus balteatus*, (**E**) *Lasioglossum* sp., (**F**) Mordellidae.

Insect flower-visiting behavior is related to the characteristics of the flower itself. *Iris* is a typical insect-pollinated flower. The species, number, and frequency of insect visits are important factors affecting the fruiting rate. Apoidea and Syrphidae were the main pollinators of *Iris*. The peak time of flower visits was from 10:00 to 14:00 every day. We did not observe any insects within flowers during morning observations on rainy or windy days. The time taken by *Apis mellifera* to visit a flower in sunny weather ranged from 3 to 17 s. The number of visits to flowers in 30 min was 112 (Table 3). When the insects entered the pollination channel between the stigma and the outer perianth to feed on pollen or to suck up nectar in the corolla tube of the upper ovary, they carried a large amount of pollen via the hairs on their thoraxes and abdomens, and a large amount of pollen mass adhered to their legs (Figure 12A–C). The pollination was accomplished by touching the stigma of another flower during the next pollen or nectar collection.

Lasioglossum sp. visited a flower for 8–28 s under clear weather conditions and visited 32 times in 30 min. They entered the pollination channel upside down when collecting pollen, and the pollen grains adhered to their head and legs could touch the stigma for pollination (Figure 12D).

Syrphus corollae visited a single flower for 45 s^{-1} min 7 s on a sunny day, and the number of flower visits was 10 in 30 min. Their heads and antennae contacted the stigma first when feeding on pollen. However, their legs did not adhere to the excess pollen mass during the visits due to having fewer downy hairs on their bodies compared with bees (Figure 12E,F).

Episyrphus balteatus visited a single flower for 24–32 s on a sunny day and six times in 30 min. They were small in size, and their heads touched the stigma first when burrowing into the pollination channel to enter the flower to feed on pollen (Figure 12G).

Megachile sp. visited a single flower for 4–16 s on a sunny day, and the number of flower visits was 106 in 30 min. Its mouthparts were longer than those of other honeybees to suck nectar directly. The leaf-cutting bees entered the pollination channel directly from the front side or by holding the styles on the reverse side during the flower visit. The tomentum on their abdomens adhered to the pollen dispersed on the outer perianth during frontal visits (Figure 12H,I). On the next visit to the flower, they entered the channel by holding the style on the opposite side, and the pollen adhering to their abdomens contacted the stigma for pollination. A few flower-visiting insects, such as flower fleas, ants, and spiders, bit the corolla tube and sucked the nectar, but no pollen was collected.

Table 3. Visitors and their behavior on flowers of *I. setosa*.

Visitor	Visitation Rates	Number (%) of Visits	Visitor Type	Reward
Apis mellifera	0.438 ± 0.080 [a]	112 (42)	Pollinator	Nectar, pollen
Megachile sp.	0.406 ± 0.087 [a]	106 (40)	Pollinator	Nectar, pollen
Lasioglossum sp.	0.154 ± 0.039 [b]	32 (12)	Pollinator	Nectar, pollen
Syrphus corollae	0.080 ± 0.058 [c]	10 (4)	Pollen thief	pollen
Episyrphus balteatus	0.066 ± 0.052 [c]	6 (2)	Pollen thief	pollen

Visits per flower per 30 min (mean ± SE). Different letters show significant differences at $p < 0.05$ (GLMs). Visits of insect species percentage of the total visits is indicated in bracket.

Figure 12. Behaviour of flower-visiting insects. (**A–C**) The pollination process of *Apis mellifera* (p. pollen ball), (**D**) Flower Visiting Behaviour of *Lasioglossum* sp. (p. pollen ball), (**E,F**) The pollination process of *Syrphus corollae*, (**G**) Flower Visiting Behaviour of *Episyrphus balteatus*, (**H,I**) The pollination process of *Megachile* sp. (p. pollen ball).

3.8. Artificial Pollination Test Results

As shown in Table 4, seven different pollination treatments had significant effects on the fruit set, seed set, and seed number of *I. setosa*. The fruit set, seed set, and seed number of T2 and T3 were all zero, indicating that *I. setosa* could not self-pollinate nor be subjected to apomixis. The seed set in the T4 group was significantly different from that in T1. The fruit set in T6 and T7 was in the range of 95% and 100%, respectively, and the number of seeds and seed set were not significantly different from those in T1. Meanwhile, the number of seeds and seed set were significantly lower in T5 than in T1, T6, and T7. The results of T6 and T7 were similar, which indicated that the breeding system of *I. setosa* is based on outcrossing with a certain degree of self-compatibility.

Table 4. Results of Artificial Pollination Trial of *I. setosa*.

Pollination Test	Number of Flowers	Fruit Setting Rate/%	Seed Number/%	Seed Setting Rate/%
T1	20	90.00	85 ± 10^{a}	63 ± 11^{a}
T2	20	0.00	0 ± 0^{d}	0 ± 0^{d}
T3	20	0.00	0 ± 0^{d}	0 ± 0^{d}
T4	20	55.00	37 ± 23^{c}	29 ± 14^{c}
T5	20	85.00	59 ± 9^{b}	43 ± 10^{b}
T6	20	95.00	88 ± 5^{a}	63 ± 9^{a}
T7	20	100.00	86 ± 7^{a}	65 ± 9^{a}

T1 group natural pollination (CK), T2 group was directly bagged, T3 group emasculation and bagged, The T4 group not bagged and emasculation, The T5 group of artificial self-inbreeding, The T6 group of artificial geitonogamy, The T7 group of artificial xenogamy, Different lowercases in the same column indicate the significant ($p < 0.05$) difference.

4. Discussion

Different species of *Iris* attract different pollinators because of their varying floral characteristics. For example, *Lapeirousia anceps* (Iridaceae), which has a bimodal distribution of floral tube lengths, and the long-proboscid fly *Moegistorhynchus longirostris* (Nemestrinidae), its main pollinating insect, have a unimodal distribution of proboscis lengths and show a preference for the long-tube phenotype [33]. The pollinators of *Iris pumila* showed a higher affinity for flowers with taller flower stems and greater brightness and size of floral organs [34]. In this study, Hymenoptera insects belonging to *Apis mellifera*, *Megachile* sp., and *Lasioglossum* sp. were common pollinators of *Iris setosa*. We did not observe any insects within flowers during morning observations on rainy or windy days.

The base of the outer perianth often has appendages called barbs or nectar guides with spots and veins, which can be used as visual signals for insects to land and enter the pollination channel [35]. The thin bluish-purple stripes that extend outward on the outer perianth of *I. setosa* are called nectar guides (Figure 1A). Most of the pollen is also dispersed on the outer perianth after the anther of *I. setosa* opens to attract insects. The structure and function of the outer perianth of *I. setosa* have been completely specialized during its long-term evolution process, which may be an important means of attracting a wide range of insect pollinators. It is hypothesized that there has been an adaptive evolutionary process between the floral features of *I. setosa* and the corresponding flower-visiting insects, which has led to a synergistic evolution.

The scent of the flower and the pollination rewards (pollen and nectar) attract different pollinating insects, which promotes pollination and cross-breeding and enhances the fertility and resilience of the offspring so that they can reproduce successfully and evolve over time [36]. The flower size and nectar guide could act as visual signals, where large flowers/patches indicate larger tunnels (where pollinators shelter), increasing the probability of fruits and seeds [37]. It was found that the nectaries of *I. setosa* were located in the area between the base of the style and the inner part of the corolla tube. Insects were most active on the day of flower opening, and the volume of nectar from a single flower was 19.6 ± 4 μL. It was hypothesized that the nectaries of *I. setosa* secrete a large amount of nectar on the day of flower opening, which attracts pollinators to visit the flowers. The frequency of flower visits and the amount of pollination reward were the main factors affecting the pollination efficiency [38]. *Apis mellifera* and *Megachile* sp. had the highest number of visits within 30 min, which were significantly higher than the other visiting insects, followed by *Lasioglossum* sp. During the observation, it was found that the bees had a large amount of pollen mass adhered to their legs, and the downy hairs on their bodies carried pollen. The amount of pollen adhered to the contact surface of the stigma was larger during the next visit to the flower, thus increasing the success rate of pollination. Furthermore, the seed-setting rate of *I. setosa* was higher under natural pollination conditions.

Spatial (herkogamy) or temporal (dichogamy) separation of sex organs is a mechanism thought to limit self-pollination and promote cross-pollination [39]. Artificial pollination and field observations have shown that only natural pollination or artificially pollinated flowers produce capsules, whereas self-pollination after bagging does not produce fruit. Flower reward (nectar and pollen) is the main means of attracting insects to plants [40]. The seed set rate obtained from group T4 (not bagged and emasculated) was significantly lower than that of group T1 (natural pollination). The main behavior of flower-visiting insects of *I. setosa* was to feed on pollen and nectar during the period of flower visitation. When pollen was not available as a pollinator's reward, only nectar could be provided, and the number of flower visits by *Syrphus corollae* and *Episyrphus balteatus* was almost zero. The higher number of flower visits by Apoidea, and the presumed pollination reward of Syrphidae, which was mainly pollen (Table 3), reduced the attraction of flower-visiting insects, thus affecting the seed set. The number of seeds and the seed set did not significantly differ between T1 (natural pollination), T6 (artificial geitonogamy), and T7 (artificial xenogamy), which is presumed to be due to the high number of species and number of pollinators; therefore, there is no pollinator restriction. According to the ratio of pollen ovules of *I. setosa* and the OCI, the breeding systems were judged to be mainly outcrossing, with some self-compatibility and need for pollinators, whereas other Mediterranean *Iris* species are largely self-incompatible [41,42]. Therefore, domesticated *I. setosa* is suitable for planting in natural habitats where insects are more abundant and widespread, easier for natural sexual reproduction, avoids inbreeding decline, and is important for maintaining species diversity.

5. Conclusions

Iris setosa presents pollen and nectar as a reward to floral visitors. Successful sexual reproduction indicates that *Apis mellifera*, *Megachile* sp., and *Lasioglossum* sp. are its major effective pollinators due to the frequency of flower visits and the number of visits. The unique flower morphology of *I. setosa* facilitates cross-pollination, which attracts a variety of flower-visiting insects while also allowing it to adapt to these insects' characteristics. In the process of flower visits, these insects complete pollen transfer and promote the effective reproduction of *I. setosa*. Moreover, the different pollination efficiencies and methods of the different insects provide a genetic diversity of *I. setosa* populations, which enables these populations to have stronger adaptive ability and survivability in the face of environmental changes.

Author Contributions: Conceptualization, X.Z. and X.L.; methodology, X.Z. and X.L.; software, X.Z.; validation, X.Z.; formal analysis, X.Z.; investigation, X.Z. and R.L.; resources, X.L. and Y.Z.; data curation, X.Z.; writing-original draft preparation, X.Z.; writing-review and editing, X.L., Y.Z., T.P. and Y.G.; visualization, X.Z.; supervision, X.L., Y.Z. and L.C.; project administration, X.L. and Y.Z.; funding acquisition, X.L. and Y.Z. All authors have read and agreed to the published version of the manuscript.

Funding: This research was funded by Jilin Provincial Science and Technology Department's Key Research and Development Projects "breeding and demonstration of good varieties (lines) of Changbai Mountain *Iris*", grant number 20240303102NC.

Institutional Review Board Statement: Not applicable.

Informed Consent Statement: Not applicable.

Data Availability Statement: Data are contained within the article.

Acknowledgments: The authors acknowledge assistance in the field, greenhouse, and lab from Liu Ruoqi, Liu Yang, Yu Jiahui, and Lu Siyu; and Lu Xi, Zhou Yunwei for their help in writing this paper.

Conflicts of Interest: The authors declare no conflict of interest.

References

1. Haig, D.; Westoby, M. On Limits to Seed Production. *Am. Nat.* **1988**, *131*, 757–759. [CrossRef]
2. García-Camacho, R.; Totland, Ø. Pollen Limitation in the Alpine: A Meta-Analysis. *Arct. Antarct. Alp. Res.* **2009**, *41*, 103–111. [CrossRef]
3. Harder, L.D.; Aizen, M.A. Floral adaptation and diversification under pollen limitation. *Philos. Trans. R. Soc. Lond. B Biol. Sci.* **2010**, *365*, 529–543. [CrossRef] [PubMed]
4. Ashman, T.L.; Knight, T.M.; Steets, J.A.; Amarasekare, P.; Burd, M.; Campbell, D.R.; Dudash, M.R.; Johnston, M.O.; Mazer, S.J.; Mitchell, R.J.; et al. Pollen limitation of plant reproduction: Ecological and evolutionary causes and consequences. *Ecology* **2004**, *85*, 2408–2421. [CrossRef]
5. Bierzychudek, P. Pollinator Limitation of Plant Reproductive Effort. *Am. Nat.* **1981**, *117*, 838–840. [CrossRef]
6. Pellegrino, G.; Noce, M.E.; Bellusci, F.; Musacchio, A. Reproductive biology and conservation genetics of *Serapias vomerace* (Orchidaceae). *Folia Geobot.* **2006**, *41*, 21–32. [CrossRef]
7. Aizen, M.A.; Harder, L.D. Expanding the limits of the pollen-limitation concept: Effects of pollen quantity and quality. *Ecology* **2007**, *88*, 271–281. [CrossRef]
8. Glaettli, M.; Barrett, S.C.H. Pollinator responses to variation in floral display and flower size in dioecious *Sagittaria latifolia* (Alismataceae). *New Phytol.* **2008**, *179*, 1193–1201. [CrossRef] [PubMed]
9. Alcaraz, M.L.; Hormaza, J.I. Inadequate Pollination Is a Key Factor Determining Low Fruit-to-Flower Ratios in Avocado. *Horticulturae* **2024**, *10*, 140. [CrossRef]
10. Mathew, B. *The Iris*; Professional Horticulture, Chartered Institute of Horticulture: New York, NY, USA, 1981; Volume 4, pp. 91–92.
11. Lê, H.G.; Hwang, B.S.; Choi, J.S.; Jeong, Y.T.; Kang, J.M.; Võ, T.C.; Oh, Y.T.; Na, B.K. *Iris setosa* Pall. ex Link Extract Reveals Amoebicidal Activity against Acanthamoeba castellanii and Acanthamoeba polyphaga with Low Toxicity to Human Corneal Cells. *Microorganisms* **2024**, *12*, 1658. [CrossRef] [PubMed]
12. Goldblatt, P. Phylogeny and classification of Iridaceae. *Ann. Mo. Bot. Gard.* **1990**, *77*, 607–627. [CrossRef]
13. Rudall, P.J.; Manning, J.C.; Goldblatt, P. Evolution of Floral Nectaries in Iridaceae. *Ann. Mo. Bot. Gard.* **2003**, *90*, 613–631. [CrossRef]
14. Crisan, I.; Vidican, R.; Plesa, A.; Mihăiescu, T. Phytoremediation Potential of *Iris* spp. *Bull. Univ. Agric. Sci. Vet. Med. Cluj-Napoca* **2021**, *78*, 1. [CrossRef]
15. Konarska, A. Morphological, anatomical, ultrastructural, and histochemical study of flowers and nectaries of *Iris sibirica* L. *Micron* **2022**, *158*, 103288. [CrossRef] [PubMed]
16. Goldblatt, P.; Manning, J.C. *The Iris Family: Natural History and Classification*; Timber Press: Portland, OR, USA, 2008.
17. Segal, B.; Sapir, Y.; Carmel, Y. Fragmentation and Pollination Crisis in the Self-Incompatible *Iris Bismarckiana* (IRIDACEAE), with Implications for Conservation. *Isr. J. Ecol. Evol.* **2006**, *52*, 111–122. [CrossRef]
18. Imbert, E.; Wang, H.; Anderson, B.; Hervouet, B.; Talavera, M.; Schatz, B. Reproductive biology and colour polymorphism in the food-deceptive *Iris lutescens* (Iridaceae). *Acta Bot. Gall.* **2014**, *161*, 117–127. [CrossRef]
19. Goldblatt, P.; Manning, J.C. Radiation of pollination systems in the Iridaceae of sub-Saharan Africa. *Ann. Bot.* **2006**, *97*, 317–344. [CrossRef] [PubMed]
20. Liu, Z.C.; Jiao, Z.J.; Dong, X.S.; Dai, J.X. Floral Syndrome and Breeding System of *Iris tectorum*. *Acta Hortic. Sin.* **2011**, *38*, 1333–1340.
21. Goldblatt, P.; Manning, J.C. The Long-Proboscid Fly Pollination System in *Gladiolus* (Iridaceae). *Ann. Mo. Bot. Gard.* **1999**, *86*, 758–774. [CrossRef]
22. Liu, L.L.; Cao, X.D.; Xu, Z.R.; Zhang, J.F.; Wu, Y.H. Studies on Floral Syndrome and Mating System of *Iris halophila*. *Acta Agrestia Sin.* **2021**, *29*, 2733–2741. [CrossRef]
23. Fan, S.R.; Fu, J.H.; Zhuang, Q.R.; Zhang, Y.X.; Wang, Y.J.; Zhang, T.; Liu, Q.Q.; Tian, S.Q.; Yuan, H.Y. Studies on Floral Syndrome and Breeding Habits of *Iris pseudacorus*. *Acta Agrestia Sin.* **2023**, *31*, 96–104. [CrossRef]
24. Liu, R.; Gao, Y.; Fan, Z.; Wang, X.; Xiao, J.; Zhang, Q. Within-day temporal isolation of two species of *Iris* (Iridaceae) sharing the same pollinator. *Biol. J. Linn. Soc.* **2020**, *130*, 447–457. [CrossRef]
25. Szőllősi, R.; Medvegy, A.; Benyes, E.; Németh, A.; Mihalik, E. Flowering phenology, floral display and reproductive success of *Iris sibirica*. *Acta Bot. Hung.* **2011**, *53*, 409–422. [CrossRef]
26. Goldblatt, P.; Manning, J.C. Systematics and biology of the African genus *Ferraria* (Iridaceae: Irideae). *Bothalia* **2011**, *41*, 1–40. [CrossRef]

27. Yuan, S.-C.; Chin, S.-W.; Lee, C.-Y.; Chen, F.-C. Phalaenopsis pollinia storage at sub-zero temperature and its pollen viability assessment. *Bot. Stud.* **2018**, *59*, 6. [CrossRef] [PubMed]
28. Li, J.J.; Ye, C.L.; Shang, X.C.; Wang, J.; Zhang, B.; Wang, Z.Z.; Zhang, G.T. Study on breeding and pollination characteristics of Gleditsia sinensis. *Zhongguo Zhong Yao Za Zhi* **2018**, *43*, 4831–4836. [CrossRef] [PubMed]
29. Dafni, A. *Pollination Ecology: A Practical Approach*; The Practical Approach Series; Oxford University Press: Oxford, UK, 1992; Volume 6, pp. 763–778, ISBN 0-19-963298-7A.
30. Cruden, R.W. Pollen-ovule ratios: A conservative indicator of breeding systems in flowering plants. *Evolution* **1977**, *31*, 32–46. [CrossRef] [PubMed]
31. Wang, J.; Liu, W.; Luo, G.; Li, Z.; Zhao, C.; Zhang, H.; Zhu, M.; Xu, Q.; Wang, X.; Zhao, C.; et al. Synergistic effect of well-defined dual sites boosting the oxygen reduction reaction. *Energy Environ. Sci.* **2018**, *11*, 3375–3379. [CrossRef]
32. Pyke, G.H.; Ren, Z.-X.; Trunschke, J.; Lunau, K.; Wang, H. Salvage of floral resources through re-absorption before flower abscission. *Sci. Rep.* **2020**, *29*, 15960. [CrossRef] [PubMed]
33. Anderson, B.; Pauw, A.; Cole, W.W.; Barrett, S.C.H. Pollination, mating and reproductive fitness in a plant population with bimodal floral-tube length. *J. Evol. Biol.* **2016**, *29*, 1631–1642. [CrossRef]
34. Budečević, S.; Hočevar, K.; Jovanović, S.M.; Vuleta, A. Phenotypic Selection on Flower Traits in Food-Deceptive Plant *Iris pumila* L. The Role of Pollinators. *Symmetry* **2023**, *15*, 1149. [CrossRef]
35. Crişan, I.; Vidican, R.; Oltean, I.; Stoie, A.; Stoian, V. *Iris* spp. flower visitors: Pollinators vs. nectar thieves. *Rom. J. Grassl. Forage Crops* **2018**, *17*, 11–20, ISSN 2068-3065.
36. Tong, Z.Y.; Wang, X.P.; Wu, L.Y.; Huang, S.Q. Nectar supplementation changes pollinator behaviour and pollination mode in *Pedicularis dichotoma*: Implications for evolutionary transitions. *Ann. Bot.* **2019**, *123*, 373–380. [CrossRef]
37. Wang, B.; Ma, G.; Lin, S.; He, X.; Chen, B.; Li, H.; Huang, L.; Yang, Y.; Wei, J. Floral Biology of *Aquilaria sinensis* (Lour.) Spreng. *Horticulturae* **2024**, *10*, 109. [CrossRef]
38. Yuan, B.; Hu, G.X.; Zhang, X.X.; Yuan, J.K.; Fan, X.M.; Yuan, D.Y. What Are the Best Pollinator Candidates for Camelia oleifera: Do Not Forget Hoverflies and Flies. *Insects* **2022**, *13*, 539. [CrossRef] [PubMed]
39. Barranco, D.; Arroyo, J.; Santos-Gally, R. Avoiding sexual interference: Herkogamy and dichogamy in style dimorphic flowers of *Narcissus broussonetii* (Amaryllidaceae). *AoB Plants* **2019**, *11*, plz038. [CrossRef] [PubMed]
40. Hicks, D.M.; Ouvrard, P.; Baldock, K.C.; Baude, M.; Goddard, M.A.; Kunin, W.E.; Mitschunas, N.; Memmott, J.; Morse, H.; Nikolitsi, M.; et al. Food for Pollinators: Quantifying the Nectar and Pollen Resources of Urban Flower Meadows. *PLoS ONE* **2016**, *11*, e0158117. [CrossRef]
41. Arafeh, R.; Sapir, Y.; Shmida, A.; Iraki, N.; Fragman, O.; Comes, H. Patterns of genetic and phenotypic variation in *Iris haynei* and *I. atrofusca* (*Iris* sect. Oncocyclus = the royal irises) along an ecogeographical gradient in Israel and the West Bank. *Mol. Ecol.* **2002**, *11*, 39–53. [CrossRef] [PubMed]
42. Sapir, Y.; Shmida, A.; Ne'eman, G. Morning floral heat as a reward to the pollinators of the Oncocyclus irises. *Oecologia* **2006**, *147*, 53–59. [CrossRef]

biology

MDPI

Article

Impact of Seasonal Atmospheric Factors and Photoperiod on Floral Biology, Plant–Pollinator Interactions, and Plant Reproduction on *Turnera ulmifolia* L. (Passifloraceae)

Ujjwal Layek [1], Nandita Das [2], Arabinda Samanta [3] and Prakash Karmakar [4,*]

1 Department of Botany, Rampurhat College, Rampurhat 731224, West Bengal, India; layekujjwal@yahoo.co.in
2 Centre for Life Sciences, Vidyasagar University, Midnapore 721102, West Bengal, India; nandita.das.mou15@gmail.com
3 Department of Botany, Jhargram Raj College, Jhargram 721507, West Bengal, India; samanta24bot@gmail.com
4 Department of Botany & Forestry, Vidyasagar University, Midnapore 721102, West Bengal, India
* Correspondence: prakashbot1973@gmail.com

Simple Summary: The study depicted the influences of atmospheric factors and photoperiod on floral biology, plant–pollinator interactions, and the reproductive success of *Turnera ulmifolia*. Temperature, light, humidity, and day length enhanced flowering intensity, pollen, and ovule production, while flower longevity decreased with higher temperature, humidity, and longer day length. Visitor traits positively correlated with flowering intensity and were negatively influenced by rainfall. Effective pollinators were *Amegilla zonata, Borbo cinnara, Halictus acrocephalus, Nomia* (*Curvinomia*) *strigata,* and *Tetragonula iridipennis*. They pollinated the flowers via the ventral thorax and abdomen and also through their legs, antennae, proboscis, and wings. *Nomia* (*Curvinomia*) *strigata* also employed pulsatory pollination. The reproductive fitness of the plant species was highest in the hot summer and lowest during the cold winter seasons.

Abstract: Reproductive traits and plant–pollinator interactions largely depend on seasonal weather conditions, which are species-specific. *Turnera ulmifolia* is an ornamental plant distributed worldwide. There is little information about plant species' reproductive ecology and environmental factors' impact on it. Here, we aimed to examine the effects of seasonal atmospheric factors (e.g., temperature, light, relative humidity, rainfall) and photoperiod on flowering, interactions with flower visitors, and the reproductive success of *Turnera ulmifolia* in West Bengal, India. Flowering intensity peaked in hot summers and dropped in cold winters, correlating positively with temperature and humidity. Flower opening and closing occurred earlier on hot days, while flower longevity increased in winter, showing a negative correlation with temperature and humidity. Pollen and ovule production were lower in cold weather, positively linked to temperature and humidity. The self-compatible plant was moderately dependent on pollinators and had no pollination deficit in open conditions. Visitor abundance, richness, and diversity varied season-wise, with higher values during spring–summer. Based on pollinating agents, the plant showed multiple pollination modes (e.g., melittophily, myophily, myrmecophily, and psychophily). Effective pollinators were *Amegilla zonata, Borbo cinnara, Halictus acrocephalus, Nomia* (*Curvinomia*) *strigata*, and *Tetragonula iridipennis*. The fruit set (%) did not differ significantly season-wise, but the seed set remained higher in the hot days of summer than in cold winter. Therefore, it can be concluded that atmospheric factors and photoperiod significantly impact floral traits, plant–pollinator interactions, and plant reproduction.

Keywords: flowering intensity; flowering phenology; melittophily; myrmecophily; psychophily; pulsatory pollination

https://doi.org/10.3390/biology14010100

1. Introduction

Flowering is an essential plant life cycle phase that strongly affects plant fitness [1,2]. Flowering traits (e.g., flowering phenology and floral morphological and functional traits) are crucial characteristics that determine the reproductive efficacy of plants and are essential for assessing biodiversity threats under changing environmental conditions [3–5]. Plants display various flowering patterns (e.g., timing, duration, and frequency) [6,7]. Flowering patterns determine the number of outcrossing, mates, near-neighbour matings, and reproductive output [8,9].

Several aspects of floral biology, including pollen vitality and stigma receptivity, play a crucial role in the reproductive success of a population and are influenced by the timing of flowering [10,11]. The floral traits (including flower longevity, pollen viability, and stigma receptivity) and the overlap periods of male and female functional traits (pollen viability and stigma receptivity) vary across various flowering times and habitats [12,13]. Studies of these floral traits are fundamental prerequisites for understanding the reproductive constraints that affect a given population.

In addition to endogenous genetic components, flowering is determined by various environmental factors, such as day length, temperature, light intensity, moisture, and stress [14,15]. Photoperiod (i.e., day length) is a prime controlling factor for flower induction for many plants [16,17]. The light intensity can affect photosynthesis, phytohormone production, and plant growth [18,19], influencing many plants' flowering [20,21]. Flowering may be reduced or delayed as irradiance is reduced [22]. Supplementary light irradiance can differentially influence flowering depending on the plant species and light quality. Some researchers (e.g., Armitage and Tsujita [23]) found that supplemental high-pressure sodium (HPS) lighting hastened flowering in some plants compared with plants grown under natural daylight conditions. Air temperature influences flowering through several modes of action. Here, it is essential to distinguish between vernalisation (i.e., the prolonged cold exposure needed for floral transition in many plants) and ambient temperature (i.e., the non-stressful temperature range for a species). In plants, ambient temperature fluctuations impact various physiological and developmental responses—including pigment formation, cell organelle development, net CO_2 assimilation rates, transpiration, flowering, and seed yield [24–26]. Atmospheric humidity also affects floral traits, including anther dehiscence and stigma receptivity [27]. Rainfall significantly affects flowering phenology and floral resource availability for visitors [28,29]. Barrett and Brown [28] showed that absent or delayed rain altered phenological events and resulted in the decreased production of flowers.

Flower-visiting insects are another factor that constitutes one of the most critical interactions for plant reproduction. Plant–pollinator interactions highly depend on floral traits [30,31] and environmental factors [32]. This interaction may drive the evolution of flowering time by influencing competition for pollinators, potentially favouring asynchronous flowering to reduce competition or encouraging synchronous flowering to enhance pollinator efficiency and service quality [1,33]. Moreover, flowering time and synchronisation shifts are linked to the regularity and behavioural patterns of insect visits to flowers, with insect activity patterns typically aligning closely with the flowering phenology of related plant species [2,34]. Changes in the behaviour patterns of flower-visiting insects severely impact plant reproduction [35]. Investigating the relationships among flowering phenology, environmental factors, and insect visitors can offer valuable insights into the selective pressures shaping the evolution of flowering traits.

Turnera ulmifolia L. (Passifloraceae) is a polymorphic complex native to the New World tropics, composed of heterostylous and homostylous forms [36]. The majority of heterostylous varieties of *Turnera ulmifolia* are ruderal weeds. In contrast, varieties *angustifolia*, *elegans*, and *intermedia* are also used as garden ornamentals, and man has played a significant role in expanding their ranges. The plant has many pharmaceutical activities, including anti-microbial [37], antioxidant [38], anti-inflammatory [39], and anti-ulcerogenic [39]. However, little information is available about its floral and pollination biology (e.g., Torres-Hernández et al. [40]). In Asian countries, there are no specific records about its reproductive fates. Knowledge of the reproductive biology of a plant species is necessary to manage its population. Therefore, research on floral biology and plant-pollinator interactions on *Turnera ulmifolia* is demanding.

Here, we aimed to provide new information about the reproductive biology of *Turnera ulmifolia* and the impacts of environmental factors on floral biology, plant-pollinator interactions, and plant reproduction. The following aspects were investigated: (a) floral biology, (b) floral visitors and pollinators, and (c) impacts of seasonal atmospheric variables on plant reproduction. We hypothesised that seasonal atmospheric factors and photoperiod might affect the floral biology of *Turnera ulmifolia* and thereby influence the plant-pollinator interactions and reproductive fitness of the plant species.

2. Materials and Methods

2.1. Study Area and Plant Species

We carried out the present works in Bolpur (23.67° N and 87.72° E) of the Birbhum district, West Bengal, India. Some flower samples were also collected from the Vidyasagar University campus's surrounding vegetation (Midnapore, Paschim Medinipur, West Bengal; 22.43° N and 87.30° E) for the microscopic study of flower morphological traits. The study area of Bolpur characterises medium-density vegetation [41]. The majority of regions cover human habitats. The study zone exhibits hot summer (April–June, day temperatures at around 35–42 °C), following monsoon (July–August, annual rainfall 1479.9 mm), pleasant autumn (September–mid-October), late autumn (mid-October–mid-November), cold winter (mid-November–January, day temperatures at around 7–15 °C), and spring (February–March).

The present works were conducted on *Turnera ulmifolia* L. var. *angustifolia* (Passifloraceae) during 2022–2023. The plant is native to the New World tropics and distributed worldwide. It is a small shrub that grows erect on roadside walls, cement blocks, and rocks. This homostylous variety is also cultivated as an ornamental.

2.2. Environmental Factors

We recorded seasonal atmospheric factors (temperature, light intensity, relative humidity, and rainfall) and photoperiod (i.e., day length) within the study area using handy instruments (thermometer, photometer, and hygrometer, for recording temperature, light intensity, and relative humidity, respectively). Data were taken throughout the day, and average values on a sampling day were considered. On a sampling day, rainfall was recorded as an ordinal number (0—no rainfall, 1—very little rainfall, 2—little rainfall, 3—moderate rainfall, 4—heavy rainfall, and 5—very heavy rainfall).

2.3. Floral Biology

Flowering phenological data (flowering frequency, time, flowering pattern) were recorded (according to Gentry [6] and Hopkins [42]). The flowering intensity was measured as the number of freshly opened flowers per day per individual. For this, we selected old, medium, and young individuals (but reached the reproductive phase). We recorded flower opening and closing times on a sampling day (n = 3 flowers per sampling day; 16 sampling days per season). Observations were conducted at one-hour intervals. Then, flower longevity (time between flower opening and closing) was estimated for each season.

Flower morphology was studied with the help of microscopes (compound light microscope: Primo Star, Zeiss; stereo microscope: Stemi 508, Zeiss; scanning electron microscope: Merlin, Zeiss). The methodologies for the SEM study are given in Table S1. All anthers of a flower were taken into a vial before anther dehiscence in the early morning (one flower on a sampling day) to determine pollen grain per flower and preserved in 70% ethanol. During counting pollen grains, anthers were crushed with a glass rod to release the pollen. After shaking, 10 μL of pollen solution was taken on a glass slide, covered with a square-shaped cover slip, and the pollen grains were counted at 10× magnification with a compound light microscope. Then, we estimated the number of pollen grains per flower by considering the total volume of pollen solution. We studied pollen morphology using light microscopy and scanning electron microscopy (see Table S1). Three flowers (one per individual) were randomly chosen on a sampling day (16 sampling days per season) to determine ovule numbers. The ovaries were dissected with a scalpel, placed in a drop of water on glass slides, and slightly pressed to spread out. Ovules were counted under a dissecting magnifying glass.

To assess pollen viability, we used two staining techniques: (i) 2,3,5-triphenyl tetrazolium chloride (TTC) [43] and (ii) iodine potassium iodide (IKI) [44]. A total of 1% TTC (0.1 g TTC and 6 g sucrose dissolved in 10 mL distilled water) was prepared, and a drop of the mixture was taken on a clean glass slide. Pollen grains (fresh pollen collected from a flower during opening, one sample per sampling day) were added to this mixture with a brush and covered with a coverslip. After two hours of incubation, the staining of pollen grains was observed under a compound microscope (counted ≥100 pollen grains). The appearance of a red colour indicated viability, and other pollens (i.e., light red, no change, black, or yellow) were regarded as non-viable. The second staining method dissolved 0.1 g potassium iodide and 0.05 g iodine in 10 mL distilled water for the IKI solution. Pollen grains were added to this solution, and after five minutes of incubation, pollen grains were observed under a compound microscope. Pollen grains stained dark (dark red or brown colour) were considered viable.

Pollen germinability was tested by in vitro germination with an agarose–sucrose medium [45]. We used 10% sucrose, 0.5% agar, and 5 ppm boric acid in the medium. A small amount of medium was taken into a Petri dish, and bulk fresh pollen grains were added to the medium and incubated for 24 h at 25 °C in darkness. Then, pollens with medium were taken on a slide and observed under a compound microscope (counted ≥100 pollen grains). We considered pollen as germinated if the pollen tube length was greater than the diameter of the pollen grains.

Stigma receptivity was checked by using hydrogen peroxide [46]. Fresh flowers (one per time interval of 1 h covering throughout flower longevity) were collected. Stigmas were excised from the flowers using razor blades and placed on a glass slide. Then, a drop of H_2O_2 solution was added. After 10–15 s, stigma receptivity was observed under a dissecting microscope. The appearance of many bubbles indicates receptivity. More bubbles indicated stronger stigma receptivity.

2.4. Mating System and Plant Reproduction

To determine the mating system, we carried out five pollination treatments: (i) open pollination, (ii) spontaneous auto-pollination, (iii) manual selfing, (iv) manual crossing, and (v) supplementary pollination. We selected flower buds in the late afternoon, and they were marked. We bagged the selected buds (for auto-pollination and manual pollination treatments) with nylon netting until the senescence of floral parts. We utilised 40 flowers for each treatment during summer, except open pollination, which we conducted for each season with 40 flowers. If butterfly larvae damaged selected flowers, we replaced them by selecting new flower buds for that particular treatment. For manual selfing, we uncovered the net at 7.00–8.00 h, added pollen grains to the stigmas of the same flowers, and immediately re-bagged them. Buds selected for cross-pollination were emasculated before anther dehiscence. For the manual crossing, pollen grains were taken from the flowers of different individuals. For supplementary pollination, we manually added pollen grains (from the same and other individuals) to the stigmas in addition to the routine pollination service provided by native pollinators. After 10 days, we recorded each treatment's fruit and seed sets.

We measured the index of self-incompatibility (ISI) according to Raduski et al. [47] as follows:

$$\text{ISI} = 1 - \frac{\text{Seed set in self pollination}}{\text{Seed set in cross pollination}}$$

Based on the ISI value, plant species can be placed among the three categories: (i) self-incompatible ($\text{ISI} \geq 0.8$), (ii) partial self-incompatible ($0.2 < \text{ISI} < 0.8$), and (iii) self-compatible ($\text{ISI} \leq 0.2$).

We calculated the index of dependency of plants on pollinators (IDP) according to Layek et al. [48] as follows:

$$\text{IDP} = 1 - \frac{\text{Re}}{\text{Rs}}$$

Re is the reproductive success (fruit set or seed set; here, we considered seed set per flower) in pollinator exclusion treatment, and Rs means reproductive success in supplementary pollination treatment. The value of IDP ranges from 0 to 1, and a higher value indicates a higher dependency on pollinators.

To assess whether plant species experience pollen transfer limitations under open field conditions, we calculated the 'coefficient of pollination deficit (D)' following the approach described by Layek et al. [48]:

$$\text{D} = 1 - \frac{\text{Ro}}{\text{Rs}}$$

Ro denotes reproductive success (here, seed set per flower) in open pollination. The value of D ranges from 0 to 1, and a higher value ($\text{D} \geq 0.1$) indicates a significant pollination deficit.

2.5. Floral Visitors

We observed the visitors at four time slots (i.e., 6.00–8.00 h, 8.00–10.00 h, 10.00–12.00 h, and 12.00–14.00 h) covering the flower longevity. Each survey (i.e., plant-based sampling) was continued for 5 min on an individual plant. On a sampling day, we conducted one observation per time slot (n = 4 observations per day; N = 4 × 16 × 6, 16 sampling days per season, 6 seasons). The encountered visitors were identified in the field or captured for later identification.

We estimated the abundance (i.e., the number of individuals of a species/plant/5 min) of each flower-visiting species. For each flower-visiting species, we calculated the relative abundance (RA) as follows (Layek et al. [31]):

$$\text{RA}\ (\%) = \frac{\text{n}i}{\text{N}} \times 100$$

The variable ni represents the number of individuals recorded for an insect species i, while N denotes the total number of individuals recorded across all flower-visiting insect species.

The richness of the flower-visiting community was estimated using the index (D) of Margalef [49] as follows:

$$D = \frac{S - 1}{ln N}$$

In this context, S represents the number of flower-visiting species, and N indicates the total number of individuals observed. The natural logarithm is expressed as *ln*. The value of D was calculated for each sample, where one sample corresponds to a single survey (a 5 min observation on an individual plant).

The diversity of flower visitors was determined using the diversity index (H') of Shannon–Weaver [50] as follows:

$$H' = -\sum_{i}^{n}(pi.ln pi)$$

Here, pi represents the proportion of each visitor species within the sample (pi = n*i*/N, where n*i* is the number of individuals recorded for species *i*, and N is the total number of individuals recorded in the sample).

We recorded the number of visits that received a flower per 5 min duration. Data were taken for all seasons covering all four time slots (N = 2 × 4 × 16 × 6 = 768 observations; n = 2 observations/time slot/sampling day, 16 sampling days per season). We documented the types of floral resources collected by the visitors, including nectar, pollen grains, and floral tissues. The flower visitation rate (VR), or foraging rate, estimated as the number of flowers visited per minute, was recorded with up to 20 observations per insect species. For the visitors with a low visitation rate (e.g., *Tetragonula iridipennis*), we counted the number of flowers visited for 5 min. Then, we converted it for a unit of time (i.e., per minute). The flower handling time (i.e., the duration spent by a visitor on a single flower on a visit) was also measured. For abundant species, N = 20 × 4 = 80 observations were conducted per species across all time slots.

2.6. Pollinating Strategies of Visitors

We documented flower visitation events and categorised them as either legitimate or illegitimate. A visit was classified as legitimate if the visitor made contact with the stigmatic surface; otherwise, it was deemed illegitimate. For legitimate visitors, we observed the different modes of pollination: (i) nototribic (pollen deposited on the dorsal side of the visitor), (ii) sternotribic (pollen deposited on the ventral side of the visitor), (iii) noto-sternotribic (pollen deposited on both dorsal and ventral sides of the visitor), and (iv) appendages mediated (i.e., through delicate parts like legs, antennae, proboscis, etc.).

We estimate the single-visit pollination efficiency of some dominant visitors. For this purpose, we bagged matured flower buds in the late afternoon (n = 10 flower buds per observation day; 20 observation days). The next morning, we uncovered the virgin flowers, and after receiving a visit, we re-bagged them immediately and tagged them with the visitor's specifications. After 10 days, we recorded fruit and seed sets. Then, we estimated the single-visit pollination efficiency index (PE*i*) (according to Spears [51]) of dominant visitors as follows:

$$PEi = \frac{Pi - Z}{U - Z}$$

P*i* represents the average number of seed sets per flower in the single-visit experiment for the visitor species *i*. Z denotes the average number of seed sets in the pollinator exclusion treatment. U indicates the average number of seed sets resulting from open pollination, where visitation is unrestricted.

We determined a combined parameter, the pollinator importance (PI), for the dominant flower visitors based on the methodology outlined by Layek et al. [52]. We considered the numeric values of relative abundance (RA), visitation rate (VR), and single-visit pollination efficiency index (PE*i*). As flowers are hermaphrodite, the flower sex selection index (FSI) would be 1, and we ignored FSI for the calculation of PI by multiplying the numeric values. Here, PI was calculated as follows:

$$PI = RA \times VR \times PEi$$

A flower visitor species with a high PI value was considered an effective pollinator for the plant species.

2.7. Statistical Analysis

The data were analysed descriptively to calculate the mean and standard deviation. The Shapiro–Wilk test was employed to assess the normality of the data distribution. For data that were not normally distributed, the non-parametric Kruskal–Wallis H test was applied (e.g., floral display size, flower longevity, ovule and pollen production, seed set, abundance, richness, diversity of visitors, number of visits per flower, flower visitation rate, and handling time). When the *p*-value was significant ($p \leq 0.05$), post hoc analyses were conducted using Dunn's test following the Kruskal–Wallis test. Spearman's rank correlation coefficient (Spearman's rho) was used to examine the relationships between environmental factors (e.g., temperature, light, relative humidity, rainfall, and photoperiod) and floral traits, reproductive success, and flower visitors. Statistical analyses were performed using SPSS version 25.0 and R (R Core Team 2022).

3. Results

3.1. Floral Biology

Turnera ulmifolia bloomed throughout the years and showed a steady-state flowering pattern. All individuals (that reached the reproductive phase) were flowering synchronously. Flowering intensity varied season-wise (Kruskal–Wallis H test: $\chi^2 = 52.78$, df = 5, $p < 0.001$), with higher during summer–monsoon and lower in winter (Table 1). The flowering intensity was highly positively correlated with temperature (Spearman's rho = 0.86, $p < 0.001$, n = 96), humidity (Spearman's rho = 0.43, $p < 0.001$, n = 96), and day length (Spearman's rho = 0.63, $p < 0.001$, n = 96) (Figure 1). Flower opening time varied season-wise. In warmer seasons (e.g., summer and monsoon), flowers opened earlier (5.00–6.00 h) than in the cold winter (8.00–9.00 h). Flower longevity also varied season-wise (Kruskal–Wallis H test: $\chi^2 = 263.52$, df = 5, $p < 0.001$). Comparatively, a higher flower longevity was recorded during cold winter (8.25 ± 0.45 h). Flower longevity was negatively correlated with temperature, humidity, and day length.

Solitary flowers were borne at the tip of a lateral branch (Figure 2). Calyx had five sepals, semi-gamosepalous (partially fused at their lower parts), a light green dorsal surface and light yellow on the ventral side. Each sepal was lanceolate (22.25 ± 1.21 mm long and 5.12 ± 0.67 mm wide), and the outer surface was hairy. The hairs were narrow and elongated (256.53 ± 97.26 µm long and 18.34 ± 2.27 µm wide at the base) and with a beaded surface (Figure 3). The corolla had five petals, polypetalous and yellow. The upper part was broad (18.60 ± 1.50 mm wide), and the base was very narrow. Five stamens were free; the filaments were 15.40 ± 1.19 mm long; anthers were 4.60 ± 0.60 mm, basifixed and dehisced longitudinally. Carpels were three; styles were free, 14.80 ± 0.83 mm long, and light yellow, each terminated with fibrous stigmas. The ovary was dome-shaped, with a greenish and hairy surface. There were many ovules within an ovary, and placentation was the parietal type. The palynomorphic study showed the pollen grains were monad, spheroidal (44.46 µm in diameter), amb triangular, trizonocolporate, exine about 2.5 µm thick, and exine ornamentation was of the reticulate type (Figure 4).

Table 1. Season-wise floral traits of *Turnera ulmifolia* in West Bengal.

Season	Flowering Intensity	Flower Opening Time	Flower Closing Time	Flower Longevity	Pollen/Flower	Ovule/Flower
Summer	8.96 ^{a} ± 5.30	5.00–6.00 h	10.00–11.00 h	4.88 ^{b} ± 0.34	14,075 ^{a} ± 323.93	60.96 ^{a} ± 11.07
Monsoon	8.48 ^{a} ± 5.31	5.00–6.00 h	10.00–11.00 h	5.00 ^{b} ± 0	14,133 ^{a} ± 370.78	61.60 ^{a} ± 13.69
Autumn	8.17 ^{a} ± 5.17	5.00–6.00 h	11.00–12.00 h	6.31 ^{ab} ± 0.48	13,685.42 ^{ab} ± 179.13	52.54 ^{ab} ± 6.26
Late autumn	7.48 ^{ab} ± 4.92	6.00–7.00 h	13.00–14.00 h	7.38 ^{a} ± 0.50	13,358.76 ^{ab} ± 292.26	51.79 ^{ab} ± 7.56
Winter	2.31 ^{b} ± 1.67	8.00–9.00 h	15.00–17.00 h	8.25 ^{a} ± 0.45	11,389.62 ^{b} ± 497.49	44.42 ^{b} ± 7.42
Spring	7.54 ^{ab} ± 4.95	7.00–8.00 h	12.00–13.00 h	5.25 ^{b} ± 0.45	13,918.31 ^{a} ± 294.78	51.46 ^{ab} ± 8.82
Throughout year	7.16 ^{ab} ± 5.20	5.00–9.00 h	10.00–17.00	6.18 ^{ab} ± 1.34	13,426.68 ^{ab} ± 1008.01	53.80 ^{ab} ± 11.12
Statistical analysis	$\chi^2 = 52.78$, df = 5, $p < 0.001$	-	-	$\chi^2 = 263.52$, df = 5, $p < 0.001$	$\chi^2 = 67.96$, df = 5, $p < 0.001$	$\chi^2 = 79.64$, df = 5, $p < 0.001$

Values are given in mean ± standard deviation. Different superscript letters within a column indicate significant differences (Dunn's post hoc test at 0.05% level).

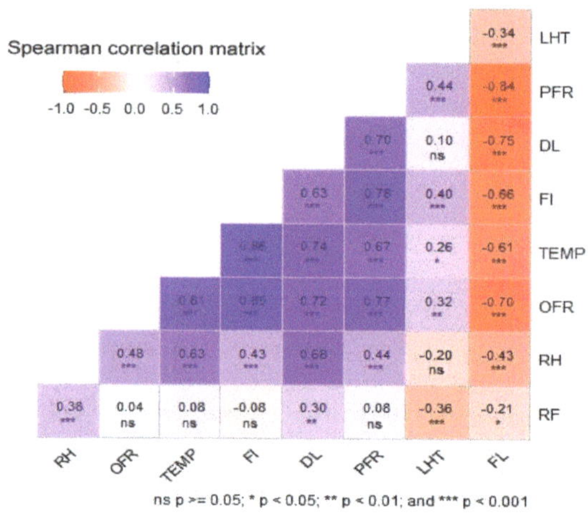

Figure 1. Correlation matrix derived from atmospheric factors and flower traits. DL: day length, FI: flowering intensity, FL: flower longevity, LTH: light intensity, OFR: ovules per flower, PFR: pollens per flower, RH: relative humidity, TEMP: temperature.

Figure 2. A flowering twig of *Turnera ulmifolia*.

Figure 3. Flower parts. (**A**) Flower, (**B**) calyx, (**C**,**D**) sepal, (**E**,**F**) a part of sepal showing hairs, (**G**) part of a hair, (**H**,**I**) petal, (**J**) stamens, (**K**) a portion of stamen showing basifixed anther, (**L**) dehisced anther showing pollen grains, (**M**) carpels, (**N**) ovary, (**O**) t.s. of ovary showing placentation, (**P**) hairs of ovary surface, and (**Q**) stigmatic surface.

Pollen production (number of pollen grains) per flower was 13,426.68 ± 1008.01 (mean ± SD, n = 96), and ovule per flower was 53.80 ± 11.12 (mean ± SD, n = 288). The number of pollens and ovules varied seasonally (Pollen: χ^2 = 67.96, df = 5, p < 0.001; ovule: χ^2 = 79.64, df = 5, p < 0.001). Pollen and ovule production remained higher during summer–monsoon and lower during winter (Table 1). The ovule-to-pollen ratio was 1:249.57. Pollen and ovule production positively correlated with temperature (pollen: Spearman's rho = 0.81; ovule: Spearman's rho = 0.75; p < 0.001, n = 96).

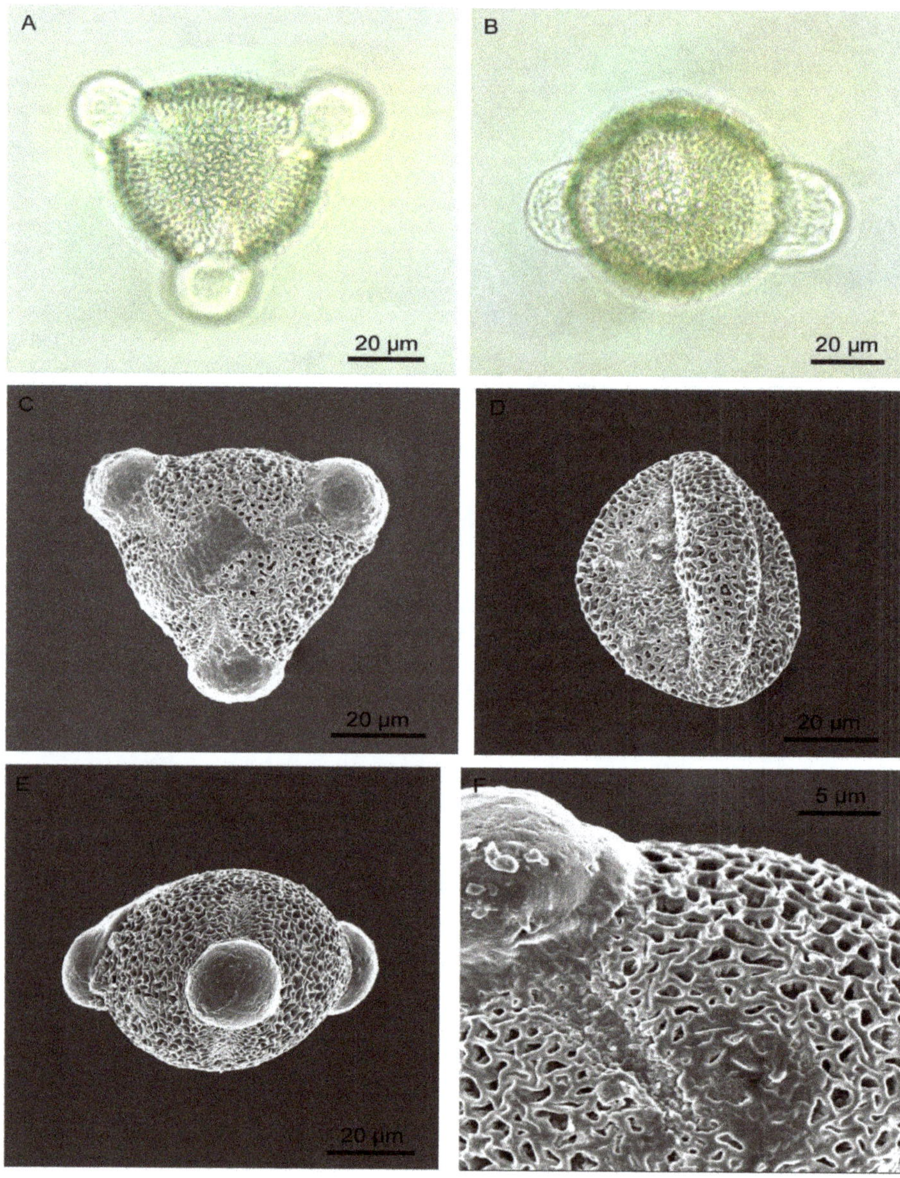

Figure 4. Pollen grains of *Turnera ulmifolia*. (**A**,**B**) light microscopy images, (**A**) polar view, (**B**) equatorial view. (**C–F**) scanning electron microscopy images, (**C**) polar view, (**D**,**E**) equatorial view, (**F**) an enlarged view showing exine ornamentations.

Anther dehiscence started before the completion of the flower opening. Anther dehisced by a longitudinal slit on the theca in each locule. At opening time, pollen viability was 77.56 ± 6.08% for the TTC test and 84.49 ± 6.33% for the IKI test (Table S2). Pollen germinability was 72.56 ± 6.17%. Pollen viability and germinability did not vary according to season. At the time of anther dehiscence, the stigma remained non-receptive, and it became receptive during the completion of the opening of a flower. Stigma remained receptive throughout the anthesis time (i.e., in opened flowers). Peak receptivity was during 8.00–10.00 h. The duration of stigma receptivity varied season-wise ($\chi^2 = 84.32$, df = 5, $p < 0.001$), with higher during winter (9.44 ± 0.51 h) and lower in summer (5.88 ± 0.34) (Figure 5 showing viable pollens and receptive stigma). The duration of stigma receptivity negatively correlated with temperature, relative humidity, and day length (temperature: Spearman's rho = –0.67; RH: Spearman's rho = –0.47; day length: Spearman's rho = –0.67; $p < 0.001$, n = 96) and positively correlated with flower longevity (Spearman's rho = 0.98; $p < 0.001$, n = 96).

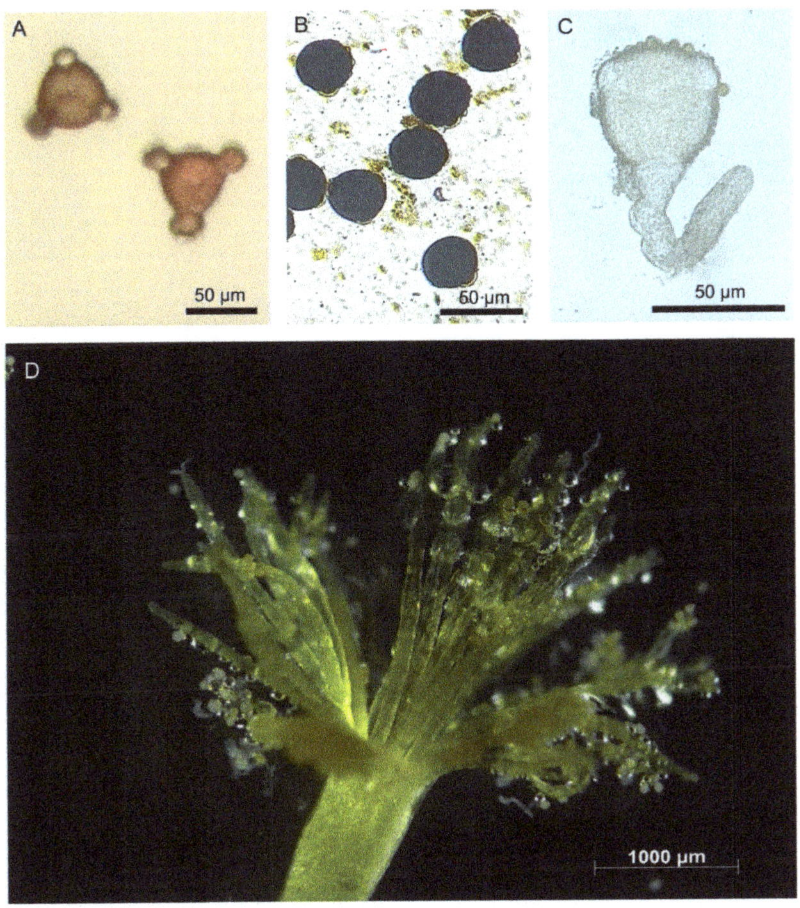

Figure 5. (**A**) Pollens stained with TTC, (**B**) pollens stained with IKI, (**C**) a germinated pollen, and (**D**) stigma showing receptivity.

3.2. Mating System and Reproduction

All five pollination treatments resulted in fruit and seed sets (Table S3). Fruit set was very high (i.e., 100%, excluding damaged flowers). Seed sets in selfing and crossing treatments were almost similar. Therefore, the plant species was fully self-compatible (index of self-compatibility, ISI = 0.02). The seed set in the pollinator exclusion treatment was comparatively lower than supplementary pollination treatments. The plant species showed moderate dependency on pollinators (IDP = 0.34). The value of the coefficient of the pollination deficit was meagre (D = 0.07), indicating there was no pollination limitation of the plant species in an open system.

The reproductive success (in terms of seed set) largely varied according to season ($\chi^2 = 31.14$, df = 5, $p < 0.001$). In summer, the plant had a greater seed set (54.72 ± 14.96 seeds/flower) and lower during winter (39.98 ± 8.92 seeds/flower) (Figure 6).

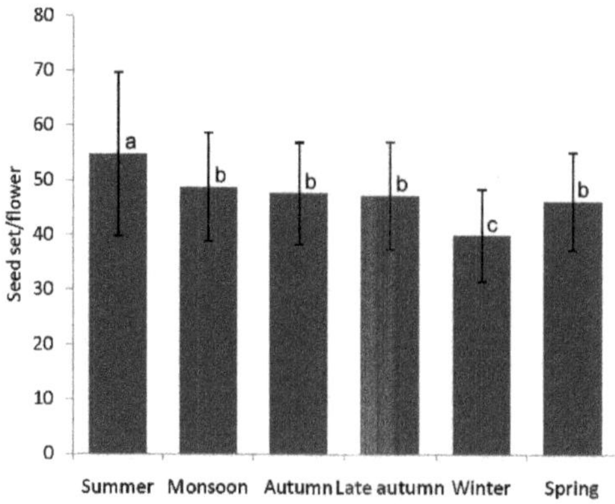

Figure 6. Season-wise seed set (number of seeds) per flower. Values are given in mean ± standard deviation. Different letters indicate significant differences (Dunn's post hoc test, $p < 0.05$).

3.3. Floral Visitors

A total of 27 insect species were documented as floral visitors of *Turnera ulmifolia* in West Bengal, India (Table 2, Figures 7 and 8). The most represented insect orders were Hymenoptera (15 species), followed by Lepidoptera (7 species), Diptera (3 species), and Coleoptera (2 species). Among the hymenopteran members, most belong to Apidae (8 species), followed by Formicidae (4 species of ants) and Halictidae (3 species). The butterflies belong to the insect families Hesperiidae (3 species), Lycaenidae (1 species), Nymphalidae (1 species), and Pieridae (2 species).

Table 2. Flower visitors of *Turnera ulmifolia* in West Bengal, India.

Floral Visitors	Abundance	Relative Abundance (%)	Flower Visitation Rate	Flower Handling Time	Floral Resources
▪ Coleoptera					
Aulacophora foveicollis	<0.01	0.21	-	-	FT
Hycleus phalarantha	<0.01	0.14	-	-	FT
▪ Diptera					
Armigeres subalbatus	0.02 ± 0.15	0.64	-	-	N
Episyrphus balteatus	0.05 ± 0.22	1.41	2.95 ± 1.00	13.12 ± 14.28	P
Helophilus peregrinus	0.04 ± 0.20	1.13	-	-	P
▪ Hymenoptera					
Amegilla zonata	0.22 ± 0.58	5.87	4.30 ± 1.56	6.27 ± 3.19	N + P
Apis cerana	0.06 ± 0.24	1.56	3.10 ± 0.91	13.50 ± 13.66	N + P
Apis dorsata	0.05 ± 0.22	1.34	3.70 ± 1.03	10.52 ± 5.23	N + P
Apis florea	0.03 ± 0.17	0.78	3.35 ± 0.93	12.26 ± 10.31	N + P
Braunsapis mixta	0.05 ± 0.24	1.34	1.16 ± 0.30	46.24 ± 16.29	N + P
Camponotus parius	0.14 ± 0.47	3.68	-	-	FT, N + P
Ceratina compacta	0.05 ± 0.22	1.27	-	-	N + P
Crematogaster laestrygon	0.14 ± 0.53	3.82	-	-	FT, N + P
Diacamma indicum	0.05 ± 0.26	1.41	-	-	FT, N + P
Halictus acrocephalus	0.48 ± 0.94	13.08	1.46 ± 0.28	39.43 ± 14.25	N + P
Lasioglossum cavernifrons	0.28 ± 0.66	7.50	1.29 ± 0.32	42.18 ± 14.62	N + P
Myrmicaria brunnea	0.13 ± 0.47	3.61	-	-	FT, N + P
Nomia strigata	0.43 ± 0.83	11.60	1.88 ± 0.43	36.24 ± 12.35	N + P
Tetragonula iridipennis	0.79 ± 1.28	21.57	0.71 ± 0.28	57.18 ± 26.44	N + P
Thyreus nitidulus	0.03 ± 0.16	0.71	3.75 ± 1.12	8.37 ± 3.71	N

Table 2. *Cont.*

Floral Visitors	Abundance	Relative Abundance (%)	Flower Visitation Rate	Flower Handling Time	Floral Resources
▪ Lepidoptera					
Acraea terpsicore	0.05 ± 0.28	1.41	-	-	FT
Borbo cinnara	0.39 ± 0.82	10.61	4.20 ± 1.47	13.47 ± 11.76	N
Catochrysops panormus	0.03 ± 0.17	0.78	1.40 ± 1.12	21.30 ± 18.29	N
Catopsilia pomona	0.02 ± 0.14	0.64	2.15 ± 1.30	8.74 ± 7.56	N
Eurema hecabe	0.02 ± 0.17	0.78	1.72 ± 1.43	17.61 ± 14.82	N
Pelopidas mathias	0.09 ± 0.34	2.40	4.05 ± 1.39	14.93 ± 12.25	N
Suastus gremius	0.03 ± 0.16	0.71	2.30 ± 2.11	18.46 ± 15.03	N

FT: floral tissue, N: nectar, P: pollen. Values are given as mean ± standard deviation.

Figure 7. Floral visitors of *Turnera ulmifolia* in West Bengal. (**A**) *Acraea terpsicore*, (**B**) *Apis cerana*, (**C**) *Apis dorsata*, (**D**) *Apis florea*, (**E**) *Armigeres subalbatus*, (**F**) *Aulacophora foveicollis*, (**G**) *Borbo cinnara*, (**H**) *Braunsapis mixta*, (**I**) *Camponotus parius*, (**J**) *Catochrysops panormus*, (**K**) *Catopsilia pumona*, and (**L**) *Crematogastor laestrygon*. Scale bars = 10 mm.

Figure 8. Floral visitors of *Turnera ulmifolia* in West Bengal. (**A**) *Diacamma indicum*, (**B**) *Episyrphus balteatus*, (**C**) *Halictus acrocephalus*, (**D**) *Helophilus peregrinus*, (**E**) *Hycleus phalarantha*, (**F**) *Lasioglossum cavernifrons*, (**G**) *Myrmicaria brunnea*, (**H**) *Nomia* (*Curvinomia*) *strigata*, (**I**) *Pelopidas mathias*, (**J**) *Saustas gremias*, (**K**) *Tetragonula iridipennis*, and (**L**) *Thyreus nitidulus*. Scale bars = 10 mm.

Visitors' abundance, richness, and diversity varied seasonally (Table 3). The highest abundance, richness, and diversity were recorded during summer (abundance: 4.89 ± 2.97 visitors/5 min/plant; richness, $D = 1.13 \pm 0.66$; the Shannon–Weaver diversity index, $H' = 0.87 \pm 0.52$) and the lowest in winter (abundance: 1.94 ± 1.71 visitors/5 min/plant; richness, $D = 0.45 \pm 0.60$; Diversity $H' = 0.32 \pm 0.43$). The visitor abundance positively correlated with flowering intensity (Spearman's rho = 0.71, $p < 0.001$, n = 96), temperature (Spearman's rho = 0.60, $p < 0.001$, n = 96), and light (Spearman's rho = 0.36, $p < 0.001$, n = 96) while negatively correlated with rainfall (Spearman's rho = −0.42, $p < 0.001$, n = 96). Visitor traits (abundance, richness, and diversity) significantly varied daytime-wise (Table S4). The abundance, richness, and diversity remained higher during 8.00–10.00 h and lower at 12.00–14.00 h.

Table 3. Abundance (number of visitors/individual/5 min), richness (Margalef's index D), and diversity (index of Shannon–Weaver, *H'*) of floral visitors on *Turnera ulmifolia* in West Bengal.

Season	Abundance	Richness	Diversity
Summer	4.89 [a] ± 2.97	1.13 [a] ± 0.66	0.87 [a] ± 0.52
Monsoon	3.23 [b] ± 2.78	0.74 [b] ± 0.70	0.56 [b] ± 0.53
Autumn	4.06 [ab] ± 2.93	0.91 [ab] ± 0.64	0.71 [ab] ± 0.52
Late autumn	3.94 [ab] ± 2.36	0.96 [ab] ± 0.62	0.72 [ab] ± 0.48
Winter	1.94 [c] ± 1.71	0.45 [c] ± 0.60	0.32 [c] ± 0.43
Spring	4.03 [ab] ± 2.54	0.94 [ab] ± 0.66	0.73 [ab] ± 0.51
Throughout year	3.68 ± 2.54	0.85 ± 0.68	0.65 ± 0.52
Statistical analysis	$\chi^2 = 44.10$, df = 5, $p < 0.001$	$\chi^2 = 36.42$, df = 5, $p < 0.001$	$\chi^2 = 42.49$, df = 5, $p < 0.001$

Values are given as mean ± standard deviation. Different superscript letters within a column indicate significant differences (Dunn's post hoc test, $p < 0.05$).

The abundant flower-visiting species were *Amegilla zonata* (abundance = 0.22 ± 0.58 individuals/plant/5 min; relative abundance = 5.87%), *Borbo cinnara* (abundance = 0.39 ± 0.82 individuals/plant/5 min; relative abundance = 10.61%), *Halictus acrocephalus* (abundance = 0.48 ± 0.94 individuals/plant/5 min; relative abundance = 13.08%), *Lasioglossum cavernifrons* (abundance = 0.28 ± 0.66 individuals/plant/5 min; relative abundance = 7.50%), *Nomia (Curvinomia) strigata* (abundance = 0.43 ± 0.83 individuals/plant/5 min; relative abundance = 11.60%), and *Tetragonula iridipennis* (abundance = 0.79 ± 1.28 individuals/plant/5 min; relative abundance = 21.57%) (Table 2).

The number of visits received by a flower per unit of time differed among the seasons ($\chi^2 = 46.98$, df = 5, $p < 0.001$). Comparatively, a higher number of visits were received during summer (2.02 ± 2.12 visits/flower/5 min) and a lower number of visits during winter (0.81 ± 0.95 visits/flower/5 min) (Figure 9). Daytime-wise, the maximum number of visits received was during 8–10 h, and the lowest number was during the afternoon (Table S5).

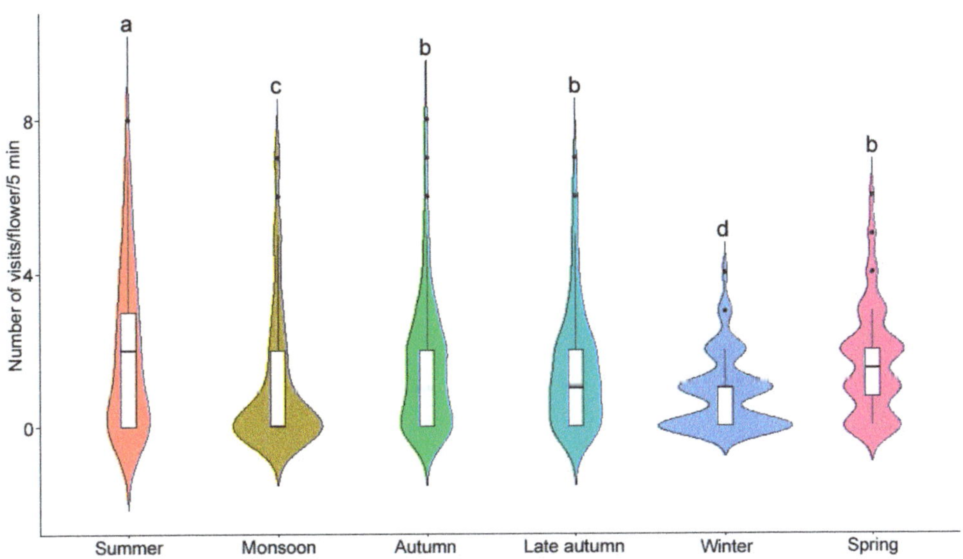

Figure 9. Violin plot showing the season-wise number of visits received by a flower. Different letters indicate significant differences (Dunn's post hoc test, $p < 0.05$).

3.4. Pollination Strategies

All flower-visiting species legitimately visited *Turnera ulmifolia* flowers (except the larvae of the tawny castor butterfly (*Acraea terpsichore*) and beetles) (Table 4). Butterflies have transferred pollens on stigma through the ventral side of the thorax and abdomen and also through their delicate body parts, including legs, antennae, proboscis, and wings (Figure 10A–D). Flies provided pollination services through their legs. Mosquitoes also touched stigmatic surfaces through their legs (Figure 10F) and provided pollination services. Sometimes, ants touched anthers and stigmatic lobes through their legs. Honeybees pollinated in a sternotribic manner (through the ventral side of the thorax and abdomen). Their legs also touched the stigmas, and occasionally, corbicular pollen loads touched the stigmatic surface. Rarely was pollination carried out through their wings. Solitary bees, when they collected nectar and pollen, touched stigmas through their legs and the ventral side of the thorax and abdomen. Sometimes, they (especially *Braunsapis mixta*, *Halictus acrocephalus*, *Lasioglossum cavernifrons*) travelled from one anther to another of a flower over the stigmatic surfaces and smeared pollens on stigmas. *Nomia* (*Curvinomia*) *strigata* showed different visitation patterns, including vertical visits with the bee's ventral side remaining facing the flower's centre (Figure 10P–Q). In most cases, their abdomens touched the flower anthers. During nectar suction from the flowers, they showed the 'pulsatory movement' of their abdomen part. They undulated the anthers through their abdominal oscillation, resulting in pollen release, and provided pollination services (i.e., pulsatory pollination). Stingless bees (*Tetragonula iridipennis*) touched stigmas through their legs during pollen collection from an anther. They also travelled over the stigmatic surfaces to move from one anther to another and provided pollination services (Figure 10T).

Table 4. Pollinating strategies of flower visitors on *Turnera ulmifolia*.

Visitors	Visitation Type	Mode of Pollination	PE*i*	PI
Ants	**	S; ventral side of thorax and abdomen, legs	-	-
Beetles	IV	-	-	-
Butterflies				
• *Acraea terpsicore*	IV	-	-	-
• *Borbo cinnara*	***	S; ventral side of thorax and abdomen, legs, antennae, proboscis, wings	0.21	9.36
• *Catochrysops panormus*	***	S; ventral side of thorax and abdomen, legs, antennae, proboscis, wings	-	-
• *Catopsilia pomona*	***	S; ventral side of thorax and abdomen, legs, antennae, proboscis, wings	-	-
• *Eurema hecabe*	***	S; ventral side of thorax and abdomen, legs, antennae, proboscis, wings	-	-
• *Pelopidas mathias*	***	S; ventral side of thorax and abdomen, legs, antennae, proboscis, wings	0.21	2.04
• *Suastus gremius*	***	S; ventral side of thorax and abdomen, legs, antennae, proboscis, wings	-	-
Flies	**	S; ventral side of thorax and abdomen, legs	-	-
Mosquitoes	*	Legs	-	-
Honeybees	***	S; ventral side of thorax and abdomen, legs, wings, corbicular pollen loads	-	-
Solitary bees				
• *Amegilla zonata*	***	S; ventral side of thorax and abdomen, legs,	0.66	16.66
• *Braunsapis mixta*	***	S; ventral side of thorax and abdomen, legs,	-	-
• *Ceratina compacta*	***	S; ventral side of thorax and abdomen, legs,	-	-
• *Halictus acrocephalus*	***	S; ventral side of thorax and abdomen, legs, scopal pollen loads	0.52	9.93
• *Lasioglossum cavernifrons*	***	S; ventral side of thorax and abdomen, legs,	0.47	4.55
• *Nomia* (*Curvinomia*) *strigata*	***	S, pulsatory pollination; ventral side of thorax and abdomen, legs,	0.50	10.90
• *Thyreus nitidulus*	**	S; ventral side of thorax and abdomen, legs,	-	-
Stingless bees	***	S; ventral side of thorax and abdomen, legs, corbicular pollen loads	0.62	9.50

PE*i*: single-visit pollination efficiency index, PI: pollinator importance, S: sternotribic, *: rarely legitimate visit, **: frequently legitimate visit, ***: mostly legitimate visit, IV: illegitimate visit.

Figure 10. Pollinating strategies of floral visitors. (**A–D**) Butterflies, legs, antennae, proboscis, and wings, respectively; (**E**) fly touching through legs and ventral side of thorax and abdomen; (**F**) mosquito touching through legs; (**G–I**) ants touching through legs and antennae; (**J–M**) honeybees touching through the ventral side of thorax and abdomen, legs, corbicular pollen loads, and wings; (**N**,**O**) *Halictus acrocephalus* touching through the ventral side of thorax and abdomen, legs, scopal pollen loads; (**P–R**) *Nomia* (*Curvinomia*) *strigata* touching through the ventral side of the abdomen and legs and showing pulsatory pollination; (**S**,**T**) stingless bees touching through the ventral side of thorax and abdomen, legs, and corbicular pollen loads.

Single-visit pollination efficiencies (PE*i*) were higher for *Amegilla zonata* (PE*i* = 0.66), *Tetragonula iridipennis* (PE*i* = 0.62), *Halictus acrocephalus* (PE*i* = 0.52), and *Nomia* (*Curvinomia*) *strigata* (PE*i* = 0.50). Pollinator importance (PI) value was the highest for *Amegilla zonata*, followed by *Nomia* (*Curvinomia*) *strigata*, *Halictus acrocephalus*, *Tetragonula iridipennis*, and *Borbo cinnara* (Table 4). These insect species provided significant pollination services to the plant and were considered effective pollinators for *Turnera ulmifolia*.

4. Discussion

Several researchers have classified the flowering patterns of plants differently. For example, Gentry [6] classified the species of Bignoniaceae into five categories (steady state, modified steady state, cornucopia, big bang, and multiple bang); Bawa [53] segregated the flowering patterns of plants into two groups (massive and extended); and Frankie et al. [54] classified them into two groups (seasonal and extended). The studied plant species showed steady-state flowering throughout the year. The steady-state flowering increases the chance of cross-pollination. It also may assure reproductive success against adverse weather conditions. All individuals bloomed synchronously. Due to the flowering synchrony, each plant can exchange genes with most plants, increasing the genetic diversity of the same population [55,56]. Flowering intensity, flower longevity, ovules, and pollen production were influenced by temperature, light, humidity, rainfall, and day length. The sensitivity of floral traits against environmental factors was well documented for many plant species (e.g., Barley [57], cocoa [58], and white yam [59]). These floral traits were favoured during summer–monsoon and declined during winter.

Regarding the maturation of reproductive traits, the plant species was protandrous, which is more common in angiosperms than protogyny [60]. This type of dichogamy is less effective for promoting outcrossing for the species. Here, most flowers may be perceived as self- and non-self-pollens, resulting in both autogamy and xenogamy. Pollen viability and germinability did not differ seasonally. The duration of stigma receptivity varied with seasons and depended on flower longevity. During winter, the longer receptive period may mitigate the low flowering frequency and effort to optimise reproductive success [61]. This feature may be an adaptation that allows biotic pollination under adverse environmental conditions. The duration of female receptivity also depends on whether pollination occurred [62,63]. In addition, long-lived stigmas allow extended cross-pollination opportunities for the plant species.

Fruit and seed sets resulted in bagged flowers, indicating that the plant species spontaneously auto-pollinated. This strategy gives the plant extra assurance of reproductive success in hazardous environments with limited pollinator activity. Seed sets in selfing remained almost equal to the output of crossing treatments, implying the self-compatibility of the plant species. Barrett and Shore [64] worked out the breeding systems of different varieties of *Turnera ulmifolia* and reported the phenomenon of self-compatibility. Self-fertilisation (the pollination referred to as 'selfing') can be advantageous in the short term for several reasons. Firstly, plants that self-fertilise gain a 50% transmission advantage over outcrossing plants, as they can both self-fertilise their own ovules and contribute outcross pollen simultaneously [65,66]. Secondly, selfing is beneficial in environments where pollinators or potential mates are scarce, providing reproductive assurance [67,68]. Additionally, selfing may enhance colonisation ability [69]. However, a key factor opposing the shift to selfing is inbreeding depression, which refers to the reduced fitness of inbred offspring compared to outcrossed progeny [70]. The seed set in the pollinator exclusion treatment was relatively lower than in the supplementary pollination treatments, indicating that the plant species relied moderately on biotic pollinators. Other species of *Turnera* also depend on pollinators for reproduction (e.g., *Turnera subuata* [71]). The value of the coefficient of pollination deficit was meagre ($D = 0.07$), indicating there was no pollination limitation of the plant species in an open system. Successful pollination of plant species depends on their flower characteristics, environmental factors, and pollinator activity [48,72].

The reproductive success of the plant species largely varied seasonally, with higher during summer and lower during winter. The seed set depends mainly on successful pollination, linked to flowering intensity, weather conditions, and pollinator activity [73,74]. In summer, some floral traits (e.g., flowering intensity, pollen and ovule production, etc.)

remained higher than in winter. More ovule production was also linked to a higher seed set. The relationship between ovule production and seed set was established by many researchers (e.g., Strelin and Aizen [75]; Cucinotta et al. [76]). In addition, high flowering intensity may attract more pollinators, which results in higher reproductive success. The seed sets in four seasons (e.g., monsoon, autumn, late autumn, and spring) did not significantly differ, though environmental factors varied. In the monsoon season, flowering intensity, pollen, and ovule production were higher than in the other three seasons. In contrast, visitor abundance was lower in the monsoon due to adverse atmospheric conditions that may have constrained plant reproduction.

Here, we first recorded detailed floral visitors of *Turnera ulmifolia*. Many insect species (here, 27) visited flowers. A similar number of insect species was also reported for another species, *Turnera subulata*, in NE Brazil [71]. Meanwhile, fewer flower-visiting species were recorded for *Turnera subulata* in Bahia, Brazil [77]. The visitor spectrum depends on plant species and varies across geographical regions [41]. Visitor abundance, richness, and diversity remained higher during summer and lower during monsoon and winter. Environmental variables highly influenced visitor traits and positively influenced flowering intensity. Many researchers (e.g., McCall and Primack [78]; Goodwin et al. [32]) also revealed that flowering patterns and environmental factors are the most important variables influencing insect visitation rates. In addition to atmospheric factors, plant–visitor interactions are also influenced by factors such as habitat fragmentation [79] and pesticide use [80]. The most abundant visitors for *Turnera ulmifolia* in West Bengal were *Borbo cinnara*, *Halictus acrocephalus*, *Nomia* (*Curvinomia*) *strigata*, and *Tetragonula iridipennis*. The visitor abundance largely depends on the plant's floral architecture, floral resources, and surrounding vegetation that deter the habitat and foods for native pollinators [81,82].

Diverse insect groups (e.g., flies, butterflies, ants, honeybees, solitary bees, and stingless bees) pollinated the plant species. Mosquitoes also legitimately visited the flowers. However, their abundance as visitors and visitation rate were very low and played a minor role in plant reproduction. Ant pollination (i.e., myrmecophily) is not very common in angiosperms, only reported for a few plant species (e.g., Vega et al. [83]; Dutton and Frederickson [84]). Here, ants were attracted by floral tissues, flower nectar, and extrafloral nectar [83], and reproductive success was enhanced by pollinating and helping in seed dispersal [85]. Many solitary bees (e.g., *Amegilla zonata*, *Halictus acrocephalus*, *Lasioglossum cavernifrons*, *Nomia* (*Curvinomia*) *strigata*, etc.) provided pollination services to the plant species. The solitary bee-mediated pollination was also recorded for other species, e.g., *Turnera subulata* [71,77]. Most visitors pollinated *Turnera ulmifolia* flowers sternotribically through the ventral surface of the thorax and abdomen. The pollination mode for a plant species largely depends on pollinator species and flower architecture [86]. Funnel-shaped flowers with exposed anthers may serve as sternotribic flowers for most visitors. The visitors transferred pollen to the stigmatic surface through their thorax and abdomen as well as delicate parts such as the legs, antennae, proboscis, and wings. The centrally placed stigmatic branches provided landing substrates for visitors, and most pollinators travelled from one anther to another over the stigmas and provided pollination services. Additionally, *Nomia* (*Curvinomia*) *strigata* performed another pollination mode, i.e., pulsatory pollination. This phenomenon was the first time we reported it for a pollinator species. In this case, the ventral surface of the abdomen was touched by anthers, and the oscillation of the abdomen caused the release of pollens from the anther and then to be deposited on stigmas. Though many insect species provided pollination services to the plant species, their efficiency (considering pollination services) varied. The pollination efficiency of pollinators depends on their abundance, foraging activity, resource-collecting behaviour, and flower traits [31]. Here, we estimated the pollinator importance (PI) of visitor species by considering relative abundance, flower visitation rate, and single-visit pollination efficiency. Based on the PI value, effective pollinators were *Amegilla zonata*, *Borbo cinnara*, *Halictus acrocephalus*, *Nomia* (*Curvinomia*) *strigata*, and *Tetragonula iridipennis*.

5. Conclusions

The study depicted the influences of atmospheric factors (e.g., temperature, light intensity, humidity, and rainfall) and photoperiod on floral biology, plant–pollinator interactions, and the reproductive success of *Turnera ulmifolia*. Temperature, light, humidity, and day length positively impacted flowering intensity, pollen, and ovule production. Temperature, humidity, and day length negatively influenced the flower longevity. Consequently, higher flowering intensity, pollen, and ovule production were recorded during the hot and humid summer and monsoon compared to the cold winter. Floral visitor abundance, richness, and diversity positively correlated with flowering intensity and were negatively influenced by rainfall. Therefore, visitor traits remained higher during summer and lower during monsoon and winter. Diverse insect groups pollinated the plant species, including ants, bees, butterflies, and flies. Effective pollinators were *Amegilla zonata*, *Borbo cinnara*, *Halictus acrocephalus*, *Nomia* (*Curvinomia*) *strigata*, and *Tetragonula iridipennis*. Most pollinators showed a sternotribic pollination mode, carried out through the ventral surface of the thorax and abdomen and through their delicate parts like legs, antennae, proboscis, and wings. *Nomia* (*Curvinomia*) *strigata* showed a unique mode of pollination, i.e., pulsatory pollination. Considering reproductive success, higher reproductive fitness was revealed during the hot summer, while comparatively lower reproductive success was revealed during the cold winter. Therefore, it can be concluded that the floral biology, plant-pollinator interactions, and reproductive success of *Turnera ulmifolia* varied seasonally, with significant influences from atmospheric factors and photoperiod.

Supplementary Materials: The following supporting information can be downloaded at: https://www.mdpi.com/article/10.3390/biology14010100/s1, Table S1: The methodologies for the microscopic study of flower parts and pollen grains; Table S2: Pollen viability, germinability, and stigma receptivity (duration in hours) of *Turnera ulmifolia*; Table S3: Reproductive success (seed set per flower) of *Turnera ulmifolia* in different pollination treatments; Table S4: Daytime-wise, visitor traits (abundance, richness, and diversity) of *Turnera ulmifolia* in West Bengal; Table S5: Daytime-wise, the number of visits received a flower (number of visits/flower/5 min) of *Turnera ulmifolia* in West Bengal.

Author Contributions: Conceptualisation: U.L. and P.K.; methodology: U.L., N.D. and A.S.; formal analysis: U.L. and A.S.; investigation: U.L. and N.D.; data curation: U.L. and N.D.; writing original draft: U.L.; writing review and editing: N.D., A.S. and P.K.; visualisation: U.L. and A.S.; supervision: P.K. All authors have read and agreed to the published version of the manuscript.

Funding: There is no funding for this research.

Institutional Review Board Statement: Not applicable.

Informed Consent Statement: Not applicable.

Data Availability Statement: The data presented in this study are available upon request from the corresponding author.

Acknowledgments: The authors thank the authorities of some institutions (e.g., IIT-Kharagpur, Rampurhat College and the Vidyasagar University of West Bengal) for providing laboratory facilities. They remain grateful to Trisha Bhandari (research scholar, Vidyasagar University) for helping with the SEM study. They are also thankful to entomologists of the Zoological Survey of India (Kolkata, West Bengal) for helping in identifying insect species.

Conflicts of Interest: The authors confirm that they have no known financial conflicts of interest or personal relationships that could have influenced the work presented in this paper.

References

1. Rathcke, B.; Lacey, E.P. Phenological patterns of terrestrial plants. *Annu. Rev. Ecol. Syst.* **1985**, *16*, 179–214. [CrossRef]
2. Hafdahl, C.E.; Craig, T.P. Flowering phenology in *Solidago altissima*: Adaptive strategies against temporal variation in temperature. *J. Plant Interact.* **2014**, *9*, 122–127. [CrossRef]
3. Fitter, A.H.; Fitter, R.S.R. Rapid changes in flowering time in British plants. *Science* **2002**, *296*, 1689–1691. [CrossRef] [PubMed]
4. Ehrlén, J.; Münzbergová, Z. Timing of flowering: Opposed selection on different fitness components and trait covariation. *Am. Nat.* **2009**, *173*, 819–830. [CrossRef]
5. Chuine, I. Why does phenology drive species distribution? *Philos. Trans. R. Soc. B Biol. Sci.* **2010**, *365*, 3149–3160. [CrossRef] [PubMed]
6. Gentry, A.H. Flowering phenology and diversity in tropical Bignoniaceae. *Biotropica* **1974**, *6*, 64–68. [CrossRef]
7. Opler, P.A.; Frankie, G.W.; Baker, H.G. Comparative phenological studies of treelet and shrub species in tropical wet and dry forests in the lowlands of Costa Rica. *J. Ecol.* **1980**, *68*, 167–188. [CrossRef]
8. Koptur, S. Outcrossing and pollinator limitation of fruit set: Breeding systems of neotropical *Inga* trees (Fabaceae: Mimosoideae). *Evolution* **1984**, *38*, 1130–1143. [CrossRef]
9. Bawa, K.S.; Kang, H.; Grayum, M.H. Relationships among time, frequency, and duration of flowering in tropical rain forest trees. *Am. J. Bot.* **2003**, *90*, 877–887. [CrossRef]
10. Thompson, J.D. How do visitation patterns vary among pollinators in relation to floral display and floral design in a generalist pollination system? *Oecologica* **2001**, *126*, 386–394. [CrossRef] [PubMed]
11. Clivati, D.; Cordeiro, G.D.; Plachno, B.J. Reproductive biology and pollination of *Utricularia reniformis* A.St.-Hil. (Lentibulariaceae). *Plant Biol.* **2014**, *16*, 677–682. [CrossRef] [PubMed]
12. Rymer, P.D.; Whelan, R.J.; Ayre, D.J.; Weston, P.H.; Russell, K.G. Reproductive success and pollinator effectiveness differ in common and rare *Persoonia* species (Proteaceae). *Biol. Conserv.* **2005**, *123*, 521–532. [CrossRef]
13. Nebot, A.; Cogoni, D.; Fenu, G.; Bacchetta, G. Floral biology and breeding system of the narrow endemic *Dianthus morisianus* Vals. (Caryophyllaceae). *Flora* **2016**, *219*, 1–7. [CrossRef]
14. Shillo, R.; Halevy, A.H. The effect of various environmental factors on flowering of gladiolus. III. Temperature and moisture. *Sci. Hortic.* **1976**, *4*, 147–155. [CrossRef]
15. Lee, Z.; Kim, S.; Choi, S.J.; Joung, E.; Kwon, M.; Park, H.J.; Shim, J.S. Regulation of flowering time by environmental factors in plants. *Plants* **2023**, *12*, 3680. [CrossRef] [PubMed]
16. Imaizumi, T.; Kay, S.A. Photoperiodic control of flowering: Not only by coincidence. *Trends Plant Sci.* **2006**, *11*, 550–558. [CrossRef] [PubMed]
17. Osnato, M.; Cota, I.; Nebhnani, P.; Cereijo, U.; Pelaz, S. Photoperiod control of plant growth: Flowering time genes beyond flowering. *Front. Plant Sci.* **2022**, *12*, 805635. [CrossRef]
18. Kami, C.; Lorrain, S.; Hornitschek, P.; Fankhauser, C. Light-regulated plant growth and development. *Curr. Top. Dev. Biol.* **2010**, *91*, 29–66. [PubMed]
19. Kurepin, L.V.; Pharis, R.P. Light signaling and the phytohormonal regulation of shoot growth. *Plant Sci.* **2014**, *229*, 280–289. [CrossRef]
20. Zhang, D.; Armitage, A.M.; Affolter, J.M.; Dirr, M.A. Environmental control of flowering and growth of *Achillea millefolium* L. 'Summer Pastels'. *HortScience* **1996**, *31*, 364–365. [CrossRef]
21. Rezazadeh, A.; Harkess, R.L.; Telmadarrehei, T. The effect of light intensity and temperature on flowering and morphology of potted red firespike. *Horticulturae* **2018**, *4*, 36. [CrossRef]
22. Cober, E.R.; Curtis, D.F.; Stewart, D.W.; Morrison, M.J. Quantifying the effects of photoperiod, temperature and daily irradiance on flowering time of soybean isolines. *Plants* **2014**, *3*, 476–497. [CrossRef]
23. Armitage, A.M.; Tsujita, M.J. The effect of supplemental light source, illumination and quantum flux density on the flowering of seed-propagated geraniums. *J. Hortic. Sci.* **1979**, *54*, 195–198. [CrossRef]
24. Smillie, R.M.; Critchley, C.; Bain, J.M.; Nott, R. Effect of growth temperature on chloroplast structure and activity in barley. *Plant Physiol.* **1978**, *62*, 191–196. [CrossRef] [PubMed]
25. Menzel, C.M.; Simpson, D.R. Effect of temperature on growth and flowering of litchi (*Litchi chinensis* Sonn.) cultivars. *J. Hortic. Sci.* **1988**, *63*, 349–360. [CrossRef]
26. Jin, B.; Wang, L.; Wang, J.; Jiang, K.Z.; Wang, Y.; Jiang, X.X.; Ni, C.Y.; Wang, Y.L.; Teng, N.J. The effect of experimental warming on leaf functional traits, leaf structure and leaf biochemistry in *Arabidopsis thaliana*. *BMC Plant Biol.* **2011**, *11*, 35. [CrossRef]
27. Lora, J.; Herrero, M.; Hormaza, J.I. Stigmatic receptivity in a dichogamous early-divergent angiosperm species, *Annona cherimola* (Annonaceae): Influence of temperature and humidity. *Am. J. Bot.* **2011**, *98*, 265–274. [CrossRef] [PubMed]
28. Barrett, A.; Brown, L. Effects of rainfall, temperature and photoperiod on the phenology of ephemeral resources for selected bushveld woody plant species in southern Africa. *PLoS ONE* **2021**, *16*, e0251421. [CrossRef] [PubMed]

29. Thuma, J.A.; Duff, C.; Pitera, M.; Januario, N.; Orians, C.M.; Starks, P.T. Nutrient enrichment and rainfall affect plant phenology and floral resource availability for pollinators. *Front. Ecol. Evol.* **2023**, *11*, 1150736. [CrossRef]
30. Morente-López, J.; Lara-Romero, C.; Ornosa, C.; Iriondo, J.M. Phenology drives species interactions and modularity in a plant-flower visitor network. *Sci. Rep.* **2018**, *8*, 9386. [CrossRef]
31. Layek, U.; Das, U.; Karmakar, P. The pollination efficiency of a pollinator depends on its foraging strategy, flowering phenology, and the flower characteristics of a plant species. *J. Asia. Pac. Entomol.* **2022**, *25*, 101882. [CrossRef]
32. Goodwin, E.K.; Rader, R.; Encinas-Viso, F.; Saunders, M.E. Weather conditions affect the visitation frequency, richness and detectability of insect flower visitors in the Australian alpine zone. *Environ. Entomol.* **2021**, *50*, 348–358. [CrossRef]
33. Cortés-Flores, K.B.; Hernández-Esquivel, A.; González-Rodríguez, G.; Ibarra-Manríquez, G. Flowering phenology, growth forms, and pollination syndromes in tropical dry forest species: Influence of phylogeny and abiotic factors. *Am. J. Bot.* **2017**, *104*, 39–49. [CrossRef] [PubMed]
34. De Jong, T.J.; Klinkhamer, P.G.L. Early flowering in *Cynoglossum officinale* L. constraint or adaptation? *Funct. Ecol.* **1991**, *5*, 750–756. [CrossRef]
35. Aguirre, A.; Dirzo, R. Effects of fragmentation on pollinator abundance and fruit set of an abundant understory palm in a Mexican tropical forest. *Biol. Conserv.* **2008**, *141*, 375–384. [CrossRef]
36. Barrett, S.C.H. Heterostyly in a tropical weed: The reproductive biology of the *Turnera ulmifolia* complex (Turneraceae). *Canadian J. Bot.* **1978**, *56*, 1713–1725. [CrossRef]
37. Coutinho, H.D.; Costa, J.G.; Lima, E.O.; Falcão-Silva, V.S.; Siqueira Júnior, J.P. Herbal therapy associated with antibiotic therapy: Potentiation of the antibiotic activity against methicillin–resistant Staphylococcus aureus by *Turnera ulmifolia* L. *BMC Complement. Altern. Med.* **2009**, *9*, 13. [CrossRef] [PubMed]
38. Nascimento, M.A.; Silva, A.K.; França, L.C.B.; Quignard, E.L.J.; López, J.A.; Almeida, M.G. *Turnera ulmifolia* L. (Turneraceae): Preliminary study of its antioxidant activity. *Bioresour. Technol.* **2006**, *97*, 1387–1391. [CrossRef]
39. Antonio, M.A.; Brito, A.S. Oral anti-inflammatory and anti-ulcerogenic activities of a hydroalcoholic extract and partitioned fractions of *Turnera ulmifolia* (Turneraceae). *J. Ethnopharmacol.* **1998**, *61*, 215–228. [CrossRef]
40. Torres-Hernández, L.; Rico-Gray, V.; Castillo-Guevara, C.; Vergara, J.A. Effect of nectar-foraging ants and wasps on the reproductive fitness of *Turnera ulmifolia* (Turneraceae) in a coastal sand dune in Mexico. *Acta Zool. Mex.* **2000**, *81*, 13–21. [CrossRef]
41. Layek, U.; Das, A.D.; Das, U.; Karmakar, P. Spatial and temporal variations in richness, diversity and abundance of floral visitors of curry plants (*Bergera koenigii* L.): Insights on plant-pollinator interactions. *Insects* **2024**, *15*, 83. [CrossRef] [PubMed]
42. Hopkins, H.C. Floral biology and pollination ecology of the neotropical species of *Parkia*. *J. Ecol.* **1984**, *72*, 1–23. [CrossRef]
43. Norton, D.J. Testing of plum pollen viability with tetrazolium salts. *J. Am. Soc. Hortic. Sci.* **1966**, *89*, 132–134.
44. Baker, H.G.; Baker, I. Starch in angiosperm pollen grains and its evolutionary significance. *Am. J. Bot.* **1979**, *66*, 591–600. [CrossRef]
45. Parfitt, E.D.; Ganeshan, S. Comparison of procedures for estimating viability of *Prunus* pollen. *HortScience* **1989**, *24*, 354–356. [CrossRef]
46. Dafni, A.; Maués, M.M. A rapid and simple procedure to determine stigma receptivity. *Sex. Plant Reprod.* **1998**, *11*, 177–180. [CrossRef]
47. Raduski, A.R.; Haney, E.B.; Igic, B. The expression of self-incompatibility in angiosperms is bimodal. *Evolution* **2012**, *66*, 1275–1283. [CrossRef]
48. Layek, U.; Bhagira, N.K.; Das, A.; Kundu, A.; Karmakar, P. Dependency of crops on pollinators and pollination deficits: An approach to measure them in considering the influence of various reproductive traits. *Agriculture* **2023**, *13*, 1563. [CrossRef]
49. Margalef, R. Temporal succession and spatial heterogeneity in phytoplankton. In *Perspectives in Marine Biology*; Buzzati-Traverso, A.A., Ed.; University of California Press: Berkeley, CA, USA, 1958; pp. 323–347.
50. Shannon, C.E.; Weaver, W. *The Mathematical Theory of Communication*; University of Illinois Press: Champaign, IL, USA, 1949; p. 117.
51. Spears, E.E., Jr. A direct measure of pollinator effectiveness. *Oecologia* **1983**, *57*, 196–199. [CrossRef] [PubMed]
52. Layek, U.; Kundu, A.; Bisui, S.; Karmakar, P. Impact of managed stingless bee and western honey bee colonies on native pollinators and yield of watermelon: A comparative study. *Ann. Agric. Sci.* **2021**, *66*, 38–45. [CrossRef]
53. Bawa, K.S. Patterns of flowering in tropical plants. In *Handbook of Experimental Pollination Biology*; Jones, C.E., Little, R.J., Eds.; Van Nostrand Reinhold: New York, NY, USA, 1983; pp. 394–410.
54. Frankie, G.W.; Baker, H.G.; Opler, P.A. Comparative phenological studies of trees in tropical wet and dry forests in the lowlands of Costa Rica. *J. Ecol.* **1974**, *62*, 881–919. [CrossRef]
55. Augspurger, C.K. Reproductive synchrony of a tropical shrub (*Hybanthus prunifolius*): Influence on pollinator attraction and movement. *Ecology* **1981**, *62*, 774–788. [CrossRef]

56. Martínez-Sánchez, J.J.; Segura, F.; Aguado, M.; Franco, J.A.; Vicente, M.J. Life history and demographic features of *Astragalus nitidiflorus*, a critically endangered species. *Flora* **2011**, *206*, 423–432. [CrossRef]
57. Karsai, I.; Kőszegi, B.; Kovács, G.; Szűcs, P.; Mészáros, K.; Bedő, Z.; Veisz, O. Effects of temperature and light intensity on flowering of barley (*Hordeum vulgare* L.). *Biol. Futura* **2008**, *59*, 205–215. [CrossRef] [PubMed]
58. Omolaja, S.S.; Aikpokpodion, P.; Oyedeji, S.; Vwioko, D.E. Rainfall and temperature effects on flowering and pollen productions in cocoa. *Afr. Crop Sci. J.* **2009**, *17*, 41–48. [CrossRef]
59. Hamadina, E.I.; Craufurd, P.Q.; Asiedu, R. Flowering intensity in white yam (*Dioscorea rotundata*). *J. Agric. Sci.* **2009**, *147*, 469–477. [CrossRef]
60. Routley, M.B.; Berlin, R.I.; Husband, B.C. Correlated evolution of dichogamy and self-incompatibility: A phylogenetic approach. *Int. J. Plant Sci.* **2004**, *165*, 983–993. [CrossRef]
61. Aronne, G.; Buonanno, M.; De Micco, V. Reproducing under a warming climate: Long winter flowering and extended flower longevity in the only Mediterranean and maritime *Primula*. *Plant Biol.* **2015**, *17*, 535–544. [CrossRef] [PubMed]
62. Vesprini, J.; Pacini, E. Temperature-dependent floral longevity in two *Helleborus* species. *Plant Syst. Evol.* **2005**, *252*, 63–70. [CrossRef]
63. Torres-Díaz, C.; Gomez-Gonzalez, S.; Stotz, G.C.; Torres-Morales, P.; Paredes, B.; Pérez-Millaqueo, M.; Gianoli, E. Extremely long-lived stigmas allow extended cross-pollination opportunities in a high Andean plant. *PLoS ONE* **2011**, *6*, e19497. [CrossRef] [PubMed]
64. Barrett, S.C.; Shore, J.S. Variation and evolution of breeding systems in the *Turnera ulmifolia* L. complex (Turneraceae). *Evolution* **1987**, *41*, 340–354. [CrossRef] [PubMed]
65. Fisher, R.A. Average excess and average effect of a gene substitution. *Ann. Eugenics* **1914**, *11*, 53–63. [CrossRef]
66. Lloyd, D. Some reproductive factors affecting the selection of self-fertilization in plants. *Am. Nat.* **1979**, *113*, 67–79. [CrossRef]
67. Pannell, J.; Barrett, S. Baker's Law revisited: Reproductive assurance in a metapopulation. *Evolution* **1998**, *52*, 657–668. [CrossRef] [PubMed]
68. Eckert, C.; Samis, K.; Dart, S. Reproductive assurance and the evolution of uniparental reproduction in flowering plants. In *Ecology and Evolution of Flowers*; Harder, L., Barrett, S., Eds.; Oxford University Press: Oxford, UK, 2006; pp. 183–200.
69. Baker, H. Self-compatibility and establishment after 'long distance' dispersal. *Evolution* **1955**, *9*, 347–349.
70. Charlesworth, D. Evolution of plant breeding systems. *Curr. Biol.* **2006**, *16*, 726–735. [CrossRef]
71. Schlindwein, C.; Medeiros, P.C. Pollination in *Turnera subulata* (Turneraceae): Unilateral reproductive dependence of the narrowly oligolectic bee *Protomeliturga turnerae* (Hymenoptera, Andrenidae). *Flora* **2006**, *201*, 178–188. [CrossRef]
72. Webber, S.M.; Garratt, M.P.; Lukac, M.; Bailey, A.P.; Huxley, T.; Potts, S.G. Quantifying crop pollinator-dependence and pollination deficits: The effects of experimental scale on yield and quality assessments. *Agric. Ecosyst. Environ.* **2020**, *304*, 107106. [CrossRef]
73. Albrecht, M.; Schmid, B.; Hautier, Y.; Müller, C.B. Diverse pollinator communities enhance plant reproductive success. *Proc. R. Soc. B Biol. Sci.* **2012**, *279*, 4845–4852. [CrossRef] [PubMed]
74. Singh, R.P.; Prasad, P.V.; Reddy, K.R. Impacts of changing climate and climate variability on seed production and seed industry. *Adv. Agron.* **2013**, *118*, 49–110.
75. Strelin, M.M.; Aizen, M.A. The interplay between ovule number, pollination and resources as determinants of seed set in a modular plant. *PeerJ* **2018**, *6*, e5384. [CrossRef] [PubMed]
76. Cucinotta, M.; Di Marzo, M.; Guazzotti, A.; de Folter, S.; Kater, M.M.; Colombo, L. Gynoecium size and ovule number are interconnected traits that impact seed yield. *J. Exp. Bot.* **2020**, *71*, 2479–2489. [CrossRef]
77. Passos, J.M.; Gimenes, M. Pollination of *Turnera subulata*: Exotic or native bees? *Iheringia Ser. Zool.* **2022**, *112*, e2022006. [CrossRef]
78. McCall, C.; Primack, R.B. Influence of flower characteristics, weather, time of day, and season on insect visitation rates in three plant communities. *Am. J. Bot.* **1992**, *79*, 434–442. [CrossRef]
79. Harris, L.F.; Johnson, S.D. The consequences of habitat fragmentation for plant–pollinator mutualisms. *Int. J. Trop. Insect Sci.* **2004**, *24*, 29–43.
80. Obregon, D.; Guerrero, O.R.; Stashenko, E.; Poveda, K. Natural habitat partially mitigates negative pesticide effects on tropical pollinator communities. *Glob. Ecol. Conserv.* **2021**, *28*, e01668. [CrossRef]
81. Vrdoljak, S.M.; Samways, M.J.; Simaika, J.P. Pollinator conservation at the local scale: Flower density, diversity and community structure increase flower visiting insect activity to mixed floral stands. *J. Insect Conserv.* **2016**, *20*, 711–721. [CrossRef]
82. Marín, L.; Martínez-Sánchez, M.E.; Sagot, P.; Navarrete, D.; Morales, H. Floral visitors in urban gardens and natural areas: Diversity and interaction networks in a neotropical urban landscape. *Basic Appl. Ecol.* **2020**, *43*, 3–15. [CrossRef]
83. Vega, C.; Arista, M.; Ortiz, P.L.; Herrera, C.M.; Talavera, S. The ant-pollination system of *Cytinus hypocistis* (Cytinaceae), a Mediterranean root holoparasite. *Ann. Bot.* **2009**, *103*, 1065–1075. [CrossRef]
84. Dutton, E.M.; Frederickson, M.E. Why ant pollination is rare: New evidence and implications of the antibiotic hypothesis. *Arthropod-Plant Inte.* **2012**, *6*, 561–569. [CrossRef]

85. Cuautle, M.; Rico-Gray, V.; Diaz-Castelazo, C. Effects of ant behaviour and presence of extrafloral nectaries on seed dispersal of the Neotropical myrmecochore *Turnera ulmifolia* L. (Turneraceae). *Biol. J. Linn. Soc.* **2005**, *86*, 67–77. [CrossRef]
86. Ramírez, M.J.; Escanilla-Jaramillo, C.; Murúa, M.M. A Specialized Combination: The Relationship between Reproductive Structure Arrangement and Breeding Systems in oil-rewarding *Calceolaria* Species (Calceolariaceae). *Evol. Biol.* **2024**, *51*, 356–365. [CrossRef]

biology

MDPI

Article

Effect of Nitrophos Fertilizer on Pollinator Dynamics and Onion Seed Yield

Syeda Fatima Bukhari [1], Mudssar Ali [1,*], Fawad Zafar Ahmad Khan [1,2] and Raimondas Mozūratis [3,4,*]

1 Institute of Plant Protection, Muhammad Nawaz Shareef University of Agriculture Multan, Multan 60000, Pakistan; fatimabukhari45@gmail.com (S.F.B.); fawad.zafar@mnsuam.edu.pk (F.Z.A.K.)
2 Department of Outreach and Continuing Education, Muhammad Nawaz Shareef University of Agriculture Multan, Multan 60000, Pakistan
3 Laboratory of Chemical and Behavioural Ecology, Institute of Ecology, Nature Research Centre, LT-08412 Vilnius, Lithuania
4 Department of Zoology, Stockholm University, SE-10691 Stockholm, Sweden
* Correspondence: mudssar.ali@mnsuam.edu.pk (M.A.); raimondas.mozuraitis@su.se (R.M.)

Simple Summary: The seed yield of flowering crops is influenced by nutrient availability and pollination. However, the combined effects of these factors have not been extensively studied. This research investigated the impact of varying insect pollination levels (0%, 25%, 50%, and 100%) and nitrophos fertilizer doses (188, 375, and 750 kg/hectare) on onion seed production. The results demonstrated that onion seed yield increased with pollination at moderate fertilizer levels. Pollinator abundance varied with fertilizer levels; honey bees were more prevalent at lower fertilizer levels, while syrphid flies were more abundant at higher levels. The umbel weight and number of seeds per umbel were highest with the intermediate fertilizer treatment. Additionally, our findings highlighted the single-visit effectiveness of the *Apis florea* bee. The data revealed a complex interplay between fertilizer treatment and pollination services, underscoring the role of insect pollination in enhancing onion seed production.

Abstract: The yield of flowering crops depends on multiple factors, with pollination and the availability of nutrients being particularly crucial. However, fewer studies have explored the combined effects of nutrients and insect-mediated pollination on onions (*Allium cepa* L.). Here, we studied how different levels of insect pollination (0%, 25%, 50%, and 100%) and nitrophos fertilizer application (188, 375, and 750 kg/hectare) affect onion seed yield. We found that pollination benefits to yield were maximum at intermediate levels of nitrophos availability. A higher pollinator abundance was recorded at a high nitrophos level, followed by intermediate and low levels. Syrphid flies and honey bees were the most abundant insect pollinators. The abundance of honey bees was higher at a low nitrophos level, while syrphid flies were more abundant at a higher nitrophos level. Moreover, the umbel weight and seed number were higher at the intermediate fertilizer level. Also, single visits by insect pollinators were observed, and the highest umbel weight was found at the intermediate fertilizer level with a single visit from *A. florea*. This finding provides evidence that resource and pollination availability have interactive, nonlinear effects on seed production. Our results also support ecological intensification as a reliable strategy for sustainable agro-ecosystem management.

Keywords: pollination ecology; honey bees; syrphid flies; umbel weight; nutrient–pollination interaction; sustainable agriculture

1. Introduction

Promoting insect pollination and improving crop diversity through non-conventional farming practices are essential to enhance food security and sustainability [1]. Even in self-compatible crops, pollination is necessary for maximum seed and food production [2]. Crops have varying pollination needs, and many depend on insect pollinators for pollen transfer, ensuring successful fertilization and fruit and seed production [3–5]. Many crops depend partially or entirely on insect pollination services.

After tomatoes, onions are the second most widely consumed vegetable globally and have the most extended history of cultivation [6]. They are consumed both raw in salads and in processed form [7]. Onion flowers are protandrous, i.e., releasing pollen before the stigma becomes receptive; therefore, cross-pollination relies mainly on biotic factors. The anther produces sticky, wet pollen grains, and, due to this factor, insect pollination contributes more than wind pollination [8]. A few studies demonstrated that effective pollination by insects leads to significantly higher seed set and yield (44 and 6 times greater, respectively) compared to self-pollinated plants [9,10]. Another study found that a lack of insect pollinators can result in a 5–14% reduction in seed quantity and a decline in seed quality [11,12].

A strong linkage between crop management practices and insect-mediated pollination has been reported [13,14]. Studies have shown that water and nutrient levels affect pollination success [13,15]. Moreover, these factors also influence pollinator health and behavior due to changes in floral resource quality and availability. For example, the maximum yield in okra, *Abelmoschus esculentus* (L.) (Moench) (Malvaceae), depends on optimal soil fertilization coupled with adequate pollinator visitation [1]. A study on common beans, *Phaseolus vulgaris* L. (Fabaceae), also showed that low nitrogen levels led to a higher abundance of insect pollinators [16]. Similar results have been reported in oil-seed rape (*Brassica napus* L. var. *oleifera*) (Brassicaceae), where low nitrogen application led to increased insect pollination [17]. Another study on sunflowers (*Helianthus annuus* L.) (Asteraceae) reported higher pollinator visitation and yield at intermediate nitrogen levels [18]. Moreover, ornamental flower Scarlet gilia *Ipomopsis aggregata* (Pursh) V.E.Grant (Polemoniaceae) and Lewis flax *Linum lewisii* Pursh (Linaceae) showed no changes in reproductive success due to fertilizer application [19].

Nutrients indirectly affect pollen quantity and quality [20], flower production, pollination, pollinator attraction, and seed setting [19]. Soil-available nutrients (nitrogen, phosphorus, and potassium) have an important influence on floral traits and pollinator attraction [21]. The insect visitation rate has been reported to correlate to nectar amino acids, pollen fatty acids, and amino acids [22]. Nitrogen and phosphorus are key nutrients that enhance crop productivity [23]. However, their use also negatively affects the environment by increasing water and soil eutrophication and greenhouse gas emissions [24,25]. In insects, bees are the most important pollinator group for plant reproduction [26]. Wild bees are essential for improving the productivity of pollinator-dependent crops [27], but their abundance and diversity are negatively affected by agricultural intensification [28]. The combined impact of pollination services and fertilizer application on onion yield has not been documented in the literature.

The current study aimed to identify the effect of soil fertilization and pollinator visitation on onion yield. To test the hypothesis that insect pollination services could be maximized at specific levels of nitrophos application, we created various insect-mediated pollination gradients along with the nitrophos application levels and assessed the onion seed yield.

2. Materials and Methods

2.1. Plant Selection

We selected the red phulkari variety of onion (*A. cepa* L.) because it has been cultivated on a large scale in Pakistan. Onions are biennial herbaceous plants grown for their edible bulbous base. Their umbels typically contain 50–2000 florets enclosed in 2–3 white-colored spathes. Being protandrous, onions require cross-pollination, and insect-mediated pollination is important in determining the final seed yield [29].

2.2. Experimental Site and Design

This study was conducted from November 2019 to April 2020 at the experimental farm of Muhammad Nawaz Shareef University of Agriculture, Multan, Pakistan. The Multan district has been categorized as a subtropical desert due to its hot summers and chilly winters, with temperatures ranging from 38 to 50 °C at maximum and from 8 to 12 °C at minimum. During summer, the average monthly rainfall is around 18 mm.

We arranged nine plots into three blocks using a randomized complete block design, with each block containing three plots. Each plot measured 15.24 m in length and width and was isolated from the surrounding plots by 2 m. Onion bulbs were sown directly into the soil at a depth of 3 cm in four rows, with a row-to-row spacing of 0.45 m and a plant-to-plant spacing of 0.15 m. Plant density was similar across the plots. After sowing, all plots were irrigated individually.

2.3. Fertilization Treatment

Fertilizer treatments began six weeks after sowing, when plants reached an average height of 8–12 cm. Three levels of nitrogen (N) and phosphorus (P) fertilizer were applied in doses of 188 kg/hectare (T1, low), 375 kg/hectare (T2, moderate), and 750 kg/hectare (T3, high). Nitrophos fertilizer was applied three times, i.e., when plants reached a height of 8–12 cm, at 20% of flowering, and 60% of flowering. To estimate the flowering percentage, we counted the total number of flowering umbels at the anthesis stage, divided them by the total number of plants, and multiplied the result by 100.

2.4. Pollination Treatment

Four levels of insect pollination treatment were used to calculate the effectiveness of pollination, i.e., 0%, 25%, 50%, and 100%. Pollination treatments were assigned before the flower opening stage. Plants of similar height and vigor were randomly selected and assigned a pollination treatment level within each plot. Pollination gradients were managed by covering umbels with fine mesh net bags. We achieved 0% pollination by completely covering the umbels, 25% pollination by removing the fine mesh bags for one day and then covering them for three days, 50% pollination by removing the fine mesh bags for two days and covering them for two days, and 100% pollination by keeping the umbels uncovered. For all the treatments with the net bags, we adjusted the bags to avoid contact with the umbels [15]. We covered and uncovered the umbels between 10:00 am and 12:00 pm, which was the peak foraging time of the pollinators [10].

2.5. Yield Parameters and Pollinator Visitation Rate

At physiological maturity, umbels were manually harvested and placed in paper bags to prevent moisture accumulation, then dried for six weeks. Seeds were harvested by shaking the umbels. We calculated the umbel weight and seed number per umbel in grams.

During the flowering stage, pollinator diversity and abundance were recorded by randomly selecting ten umbels per plot and observing each umbel for one minute, counting the visiting insect pollinator species in each plot. The foraging behavior of abundant insect pollinators was recorded in terms of visit duration (time spent on an umbel) and the number of umbellules (flowers within an umbel) visited per visit (visitation rate). Observations were made between 10:00 a.m. and 12:00 p.m.

2.6. Effectiveness of Pollinators in Fertilizer Treatment

The effectiveness of pollinators was measured by recording pollen deposition from a single visit. Nine umbels of the same age, size, and vigor were visually selected from each fertilizer treatment plot. Umbels were covered with nylon mesh bags before flowering. During peak activity time (10:00 a.m.–12:00 p.m.) [10], the bags were removed when 50% of the flowers had opened. One pollinator (bee or fly) was allowed to visit the umbel, which was covered again after the pollinator visit [10]. Open-pollinated plants and caged plants (with no insect visitation) were also kept for yield comparison. Yield parameters, including umbel weight and number of seeds per umbel, were recorded using a digital weighing balance.

2.7. Statistical Analysis

Statistical analyses were performed using XL STAT and Statistics 8.1. Data for abundance, visit duration, and visitation rate of insect pollinators in each treatment were transformed using log transformation to ensure a normal distribution. After that, a two-way analysis of variance (ANOVA) was performed. When the interaction between factors was not significant, the main effects of each factor were evaluated independently within the two-way ANOVA. Similarly, yield parameters, i.e., umbel weight and seeds per umbel were analyzed. The means were separated using Tukey's HSD all-pairwise comparisons test. The abundance, visit duration, and visitation rate of bees and flies were subjected to a paired-sample *t*-test.

3. Results

The pollinators visiting onion umbels included three bee species, twelve dipteran fly species, and two wasp species (Figure 1). Among the bees observed, *A. florea* was the most abundant species, followed by *Xylocopa* sp. We also observed twelve dipteran fly species belonging to five families: Calliphoridae, Muscidae, Sarcophagidae, Syrphidae, and Stratiomyidae. The abundance of hoverfly *Eristalinus aeneus* was the highest, followed by *Episyrphus balteatus* and *Sphaerophoria bengalensis* (Table 1).

Table 1. List of insect pollinators foraging on onion flowers in plots with different levels of fertilizer.

Pollinator Group	Order	Family	Scientific Name
Honey bees and wild bees	Hymenoptera	Apidae	*Apis florea* *Apis dorsata* *Xylocopa* sp.
Wasps		Vespidae	*Vespa orientalis* *Polistes* sp.
Flies	Diptera	Calliphoridae	*Calliphoridae* sp.
		Muscidae	*Musca domestica*
		Sarcophagidae	*Sarcophaga* sp.
		Syrphidae	*Eristalinus aeneus* *Eupeodes corollae* *Sphaerophoria scripta* *Syrphus ribesii* *Episyrphus balteatus* *Eristalis tenax* *Mesembrius* sp. *Melanostoma* sp
		Stratiomyidae	*Hedriodiscus* sp.

Figure 1. Different insect pollinators forage on onion crops: (**A**) *Apis florea*; (**B**) *Apis dorsata*; (**C**) *Xylocopa* sp.; (**D**) *Polistes* sp.; (**E**) *Vespa* sp.; (**F**) *Calliphora* sp.; (**G**) *Stratiomyidae* sp.; (**H**) *Eristalinus aeneus*; (**I**) *Eristalinus laetus*; (**J**) *Eupeodes corollae*; (**K**) *Musca domestica*; (**L**) *Episyrphus balteatus*; (**M**) *Eristalinus megacephalus*; (**N**) *Sphaerophoria bengalensis*; (**O**) *Eristalis tenax*; (**P**) *Sarcophaga* sp.; (**Q**) *Chrysoperla carnea*; and (**R**) *Vanessa cardui*.

Other insects, such as *Chrysoperla carnea*, and a butterfly species, *Vanessa cardui*, were occasionally found visiting the onion umbels. No significant differences were found for the effect of fertilizer on the total abundance of major pollinator species, including bees and flies ($p = 0.6784$) (Table 2). Overall, the abundance of bees was significantly higher compared to flies (Figure 2).

Table 2. The abundance of major insect pollinators at different levels of nitrophos fertilizer.

Pollinator	Nitrophos Levels	Abundance
Bees	Low	14.72 ± 9.22
	Moderate	8.0 ± 5.09
	High	7.8 ± 5.2
Flies	Low	18.0 ± 10.21
	Moderate	22.25 ± 11.71
	High	24.33 ± 12.52
		$F = 0.39$, df = 2
		$p = 0.6784$

Overall, bees demonstrated a higher number of visits per umbel and a higher stay time per umbel compared to the flies (Figure 3). Furthermore, the visitation rate of primary insect pollinators on onions revealed differences across low-, medium-, and high-fertilizer plots. Bees had the highest visitation rate in medium-level fertilizer plots (T2), while flies had the highest visitation rate in high-level fertilizer plots (T3). The impact of fertilizer on the visit duration of bees and flies was not significant ($p = 0.1509$) (Table 3). Overall, moderate and high levels of nitrophos application led to significantly higher stay times of pollinators (bees and flies) compared to lower application levels. However, no significant differences were recorded for several umbels visited by pollinators (Figure 4).

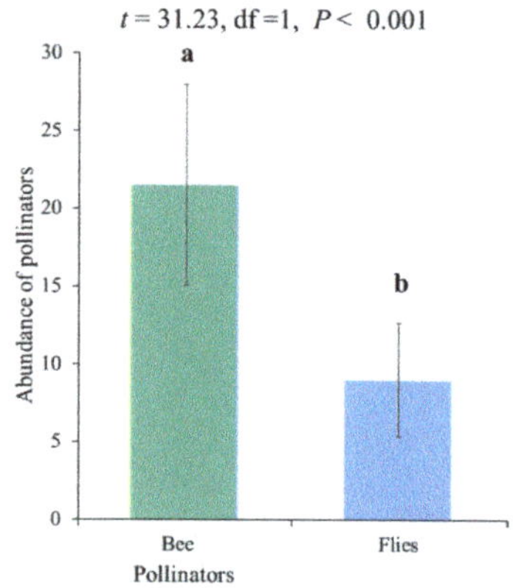

Figure 2. Overall abundance of onion crop pollinators (bees and flies) at all levels of fertilizer. Bars having different letters (a, b) show a statistically significant difference between groups based on a *t*-test ($p < 0.05$).

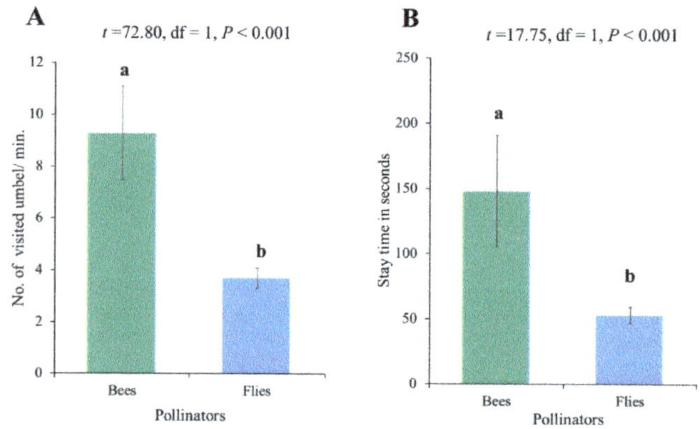

Figure 3. Foraging behavior of onion crop pollinators (bees and flies) at all levels of fertilizer: (**A**) no. of visited onion umbel/minute by bees and flies and (**B**) visit duration of bees and flies on the umbel. Bars having different letters (a, b) show a statistically significant difference between groups based on a *t*-test ($p < 0.05$).

Table 3. Yield parameters under three levels of fertilizer application and different pollination treatments.

Pollination Treatment	Nitrophos Levels	Umbel Weight (Grams)	No. Seeds per Umbel
	Low	7.64 ± 0.73 a	37.00 ± 4.85
100%	Moderate	8.04 ± 0.79 a	45.22 ± 5.49
	High	7.50 ± 0.47 a	34.55 ± 2.88
	Low	4.68 ± 0.42 cd	13.55 ± 2.59
50%	Moderate	6.73 ± 0.67 ab	20.88 ± 2.18
	High	4.91 ± 0.74 cd	16.33 ± 3.94
	Low	3.98 ± 0.49 d	6.11 ± 1.24
25%	Moderate	4.29 ± 0.32 cd	9.77 ± 1.16
	High	5.44 ± 0.48 cd	8.00 ± 1.93
	Low	2.47 ± 0.29 e	2.11 ± 0.77
0%	Moderate	1.93 ± 0.26 e	1.22 ± 0.49
	High	2.36 ± 0.30 e	1.77 ± 0.68
		$F = 2.13$, df = 6	$F = 0.81$, df = 6
		$p = 0.0567$	$p = 0.5677$

Means within a column followed by different letters show significant differences, as determined by Tukey's HSD all-pairwise comparisons test. The column without letters shows no differences between means.

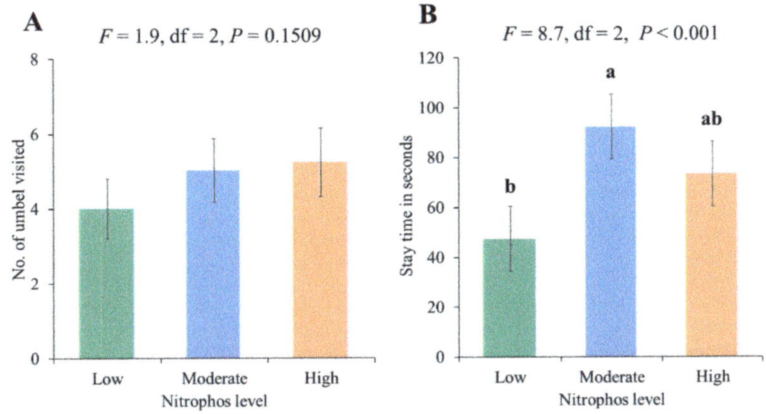

Figure 4. Foraging behavior of onion crop pollinators (bees and flies) at different levels of fertilizer application: (**A**) no. of visited onion umbel/minute by bees and flies and (**B**) visit duration of bees and flies on umbel. Bars with different letters are significantly different (Tukey's HSD all-pairwise comparisons test). Bars without letters are not significantly different.

The results of pollination treatments at low, medium, and high fertilizer levels showed that the maximum umbel weight and number of seeds were recorded in the 100% pollination treatment, followed by the 50% pollination treatment (Table 4). The highest umbel weight (8.04 ± 0.79) and number of seeds (45.22 ± 5.49) were observed at moderate fertilizer levels with the 100% pollination treatment. However, there was a decline in seed yield at high fertilizer levels (T3) with 100% pollination. The least reproductive success was recorded for 0% pollination across all fertilizer levels. The effects of fertilization and pollination treatments on onion umbel weight were insignificant ($p = 0.567$). Overall, the moderate fertilizer level (375 kg/hectare) combined with 100% pollination provided the highest seed yield in onions (Table 4).

Table 4. Visitation rate and visit duration of major pollinator groups on onion flowers at different nitrophos fertilizer levels.

Pollinator	Nitrophos Levels	Visitation Rate (No. of Visited Umbel/min.)	Visit Duration (Seconds)
Bees	Low	7.33 ± 2.52 b	93.37 ± 40.38
	Moderate	11.11 ± 3.59 a	203.86 ± 78.54
	High	9.44 ± 3.47 b	146.33 ± 96.33
Flies	Low	3.23 ± 0.75 c	36.77 ± 8.35
	Moderate	3.61 ± 0.44 bc	66.65 ± 11.61
	High	4.26 ± 0.76 bc	56.45 ± 11.71
		$F = 8.68$, df = 2	$F = 1.91$, df = 2
		$p = 0.0002$	$p = 0.1509$

Means within a column followed by different letters show significant differences, as determined by Tukey's HSD all-pairwise comparisons test. The column without letters shows no differences between means.

The effectiveness of a single visit by bees and flies differed significantly in terms of umbel weight. Moreover, the effect on seed number was not significantly different ($p = 0.615$). The highest umbel weight was achieved with a single visit from bees at the low fertilizer level, followed by the high fertilizer level, with a slight difference (Table 5). Overall, a significantly higher number of seeds and a higher umbel weight were recorded for moderate nitrophos application (Figure 5). Moreover, the 100% pollination treatment led to higher umbel weight and number of seeds (Figure 6).

Table 5. Single-visit efficacy at different levels of fertilizer.

Pollinators	Nitrophos Levels	Umbel Weight (Grams)	No. Seeds per Umbel
	Low	9.27 ± 0.27 a	35.00 ± 0.82
Bees	Moderate	6.98 ± 0.17 ab	32.00 ± 0.41
	High	7.45 ± 0.29 a	28.00 ± 1.08
	Low	3.05 ± 0.28 c	16.60 ± 3.74
Flies	Moderate	4.91 ± 0.81 bc	18.80 ± 4.20
	High	0.91 ± 0.48 c	16.80 ± 2.63
		$F = 7.0$, df = 2	$F = 0.5$, df = 2
		$p = 0.0046$	$p = 0.6149$

Means within a column followed by different letters show significant differences, as determined by Tukey's HSD all-pairwise comparisons test. The column without letters shows no differences between means.

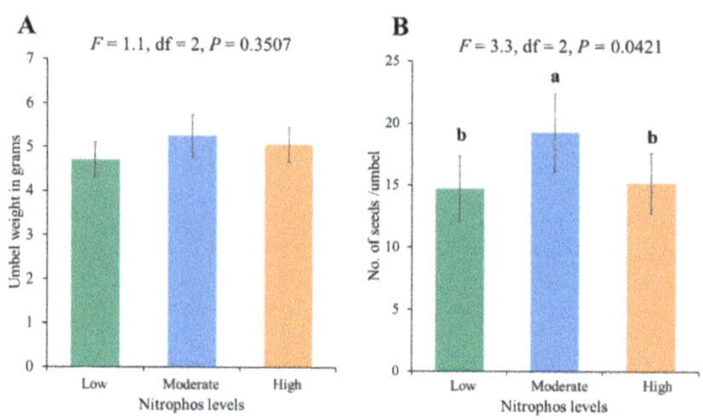

Figure 5. Yield parameters under three levels of fertilizer application: (**A**) weight of umbel in grams and (**B**) no. of seeds/umbel. Bars with different letters are significantly different (Tukey's HSD all-pairwise comparisons test). Bars without letters are not significantly different.

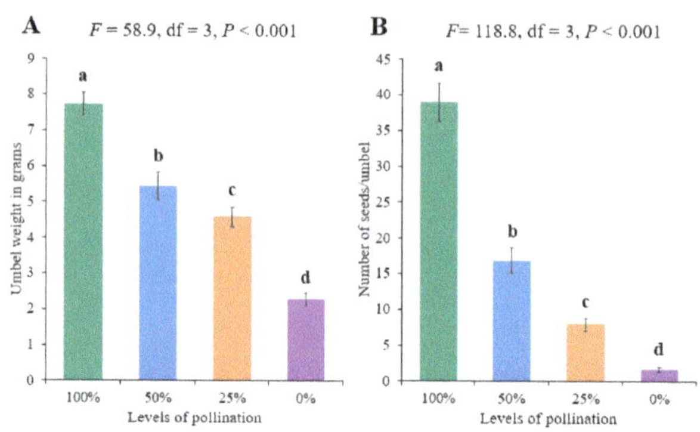

Figure 6. Reproductive parameters under different pollination treatments: (**A**) weight of umbel in grams and (**B**) no. of seeds/umbel. Bars with different letters are significantly different (Tukey's HSD all-pairwise comparisons test).

4. Discussion

The current study revealed that hymenopterans (71%) and dipterans (29%) were the most abundant visitors of onion flowers, while neuropterans and lepidopterans were occasionally found visiting the onion umbels. These findings aligned well with other studies that identified honey bees as the most abundant pollinators of onions [12,30–32]. Additionally, some studies have shown that syrphid flies played a vital role in onion pollination [10,33]. In our study, dipterans were the most species-rich order of onion pollinators.

Our research established the link between nitrophos fertilizer application and pollinator visitation. Hymenopterans (bees) preferred low- and medium-level fertilizer plots, while dipterans (flies) preferred to visit high-level fertilizer plots. The current results differed from the published findings, which showed that the bumble bee visitation rate was significantly higher for *Impatiens capensis* Meerb plants treated with high fertilizer amounts compared to those which had received low and no fertilizer treatments [34].

Published data showed that high fertilizer application altered the quantitative floral nutrition composition by changing the quality of nectar and pollen, which, in turn, influenced pollinator behavior [16,22,35–37]. It was shown that high fertilizer application decreased the quantity of essential amino acids in the plant and negatively affected bumble bees' behavior [16]. The decreased visitation rate of onion flowers by bees in high-fertilization plots could be due to a similar effect of suboptimal fertilizer levels. The higher attraction of dipterans could be explained by their preferences for high levels of lipids, sugars, fructose, glycogen, and carbohydrates necessary for survival [38]. This finding suggests that high fertilizer levels can change the abundance and species composition of onion pollinators. On the other hand, another study reported that soil fertilizer application in okra, up to its optimal level, played an important role in improving the floral cues and attracting more pollinator fauna [1]. The impact of high fertilization levels on flower-visiting insects is pollinator-flowering plant species-specific and widely variable [39].

Our study showed that the benefits of pollination on onion yield could be maximized at moderate or recommended nitrophos applications, with a decrease in yield observed at higher nitrophos applications. Fertilizer application had a direct effect on the mature umbel weight [40]. We found that the combination of soil fertilization and insect pollination led to heavier umbel/plant in plants that were 100% pollinated compared to those with 0% pollination. Plants with 0% pollination, representing self-pollination, exhibited the lowest yield parameters due to suboptimal soil nutrients during both vegetative and reproductive growth phases, decreasing plant biomass production and yield parameters throughout the cropping seasons. This suggests that, while self-compatible plants can produce seeds without cross-pollination, they may experience reduced seed production in the absence of pollinators [10]. Our findings demonstrate that pollinators enhance onion seed production by increasing flower attractiveness and boosting seed production. We observed that exposure to 25% pollination had a minimal impact on onion yield, similar to when pollinators were entirely absent, while successful pollination exposures occurred at 50% and 100% under minimal and optimal soil fertilization. The variability in insect pollinator abundance suggests a potential influence on pollinator efficiency. The low yields in plants with 0% and 25% pollination were probably due to insufficient insect visitation, resulting in limited or no pollen transfer between plants, thus reducing the number and weight of seeds. This aligns with previous evidence, highlighting the positive interaction between soil fertilization and sufficient pollinators to improve sunflower and okra production parameters [1]. Another study reported that insect pollination enhances the average crop production by 18% to 71% (depending on the crop type) [41].

In addition, this study is the first example of single visits by insect pollinators, aiming to evaluate the effectiveness of pollinators at different nitrophos application levels. The mature umbel weight was highest where bees visited both at low and high levels of fertilizer, while a higher umbel weight was recorded at intermediate levels of fertilizer application. On the other hand, a decrease in umbel weight was seen when fertilizer levels increased. Interestingly, when the fertilizer was combined with a single pollinator visit, the amount of seeds produced per umbel was not significantly different. Furthermore, similar results were observed in fly single visits for all treatment levels, supporting the hypothesis that higher fertilizer concentrations reduce onion crop productivity and pollination efficiency.

Our results highlight the complex interlinkage between fertilizer application, crop productivity, and pollinator activity, emphasizing the importance of management strategies in agricultural landscapes. Understanding these mechanisms can help conserve the pollinator population. Further studies are needed to understand the physiological mechanisms associated with higher fertilizer application levels and low pollinator visits. Moreover, field trials should be conducted in different locations with different pollinator networks. Long-term studies are needed to evaluate the effects of different fertilizers on plant–pollinator interactions.

5. Conclusions

This study demonstrated the significant impact of nitrophos fertilizer application and insect pollinator visitation on onion (*Allium cepa* L.) seed production. Moderate fertilizer levels enhanced pollinator activity and onion umbel weight. Specific patterns of insect pollinator preferences emerged, with bees, including *Apis* species, favoring low-to-moderate fertilizer levels, while syrphid flies preferred higher levels. The highest umbel weight was achieved with the moderate fertilizer treatment, highlighting the bottom–up effects of moderate fertilizer use on pollinators and, consequently, onion seed yield. Future research could explore the use of other fertilizers, including micronutrients, and consider conducting long-term trials under diverse field conditions and across various crops to understand better fertilizer-mediated changes in pollinator behavior and their impact on crop yield.

Author Contributions: Conceptualization, M.A. and S.F.B.; methodology, M.A.; software, F.Z.A.K.; validation, M.A.; formal analysis, M.A. and F.Z.A.K.; investigation, S.F.B.; resources, M.A., F.Z.A.K. and R.M.; data curation, S.F.B.; writing—original draft preparation, F.Z.A.K. and S.F.B.; writing—review and editing, F.Z.A.K. and M.A.; visualization, F.Z.A.K.; supervision, M.A.; project administration, M.A. and F.Z.A.K.; and funding acquisition, R.M. All authors have read and agreed to the published version of the manuscript.

Funding: This research was funded by the MNS University of Agriculture, Multan, Pakistan, and a Lithuanian state grant awarded to R.M. through the Nature Research Centre, Program 2: Climate and Ecosystems, Vilnius, Lithuania.

Institutional Review Board Statement: Not applicable.

Informed Consent Statement: Not applicable.

Data Availability Statement: The data used in this study can be provided by the corresponding author upon request.

Acknowledgments: The authors thank the farm management team at the MNS University of Agriculture, Multan, Pakistan. The authors are thankful to the three anonymous reviewers for their valuable contributions to improving the manuscript.

Conflicts of Interest: The authors declare no conflicts of interest. The funders had no role in the design of the study; in the collection, analyses, or interpretation of data; in the writing of the manuscript; or in the decision to publish the results.

References

1. Adelabu, D.B.; Franke, A.C. Response of Soil Fertilization and Insect Pollination on Okra Production: Prospect for Optimizing Underutilized Crop Management. *J. Agric. Food Res.* **2023**, *14*, 100869. [CrossRef]
2. Ali, Q.; Ali, M.; Khan, F.Z.A.; Noureldeen, A.; Alghamdi, A.; Darwish, H.; Fatima, A.; Jalali, A.I.; Prendergast, K.; Saeed, S. Water Deprivation and Sowing Times Alter Plant–Pollination Interactions and Seed Yield in Sunflower, *Helianthus annuus* L. (Asteraceae). *Plants* **2024**, *13*, 3194. [CrossRef] [PubMed]
3. Aizen, M.A.; Garibaldi, L.A.; Cunningham, S.A.; Klein, A.M. How Much Does Agriculture Depend on Pollinators? Lessons from Long-Term Trends in Crop Production. *Ann. Bot.* **2009**, *103*, 1579–1588. [CrossRef] [PubMed]
4. Abdullah, S.; Ali, M.; Khan, F.Z.A.; Sajjad, A.; Qayyum, M.A.; Ahmad, N. Solitary Bees Are More Efficient Pollinators of Sponge Gourd than Giant Honeybees and Syrphid Flies. *Sociobiology* **2024**, *71*, e10279. [CrossRef]

5. Ali, Q.; Ali, M.; Khan, F.Z.A.; Awan, T.H. Comparing the Efficacy of Single and Multiple Visits by Honey and Solitary Bees on Sunflower Seed Production. *Sociobiology* **2024**, *71*, e10425. [CrossRef]
6. Benítez, V.; Mollá, E.; Martín-Cabrejas, M.A.; Aguilera, Y.; López-Andréu, F.J.; Esteban, R.M. Effect of Sterilisation on Dietary Fibre and Physicochemical Properties of Onion By-Products. *Food Chem.* **2011**, *127*, 501–507. [CrossRef]
7. Lawande, K.E. Onion. In *Handbook of Herbs and Spices*; Elsevier: Amsterdam, The Netherlands, 2012; pp. 417–429. ISBN 978-0-85709-039-3.
8. Shemesh, E.; Scholten, O.; Rabinowitch, H.D.; Kamenetsky, R. Unlocking Variability: Inherent Variation and Developmental Traits of Garlic Plants Originated from Sexual Reproduction. *Planta* **2008**, *227*, 1013–1024. [CrossRef]
9. Bukhari, S.F.; Ali, M.; Khan, F.Z.A.; Awan, T.H. From Petals to Seeds: Understanding the Role of Hymenopteran and Dipteran Pollinators in the Reproductive Success of Onion (*Allium cepa* L.). *Plant Bull.* **2024**, *3*, 19–25. [CrossRef]
10. Sajjad, A.; Saeed, S.; Masood, A. Pollinator Community of Onion (*Allium cepa* L.) and Its Role in Crop Reproductive Success. *Pak. J. Zool.* **2008**, *40*, 451–456.
11. Walker, M.K.; Howlett, B.G.; Wallace, A.R.; Mccallum, J.A.; Teulon, D.A.J. The Diversity and Abundance of Small Arthropods in Onion, *Allium cepa*, Seed Crops, and Their Potential Role in Pollination. *J. Insect Sci.* **2011**, *11*, 98. [CrossRef]
12. Devi, S.; Gulati, R.; Tehri, K.; Poonia, A. The Pollination Biology of Onion (*Allium cepa* L.)—A Review. *Agric. Rev.* **2015**, *36*, 1–13. [CrossRef]
13. Tamburini, G.; Lami, F.; Marini, L. Pollination Benefits Are Maximized at Intermediate Nutrient Levels. *Proc. R. Soc. B* **2017**, *284*, 20170729. [CrossRef] [PubMed]
14. Sawe, T.; Nielsen, A.; Eldegard, K. Crop Pollination in Small-Scale Agriculture in Tanzania: Household Dependence, Awareness and Conservation. *Sustainability* **2020**, *12*, 2228. [CrossRef]
15. Tamburini, G.; Berti, A.; Morari, F.; Marini, L. Degradation of Soil Fertility Can Cancel Pollination Benefits in Sunflower. *Oecologia* **2016**, *180*, 581–587. [CrossRef]
16. Ramos, D.D.L.; Bustamante, M.M.C.; da Silva e Silva, F.D.; Carvalheiro, L.G. Crop Fertilization Affects Pollination Service Provision—Common Bean as a Case Study. *PLoS ONE* **2018**, *13*, e0204460. [CrossRef]
17. Marini, L.; Tamburini, G.; Petrucco-Toffolo, E.; Lindström, S.A.M.; Zanetti, F.; Mosca, G.; Bommarco, R. Crop Management Modifies the Benefits of Insect Pollination in Oilseed Rape. *Agric. Ecosyst. Environ.* **2015**, *207*, 61–66. [CrossRef]
18. Haider, S.; Khan, F.Z.A.; Gul, H.T.; Ali, M.; Iqbal, S. Assessing the Role of Conservation Strips in Enhancing Beneficial Fauna in the Wheat-Cotton Agricultural System in Punjab, Pakistan. *Pak. J. Zool.* **2024**, *56*, 1–9. [CrossRef]
19. Burkle, L.A.; Irwin, R.E. The Effects of Nutrient Addition on Floral Characters and Pollination in Two Subalpine Plants, Ipomopsis Aggregata and Linum Lewisii. *Plant Ecol.* **2009**, *203*, 83–98. [CrossRef]
20. Lau, T.; Stephenson, A.G. Effects of Soil Nitrogen on Pollen Production, Pollen Grain Size, and Pollen Performance in Cucurbita Pepo (Cucurbitaceae). *Am. J. Bot.* **1993**, *80*, 763–768. [CrossRef]
21. Vaudo, A.D.; Erickson, E.; Patch, H.M.; Grozinger, C.M.; Mu, J. Impacts of Soil Nutrition on Floral Traits, Pollinator Attraction, and Fitness in Cucumbers (*Cucumis sativus* L.). *Sci. Rep.* **2022**, *12*, 21802. [CrossRef]
22. Russo, L.; Ruedenauer, F.; Gronert, A.; Van De Vreken, I.; Vanderplanck, M.; Michez, D.; Klein, A.; Leonhardt, S.; Stout, J.C. Fertilizer and Herbicide Alter Nectar and Pollen Quality with Consequences for Pollinator Floral Choices. *PeerJ* **2023**, *11*, e15452. [CrossRef]
23. Vitousek, P.M.; Porder, S.; Houlton, B.Z.; Chadwick, O.A. Terrestrial Phosphorus Limitation: Mechanisms, Implications, and Nitrogen–Phosphorus Interactions. *Ecol. Appl.* **2010**, *20*, 5–15. [CrossRef] [PubMed]
24. Tilman, D.; Balzer, C.; Hill, J.; Befort, B.L. Global Food Demand and the Sustainable Intensification of Agriculture. *Proc. Natl. Acad. Sci. USA* **2011**, *108*, 20260–20264. [CrossRef]
25. Burney, J.A.; Davis, S.J.; Lobell, D.B. Greenhouse Gas Mitigation by Agricultural Intensification. *Proc. Natl. Acad. Sci. USA* **2010**, *107*, 12052–12057. [CrossRef] [PubMed]
26. Knight, T.M.; Steets, J.A.; Vamosi, J.C.; Mazer, S.J.; Burd, M.; Campbell, D.R.; Dudash, M.R.; Johnston, M.O.; Mitchell, R.J.; Ashman, T.-L. Pollen Limitation of Plant Reproduction: Pattern and Process. *Annu. Rev. Ecol. Evol. Syst.* **2005**, *36*, 467–497. [CrossRef]
27. Dainese, M.; Martin, E.A.; Aizen, M.A.; Albrecht, M.; Bartomeus, I.; Bommarco, R.; Carvalheiro, L.G.; Chaplin-Kramer, R.; Gagic, V.; Garibaldi, L.A.; et al. A Global Synthesis Reveals Biodiversity-Mediated Benefits for Crop Production. *Sci. Adv.* **2019**, *5*, eaax0121. [CrossRef]
28. Grab, H.; Danforth, B.; Poveda, K.; Loeb, G. Landscape Simplification Reduces Classical Biological Control and Crop Yield. *Ecol. Appl.* **2018**, *28*, 348–355. [CrossRef]
29. Soto, V.C.; Maldonado, I.B.; Gil, R.A.; Peralta, I.E.; Silva, M.F.; Galmarini, C.R. Nectar and Flower Traits of Different Onion Male Sterile Lines Related to Pollination Efficiency and Seed Yield of F1 Hybrids. *J. Econ. Entomol.* **2013**, *106*, 1386–1394. [CrossRef]
30. Mazeed, A.R.; Marey, R.A. Pollinators Activity on Onion Flowers and Its Effect on Seeds Yield at Sohag Governorate, Egypt. *Egypt. J. Agric. Res.* **2018**, *96*, 465–475. [CrossRef]

31. Chandel, R.S.; Thakur, R.K.; Bhardwaj, N.R.; Pathania, N. Onion Seed Crop Pollination: A Missing Dimension in Mountain Horticulture. *Acta Hortic.* **2004**, *631*, 79–86. [CrossRef]
32. Chaudhary, N.; Sihag, R. Diversity, Foraging Behaviour and Foraging Efficiency of Different Pollinators Visiting Onion (*Allium cepa* L.) Blossoms. *Korean J. Apic.* **2003**, *18*, 103–108.
33. Saeed, S.; Sajjad, A.; Kwon, O.; Kwon, Y.J. Fidelity of Hymenoptera and Diptera Pollinators in Onion (*Allium cepa* L.) Pollination. *Entomol. Res.* **2008**, *38*, 276–280. [CrossRef]
34. Leach, M.; Dibble, A.C.; Stack, L.B.; Perkins, L.B.; Drummond, F.A. The Effect of Plant Nutrition on Bee Flower Visitation. *J. Kans. Entomol. Soc.* **2023**, *94*, 277–300. [CrossRef]
35. Gijbels, P.; Van Den Ende, W.; Honnay, O. Landscape Scale Variation in Nectar Amino Acid and Sugar Composition in a Lepidoptera Pollinated Orchid Species and Its Relation with Fruit Set. *J. Ecol.* **2014**, *102*, 136–144. [CrossRef]
36. Boff, S.; Keller, A.; Raizer, J.; Lupi, D. Decreased Efficiency of Pollen Collection Due to Sulfoxaflor Exposure Leads to a Reduction in the Size of Bumble Bee Workers in Late European Summer. *Front. Ecol. Evol.* **2022**, *10*, 842563. [CrossRef]
37. Ceulemans, T.; Hulsmans, E.; Vanden Ende, W.; Honnay, O. Nutrient Enrichment Is Associated with Altered Nectar and Pollen Chemical Composition in Succisa Pratensis Moench and Increased Larval Mortality of Its Pollinator *Bombus terrestris* L. *PLoS ONE* **2017**, *12*, e0175160. [CrossRef]
38. Pinheiro, L.A.; Torres, L.M.; Raimundo, J.; Santos, S.A.P. Effects of Pollen, Sugars and Honeydew on Lifespan and Nutrient Levels of Episyrphus Balteatus. *BioControl* **2015**, *60*, 47–57. [CrossRef]
39. Stevens, C.J.; David, T.I.; Storkey, J. Atmospheric Nitrogen Deposition in Terrestrial Ecosystems: Its Impact on Plant Communities and Consequences across Trophic Levels. *Funct. Ecol.* **2018**, *32*, 1757–1769. [CrossRef]
40. Qasem, J.R. Response of Onion (*Allium cepa* L.) Plants to Fertilizers, Weed Competition Duration, and Planting Times in the Central Jordan Valley. *Weed Biol. Manag.* **2006**, *6*, 212–220. [CrossRef]
41. Bartomeus, I.; Gagic, V.; Bommarco, R. Pollinators, Pests and Soil Properties Interactively Shape Oilseed Rape Yield. *Basic Appl. Ecol.* **2015**, *16*, 737–745. [CrossRef]

biology

MDPI

Review

Reducing Risks to Native Pollinators by Introduced Bees: A Review of Canada's Legislation with Recommendations for Yukon Territory

Maria Leung [1,*] and Donald Reid [2]

[1] Wild Tracks Ecological Consulting, Whitehorse, YT Y1A 5T2, Canada
[2] Wildlife Conservation Society Canada, Toronto, ON M5S 3A7, Canada; dreid@wcs.org
* Correspondence: wildtracks@northwestel.net

Simple Summary: Natural ecosystems support a rich variety of native bees, many of which are important economically. Honey bees, which are not native to North America, and domesticated bumble bees spread pathogens (diseases and parasites) and outcompete native bees for nectar and pollen. We aimed to reduce these threats by finding Canadian provincial legislation that could direct legislation in Yukon Territory, where native pollinator communities are still healthy and essential for pollinating agricultural crops. We classified the legislative requirements as follows: tracking numbers and locations of honey bee hives (registry); controlling the spread of pathogens (inspections, quarantines, and cleaning); controlling the competition with native pollinators (limiting shared use of space); and applying regulations to all domesticated bee species. There is little Canadian legislation controlling the interactions of honey bees and native pollinators; the competition problem is poorly addressed. We recommend controls on the numbers, locations, and timing of honey bee hives on public lands in the form of permitted "foraging leases" (for domestic insect livestock), similar to "grazing leases" (for domestic mammalian livestock). We detail changes to Yukon legislation (Animal Health, Animal Protection and Control, Wildlife, and Public Lands Acts) to deal with pathogen spread and competition. Protecting native pollinators will conserve the successful pollination of flowers producing foods and medicines for people.

Abstract: Imported, non-native honey bees and bumble bees threaten native pollinators by spreading pathogens (disease and parasites) and outcompeting native pollinators for nectar and pollen. We reviewed Canadian federal, provincial/territorial, and municipal legislation to find governance requirements that potentially reduce these threats. We classified the requirements as follows: tracking the number and location of honey bee hives (registry); controlling the spread of pathogens (registry with inspections, quarantines, and cleaning regimes); controlling the competition with native pollinators (limiting shared use of space); and making regulations applicable to all domesticated bees in addition to honey bees. Policies and regulations to control the competition from honey bees are generally lacking. So, we propose the concept of "foraging leases" to manage the location and duration of honey bee apiaries on public lands. Based on the identified requirements, we recommend amendments to the legislation in Yukon Territory, a jurisdiction that still has healthy native pollinator communities that pollinate various human food and medicine crops. Recommendations include tracking hives and their pathogen status with beekeeping regulations enabled by the Animal Health Act, controlling the use of imported bumble bees with changes to the Animal Protection and Control Act and/or the Wildlife Act, and restricting use of public lands for apiaries with the Public Lands Act.

 https://doi.org/10.3390/biology14030282

Keywords: boreal agriculture; native pollinator conservation; bumble bee; foraging lease; honey bee; recommended legislation; pathogens; Canada; Yukon

1. Introduction

Pollinators are necessary for the functioning and stability of terrestrial ecosystems, and natural ecosystems function well with native pollinator species [1]. The most efficient pollinators, the bees, are experiencing a disproportionate amount of global insect decline [2]. Two factors contributing to this are (i) novel pathogens transmitted by, and (ii) competition with, honey bees and managed bumble bees.

The honey bee (*Apis mellifera*), a domesticated species not native to North America, can infect native bees and other arthropods with various pathogens (e.g., deformed wing virus, sac brood virus, and small hive beetle) [3–5]. Commercially raised bumble bee species (*Bombus* sp.) carrying *Nosema bombi*, a microsporidian parasite, are implicated in the decline of several native bumble bee species [6,7]. Less understood are the impacts of imported alfalfa leaf-cutter bees (*Megachile rotundata*) and blue orchard mason bees (*Osmia lignaria*) [8] that are closely related to native *Megachile* and *Osmia* species in northern Canada.

Honey bees and introduced bumble bees also outcompete and displace native pollinators and disrupt plant–pollinator networks [9,10]. Honey bee hives typically begin with 10,000 worker bees in spring and grow throughout summer. The displacement of native bumble bees by introduced bumble bees has occurred in Canada. The common eastern bumble bee (*Bombus impatiens*), which is used in commercial greenhouses, has expanded its range into southern British Columbia and Washington State [11] and is the most recorded bumble bee in the Greater Vancouver Area of British Columbia [12].

Yukon Territory, Canada, is mostly in the northern boreal biome, with a relatively low human population density, low extent of private agricultural land, and high extent of wild lands under public (i.e., territorial government) jurisdiction compared with many parts of Canada. It is a jurisdiction where native pollinator populations are still largely intact and perform a vital pollination service for human food sources in both wild and agricultural ecosystems, including various berry crops [13–15]. Native pollinator species (e.g., *Bombus* spp.) are more cold-tolerant than imported honey bees, and thus, are better adapted to provide this pollination service [15]. Imported honey bees and bumble bees have limited agricultural functions in pollinating field and greenhouse crops and in honey production [16]. Pro-active management of these imported species has the best chance of conserving native pollinators in this context.

Four bumble bee species that occur in Yukon have been assessed by the Committee on the Status of Endangered Wildlife in Canada [17–20]. The main threats identified for these bumble bees are pathogens, pesticides, climate change, invasive/problematic species, and habitat loss. These threats also pose risks to the native pollinator community in general. In the Recovery Strategies and Management Plans for the four at-risk bumble bees, a high priority is placed on addressing the threats by creating, amending, or influencing environment-related provincial or territorial laws and/or regulations, policies, and guidelines. This legislative approach makes sense for Yukon, which already has some legislation resembling that of the provinces where beekeeping is more tightly controlled to reduce the risk of pathogen spread. Internationally, this legislative approach aligns with the Kunming–Montreal Global Biodiversity Framework's targets 6, 11, and 14 [21], which aim to mitigate the impacts of alien species, maintain pollination services, and integrate the valued contributions of biodiversity into legislation. It would also help fulfil Canada's 2030 Nature Strategy that aims to address these targets [22].

Our aim was to review legislation and regulations from federal, provincial/territorial, and municipal governments across Canada for examples of how the threats (pathogen introduction; competition) posed by non-native pollinators (honey bees, domesticated bumble bees) can be reduced and managed so that we can (i) recommend changes to current legislation and regulations in Yukon Territory and (ii) provide a review and general recommendations for improvements to legislation in other Canadian and North American jurisdictions.

We found that the most useful regulatory and management approaches include the mandatory registration and monitoring of honey bee apiaries, prescriptive rather than permissive approaches to commercial extractive uses of public lands, the concept of foraging leases for honey bees on public lands, and seasonal limitations on the use of domesticated bumble bees in greenhouses.

2. Materials and Methods

To understand how Canadian governments currently aim to reduce the threat of pathogen spillover to, and competition with, native pollinators, we examined the legislation governing beekeeping and the use of other managed bees at the federal and provincial/territorial levels.

We searched Canadian federal legislation for regulations dealing with bees, which led to the work of the Canadian Food Inspection Agency (CFIA) under the Health of Animals Act. The 10 provinces and 3 territories in Canada have jurisdiction over most animals and lands, so we searched their legislation for the governance of honey bees, other domesticated and imported bees, animal pathogens (particularly for honey bees), and beekeeping on public lands. This involved searches through the consolidated statutes for each jurisdiction.

The "legislation" consists of acts, regulations, and enabling statutes supplemented by other legally enforceable policies, such as protocols. This legislation can be implemented by the federal, provincial/territorial, or municipal levels of government (see the Results Section), though the bulk of it is provincial (Table 1).

Though we recognize that the main purpose of honey bee legislation and regulation is to protect the health of honey bees and the honey bee industry (including by reducing the risks of pathogen transmission between honey bee hives), many of these precautions also reduce the risk to wild pollinators by averting the introduction and spread of diseases and pests that potentially affect wild pollinators.

We categorized the legislative requirements by general function or intent (termed "requirement") with respect to their benefit to native pollinators. We considered five categories: tracking the number and location of honey bee hives; controlling the spread of pathogens; controlling the competition with native pollinators; reducing the lethal and sub-lethal effects of pesticides; and making regulations applicable to domesticated bees other than honey bees. Within each category of requirement, we itemized the various actions required to satisfy the intent. For each requirement and action, we documented the example wording and noted additional clauses that could potentially abate threats. No dedicated legislation exists for beekeeping in Yukon, but what other legislation does exist was examined for its applicability to the categories identified from the provincial pieces of legislation.

We also tabulated the pathogens identified by each province and whether they were considered reportable in each jurisdiction. The intent here was to understand the scope and sufficiency of the listing processes.

Table 1. The statutes and regulations governing beekeeping in each Canadian province.

Province	Statute	Regulations
British Columbia	Animal Health Act, Statutes of British Columbia (2014 c. 16) Lands Act, Revised Statutes of British Columbia (1996, c. 245)	Bee Regulation, British Columbia Regulation 3/2015 Reportable and Notifiable Disease Regulation 7/2015 O.C. 14/2025 British Columbia Land Use Policy Permission, 2023
Alberta	Bee Act, Revised Statutes of Alberta (2000, c. B-2)	Bee Regulation, Alberta Regulation 194/2003
Saskatchewan	The Apiaries Act, Statutes of Saskatchewan (2005, c. A-22.01) The Animal Health Act, Statutes of Saskatchewan (2019, c. A-20.01)	The Apiaries Regulations, Saskatchewan Regulations, 2005 The Animal Health Regulations, 2019
Manitoba	The Bee Act, Continuing Consolidation of the Statutes of Manitoba (1988, c. B15) Animal Diseases Act, Continuing Consolidation of the Statutes of Manitoba (2022, c. A85)	Diseases of Bees Designation Regulation, 2024 Reportable Diseases Regulation, 2007
Ontario	Bees Act, Revised Statutes of Ontario (1990, c. B-6)	Regulation 57 General, Revised Regulations of Ontario 1990
Quebec	Animal Health Protection Act, Revised Statutes of Quebec (1964, P-42)	Regulations Respecting the Registration of Beekeepers, 2024 Regulation Respecting the Inscription Affixed on Hives, 2024 Regulations to Designate Contagious or Parasitic Diseases, Infectious Agents and Syndromes, 2020
New Brunswick	Bee Act, Statute of New Brunswick (2021, c. 21)	New Brunswick Regulation 13/2023
Nova Scotia	Bee Industry Act, Statutes of Nova Scotia (2005, c. 3)	Bee Industry Regulations, Nova Scotia Regulation 133/2012
Prince Edward Island	Animal Health Act, Revised Statutes of Prince Edward Island (1988, c. A-11.1) [1]	Bee Health Regulations, 2019
Newfoundland	Animal Health and Protection Act, Statutes of Newfoundland and Labrador (2013, c. 1)	Animal Health Regulations, 2012 Animal Reportable Diseases Regulations, 2012

[1] This is under revision (An Act to Amend the Animal Health Act Bill No. 45, 2024).

We then identified the requirements within municipal bylaws. Many municipalities have bylaws regulating beekeeping in urban areas. These are often less prominent than provincial/territorial governance but can still potentially reduce the risk of pathogen spillover to, and competition with, native pollinators. An exhaustive search of all municipalities across Canada was beyond the scope of this study. Instead, municipal bylaws and regulations were collected for the northern territories where available, along with a selection of representative ones from each province, if available.

To understand what legislation exists to reduce the risks associated with introduced bumble bees competing with, or spreading pathogens to, native bumble bees, we identified provincial and territorial legislation that complements the federal Species at Risk Act (SARA) and examined it for its relevance to pollinator species at risk. Other federal legislation relevant to the use of managed bumble bees was also identified, which was mainly legislation related to the importation of live organisms into Canada or into individual provinces.

3. Results

Three levels of government—federal, provincial/territorial, and municipal—have legislation relevant to honey bees, managed bumble bees, or other introduced bees. The federal government is responsible for regulating the import of organisms into Canada and the reporting of pathogens listed by the World Organisation for Animal Health. It is also responsible for wildlife species at risk on federal lands. The provinces and territories have the most jurisdiction over domestic livestock, such as honey bees, and animal husbandry, such as beekeeping. They also have jurisdiction over what may enter the province or territory, and over management of non-federal public lands and species at risk on non-federal lands. Municipalities govern activities within urban boundaries, including activities that have traditionally been carried out on rural lands, such as the raising of chickens or pigs and beekeeping.

3.1. Federal

For all of Canada, the federal Health of Animals Regulation (Consolidated Regulations of Canada, c. 296), enabled by the Health of Animals Act (Statutes of Canada, 1990, c.21), applies. The Canadian Food and Inspection Agency (CFIA) enforces the regulations and issues import permits for honey bees that meet the requirements of permissible source countries and any other measures intended to reduce the spread of pathogens [23]. Applications to import other species of bees, such as blue orchard mason bees and bumble bees, into Canada are assessed on a case-by-case basis under the authority of the Plant Protection Act (Statute of Canada, 1990, c.22) and CFIA Directive 12-02 [24]. The CFIA prohibits importation of alfalfa leaf-cutting bees from outside Canada due to the risk of pathogens, including chalkbrood [25]. Only one provincial region, Newfoundland and Labrador, also prohibits the importation of alfalfa leaf-cutting bees (which could come from other provinces), as well as all other managed bees (e.g., bumble bees, blue orchard mason bees) aside from honey bees; this is regulated by the Wild Life Regulations (Consolidated Newfoundland and Labrador Regulation 1156/96 s.83).

3.2. Species at Risk Legislation

Species at risk legislation governs the assessment process for identifying wildlife species at risk and implementing conservation actions. Federally identified wildlife species at risk do include some native pollinators, such as the Gypsy Cuckoo Bumble Bee (*Bombus bohemicus*) and McKay's Bumble Bee (*Bombus mckayi*). However, Canada's federal Species at Risk Act only applies to federal lands, such as National Parks. In Yukon, there is no complementary territorial Species at Risk Act for territorial public lands, which account for approximately 85% of the Yukon's land base [26]. Six provinces and one territory have some form of a species at risk act (Table 2) [27]. All the provinces with species at risk legislation have included at least some insect pollinators in their lists of species at risk. These pollinators include bees, butterflies, and moths. Northwest Territories' Species at Risk Act has enabled the assessments of three bumble bees that are listed federally and in some other provinces but has not included them in the resulting list of species at risk (Table 2).

Table 2. Provincial and territorial statutes for species at risk and a list of any pollinator species to which the legislation applies.

Province	Statute	Listed Pollinator Species
British Columbia	None	
Alberta	None	
Saskatchewan	None	
Manitoba	The Endangered Species and Ecosystems Act, Continuing Consolidation of the Statutes of Manitoba (1989–1990, c. 39)	Dakota skipper (*Hesperia dacotae*) Ottoe skipper (*Hesperia ottoe*) Uncas skipper (*Hesperia uncas*) Dusky dune moth (*Copablepharon longipenne*) Pale yellow dune moth (*Copablepharon grandis*) Poweshiek skipperling (*Oarisma poweshiek*) Verna's flower moth (*Schinia verna*) White flower moth (*Schinia bimatris*)
Ontario	Endangered Species Act, Statutes of Ontario (2007, c. 6)	Bogbean buckmoth (*Hemileuca* sp.) False-foxglove sun moth (*Pyrrhia aurantiago*) Gypsy cuckoo bumble bee (*Bombus bohemicus*) Mottled duskywing (*Erynnis martialis*) Rusty-patched bumble bee (*Bombus affinis*) Suckley's cuckoo bumble bee (*Bombus suckleyi*) Northern oak hairstreak (*Satyrium favonius ontario*) Reversed haploa moth (*Haploa reversa*) American bumble bee (*Bombus pensylvanicus*) Dukes' skipper (*Euphyes dukesi*) Monarch (*Danaus plexippus*) Red-tailed leafhopper (*Aflexia rubranura*) West Virginia white (*Pieris virginiensis*) Yellow-banded bumble bee (*Bombus terricola*)
Quebec	Act Respecting Threatened or Vulnerable Species, Revised Statutes of Quebec (1989, c. E-12.01)	Rusty-patched bumble bee (*Bombus affinis*) Salt marsh copper (*Lycaena dospassosi*) Maritime ringlet (*Coenonympha nipisiquit*)
New Brunswick	Endangered Species Act, Revised Statutes of New Brunswick (2012, c. 6)	Bumble bee, Bohemian cuckoo (*Bombus bohemicus*) Bumble bee, rusty-patched (*Bombus affinis*) Bumble bee, Suckley's cuckoo (*Bombus suckleyi*) Monarch (*Danaus plexippus*) Ringlet, maritime (*Coenonympha nipisiquit*)
Nova Scotia	Endangered Species Act, Statutes of Nova Scotia (1998, c. 11. s. 1)	Gypsy cuckoo bumble bee (*Bombus bohemicus*) Macropis buckoo bee (*Epeoloides pilosulus*) Monarch (*Danaus plexippus*) Sable Island sweat bee (*Lasioglossum sablense*) Yellow-banded bumble bee (*Bombus terricola*)
Prince Edward Island	None	
Newfoundland	Endangered Species Act, Statutes of Newfoundland and Labrador (2001, c. E-10.1)	Gypsy cuckoo bumble bee (*Bombus bohemicus*) Suckley's cuckoo bumble bee (*Bombus suckleyi*) Yellow-banded bumble bee (*Bombus terricola*)
Yukon	None	
Northwest Territories	Species at Risk (NWT) Act, Statutes of Northwest Territories (2009, c. 16)	No pollinators listed [1]
Nunavut	None	

[1] NWT Species at Risk Committee (SARC) considers gypsy cuckoo bumble bee and McKay's bumble bee to be data deficient and yellow-banded bumble bee to be not at risk; Suckley's cuckoo bumble bee is not assessed. SARC assessments enabled by Species at Risk (NWT) Act.

3.3. Provincial and Territorial Legislation

All provinces have legislation that actively governs beekeeping. Six provinces have statutes specific to bees or apiaries (apiary: a collection of honey bee hives). The other four provinces—British Columbia, Quebec, Newfoundland, and Prince Edward Island—use their statutes on animal health to enable regulations pertaining to beekeeping (Table 1). Additional statutes also support beekeeping in some provinces. For example, despite Saskatchewan having an Apiaries Act, its Animal Health Act and Regulations control the reporting of bee pathogens. None of the three territories—Yukon, Northwest Territories, or Nunavut—has territorial statutes specific to beekeeping or beekeeping regulations. However, the Yukon's Animal Health Regulation (O.I.C. 2018/159) lists mandatory reporting of suspected or confirmed cases of two honey bee "transmissible hazards": American foulbrood (*Paenibacillus* larvae) and the small hive beetle (*Aethina tumida*).

From the statutes and regulations overseeing beekeeping, we identified sets of requirements that are of benefit to native pollinators (Table 3) while facilitating the management of the honey bee industry. These legislative requirements, and associated actions, comprise a systematic assessment of governance functions.

Table 3. Requirements within provincial legislation pertaining to beekeeping that potentially reduce risk to native pollinators. Requirements are organized first by general functions, and then by actions (identified in italics). See Table S1 in the Supplementary Materials for example legislative wording for each action identified here.

1. Requirements that track the number and location of honey bee hives.
Registering hives: Bee Regulation (Alberta Regulation 194/2003; 221/2004, s.3)
Identification of hives: New Brunswick Regulation 2023-13, s.18(1)
Registering specific location of hives: Bee Industry Regulations (Nova Scotia Regulation 319/2007, s.5)
Mandatory reporting of movement of honey bee hives/colonies: Bee Health Regulations (Pursuant to Section 4 of the Animal Health and Protection Act, Revised Statutes of Prince Edward Island. 1988, Cap. A-11.1, s.6(1), s.6(2))
2. Requirements that control the spread of pathogens
Mandatory inspection of bees prior to importation: Animal Health Regulations (Newfoundland and Labrador Regulation 33/2012, s.7(1), s.7(2))
Mandatory quarantine period: Animal Health Regulations (Newfoundland and Labrador Regulation 33/2012, s.7(3))
Mandatory inspection of used honey bee equipment prior to importation: Bee Industry Regulations (Nova Scotia Regulation 319/2007, s.6(1), s.6(2a), s.6(2b))
Mandatory sanitization of equipment prior to importation: 2024 Nova Scotia Honey Bee Health Importation Protocol
Mandatory reporting of disease and pests: Bees Act (Revised Statutes of Ontario, 1990, c. B.6, s.10)
Listing of specific pathogens: refer to Table 4 for a list by province
Restrictions on the source of honey bees and honey bee equipment: The Apiaries Regulation (Saskatchewan 2005 cA-22.01 Regulation 1, s.4)

Table 3. *Cont.*

3. Requirements that can control competition with native pollinators
Regulations controlling apiaries on public land: British Columbia Land Use Policy Permission, 2023. Appendix 4. Conditions for temporary apiaries (issued under authority of Lands Act (C.245, Revised Statutes of British Columbia, 1996, s.60))
Restricting number of hives: The City of Dawson Bylaw #12-28, s.6.08
4. Requirements that reduce lethal and sub-lethal effects of pesticides on native pollinators [1]
Restricting pesticides while flowers are in bloom: Animal Health Protection Act (Revised Statutes of Quebec, 1964 Ch. P-42, s. 11.12)
5. Requirements that make regulations applicable to domesticated bees in addition to honey bees
Defining imported bee species (e.g., *honey bees, bumble bees, alfalfa leaf-cutter bees, blue orchard bees*): The Bee Act (Continuing Consolidation of the Statutes of Manitoba c. B15, s.1)

[1] Although we note this as an existing legislative requirement in some jurisdictions, we do not discuss this topic further because of its complexity. Substantial research has been performed, often on species other than bees, leading to diverse governance outcomes beyond the scope of this paper.

Table 4. Pathogens identified in provincial and territorial statutes and regulations: X—identified only; R—reportable; N—notifiable. "Reportable" and "Notifiable" are defined in the text. Two-letter codes for jurisdictions are as follows: YT—Yukon Territory; BC—British Columbia; AB—Alberta; SK—Saskatchewan; MB—Manitoba; ON—Ontario; QC—Québec; NB—New Brunswick; NS—Nova Scotia; PEI—Prince Edward Island; NL—Newfoundland and Labrador.

Common Name	Scientific Name	YT	BC	AB	SK	MB	ON	QC	NB	NS	PEI	NL
Mites												
Asian bee mite sp. 1	*Tropilaelaps mercedesae*			X		X	R	R	R	R		
Asian bee mite sp. 2	*Tropilaelaps clareae*		R	X	X	X	R	R	R	R		
Euvarroa genus	*Euvarroa*					X	R		R			
Tracheal mites	*Acarapis woodi*		N	X	X	R	R		X	R	R	R
Varroa mites	*Varroa destructor*		N	X	X	X	R	X	R [1]	R [1]	R	R
Bees, Wasps, and Hornets												
Africanized honey bee	*Apis mellifera scutellata* and hybrids		R	X	X	X	R	R	R	R	X	
Asian honey bee	*Apis cerana* and species complex		R	X		X	R		R	R	X	
Cape honey bee	*Apis mellifera capensis*		R	X		X	R		R	R	X	
Dwarf honey bee	*Apis florea* and species complex					X	R		R			
Giant honey bee	*Apis dorsata*			X		X	R		R			
Asian predatory wasp	*Apis vilutina*		R									
Northern giant hornet	*Vespa mandarinia*		R	X		X	R		R	R		
Southern giant hornet	*Vespa soror*			X		X			R			
Yellow-legged hornet	*Vespa velutina*			X		X			R		X	
Beetles												
Large African hive beetle sp. 1	*Oplostomus fuligineus*					X	R		R			
Large African hive beetle sp. 2	*Oplostomus haroldi*					X	R		R			
Small hive beetle	*Aethina tumida*	R	R	X	X	R	R	R	X	R	R	X
Moths												
Greater wax moth	*Galleria mellonella*		N			X			X			R
Lesser wax moth	*Achoia grisella*		N			X			X			
Other insects												
Bee louse	*Braula coeca*					X			X			

Table 4. *Cont.*

Common Name	Scientific Name	YT	BC	AB	SK	MB	ON	QC	NB	NS	PEI	NL
Fungi												
Chalkbrood	*Ascosphaera apis*		N	X		X			X	X		
Nosema sp. 1	*Nosema apis* (*Vairimorpha apis*)		N	X		R	R		X	X		R
Nosema sp. 2	*Nosema ceranae* (*Vairimorpha ceranae*)		N	X		X	R		X	R		
Bacteria												
American foulbrood	*Paenibacillus* larvae	R	R	X	N	R	R	R	R	R	R	R
European foulbrood	*Melissococcus plutonius* and bacteria		N	X		R	R		R	R		R
Viruses												
Acute bee paralysis virus			N			X						
Black queen cell virus			N			X						
Chronic bee paralysis virus						X						
Deformed wing virus			N			X						
Israel acute paralysis virus			N			X						
Kashmir bee virus			N			X						
Sacbrood virus	Family Iflaviridae		N	X		X			X	X		
Varroa destructor virus			N			X						

[1] Reportable for miticide-resistant types only.

The first requirement is the obligation to track the number and location of honey bee hives, including the following specific actions: register hives; identify individual hives; register the specific location of each hive; and report all movements of hives or colonies (Table 3). Except for Prince Edward Island, all provinces require beekeepers to register their hives. Information collected during registration usually includes the number and location of hives, which is essential in tracking and containing pathogens, and may provide a density estimate of honey bees on the landscape. For the adequate tracking and containment of the threats of pathogen spread and competition, we consider all four of these actions to be necessary.

The second requirement is to control the spread of diseases and pests (pathogens). This category includes various legislated actions, not all of which are mandatory or enacted in every province. These actions include restrictions on the sources of honey bees, quarantines, inspections of colonies, sanitization of equipment, and mandatory reporting of diseases and pests. All provinces control the importation of honey bees and beekeeping equipment from other provinces, and in the case of British Columbia, the movement of hives within the province is closely regulated, allowing for quarantine zones when pathogen outbreaks occur.

The third requirement, to control competition between honey bees and native pollinators, has had very little legislative attention (Table 3). British Columbia is the only province to have specific regulations for apiaries on public ("Crown") land. These allow the province to control the locations, densities, and duration of privately owned hives on public lands, all of which are necessary tools to influence the competition between honey bees and the already established native pollinator community. This approach is easier with public lands legislation that is prescriptive (i.e., prohibiting any private use of public lands unless that use is explicitly allowed with conditions detailed in the legislation) as compared with permissive (i.e., allowing all private uses of public land with controls and prohibitions only occurring when specifically addressed in the legislation). Restrictions on the number of hives on any piece of land were generally absent in provincial legislation, but common at the municipal level, which is discussed later.

The fourth requirement, to reduce the negative effects of pesticides on bees, is generally addressed in legislation other than what we reviewed and was beyond the scope of this study. However, it is worth mentioning that Ontario's Bee Act and Quebec's Animal Health Protection Act both prohibit the spraying of fruit trees with any substance toxic to bees while the trees are in bloom.

The fifth requirement is to make regulations applicable to all domesticated bee species, including honey bees (Table 3). This has had relatively little legislative attention, partly because importation into Canada is controlled by federal legislation (see above). Some provinces list various domesticated species, as well as honey bees, giving the government the option to act on threats these domesticated species pose.

Returning to the question of pathogen reporting and control (the second requirement), all provinces have identified some pathogens associated with beekeeping. These pathogens are listed in provincial legislation used to facilitate their control and eradication. The pathogens, including disease organisms and pest species, listed from provinces across Canada comprise the following numbers of species: five mites, five honey bees, four wasps or hornets, three beetles, two wax moths, one louse, three fungi, two bacteria, and eight types of virus (Table 4). All 10 provinces list the small hive beetle, Varroa mite, and American foulbrood. Only British Columbia and Manitoba identify viruses in their legislation. Some provinces specifically list "reportable" and "notifiable" pathogens. "Reportable" pathogens require immediate action to contain their spread. "Notifiable" pathogens are monitored to help understand their presence and impacts. In provinces where notifiable and reportable pathogens are named, the beekeeper is obligated to report suspected cases within a limited time.

At 23, British Columbia identified the most pathogens in its legislation, followed by Alberta and Ontario at 18. All provinces have at least one reportable or notifiable pathogen, with the exception of Alberta, which has no honey bee pathogens listed in its Reportable and Notifiable Diseases Regulation (Alberta Regulation 129/2014). The reporting of some pathogens is still recommended in Alberta [28]. The Canadian Federal Inspection Agency also monitors tracheal mites, the small hive beetle, American foulbrood, and European foulbrood (Health of Animals Regulations, Consolidated Regulations of Canada, 2024, c. 296). Mandatory reporting by laboratories across Canada contributes to this monitoring effort.

3.4. Municipal Legislation

Municipal legislation pertaining to beekeeping is generally more developed in western Canada than in central or eastern Canada, with British Columbia having upward of 20 municipalities permitting and regulating beekeeping. There were considerable similarities between bylaws, and the sample (see Table S2 in the Supplementary Materials) appeared sufficient to identify the main requirements. Regulations are usually enabled by animal control bylaws (e.g., Edmonton Animal Licensing and Control Bylaw No. 13145), zoning bylaws (e.g., Winnipeg Zoning Bylaw), or dedicated urban beekeeping bylaws (e.g., Chilliwack Urban Beekeeping Bylaw 2019, No. 4680).

The focus of the regulations is on keeping the public safe from the defensive behavior of honey bees, especially bee stings. Regulations usually stipulate the maximum number of beehives for the size of the property and the placement of beehives relative to property boundaries, and often mention having the hive entrance directed away from adjacent properties and/or requirements for a physical barrier, such as a fence or hedge to discourage the direct flight of honey bees to adjacent properties (e.g., Municipal Planning Guide to Zoning Bylaws in Manitoba [29]). A few municipalities require proof that the applicant has taken a beekeeping course or is being mentored by an experienced beekeeper (e.g., Mount Pearl Development Regulations 2010).

The two municipalities in Yukon that explicitly allow beekeeping use different bylaws to regulate the activity. The City of Whitehorse uses its Zoning Bylaw 2012–20 to specify where, within the city limits, beekeeping is permitted. Beyond this, there are no published regulations on the placement of beehives within properties or the permissible number of beehives and no permitting process to keep track of beekeeping activity. The beekeeping regulations in Dawson City are enabled by its Animal Control Bylaw and are similar to the municipalities in southern Canada, with well-developed beekeeping regulations specifying the maximum number of beehives, placement, and management practices to reduce risks to neighbouring properties. Dawson City has a permitting process in place that provides a record of the beekeeper, where hives are located, and from where they were moved.

4. Discussion

4.1. The Yukon Context

Having a diversity of pollinators for the large diversity of plants native to Yukon creates a complex plant–pollinator network. The production of northern wild foods, such as blueberries, lingonberries (*Vaccinium* spp.), and cloudberries (*Rubus chamaemorus*), as well as some domestic crops (e.g., haskap (*Lonicera caerulea*), beans (*Phaseolus* spp.), and squash (*Cucurbita* spp.)), depends on these pollinators [13–15]. In Yukon, there are over a hundred species of bees; over a hundred species of butterflies; and many more moths, flies, and other pollinator insects [30]. This native pollinator network is thought to be relatively intact compared with networks farther south in Canada and North America because Yukon has a relatively low human footprint. The bumble bees listed in Canada's Species at Risk Act are believed to be closer to historic levels in Yukon than in other parts of Canada because less of the landscape has been subjected to alteration by agriculture, resource extraction, or other uses. So, now is an appropriate time to act to strengthen the conservation of native pollinator species in Yukon before very many become threatened.

The Canadian legislative framework for reducing threats to native pollinators has improved in recent years. A 2007 review of federal and provincial legislation [31] concluded that Canadian federal and provincial legislation lacked explicit provisions for pollinator conservation, but did have provisions related to pollinator conservation. Since that study, there have been numerous amendments to existing legislation and the introduction of new legislation. Although provisions specific to the conservation of native pollinators are still generally lacking across Canada, there are many examples of legislation that is potentially useful in reducing the risks to native pollinators. In the context of legislation across the country, we examined current Yukon legislation to evaluate its ability to address the threat of pathogen spillover to native pollinators from introduced bees and the threat of competition with native pollinators by introduced bees.

4.2. Pathogen Spillover

The main sources of pathogens affecting native bees are imported honey bees, a species not native to Canada, and commercially raised and imported bumble bees (mainly *Bombus impatiens*) used in greenhouse operations [32,33]. Pathogens and parasites weaken native pollinators, and when pervasive, or combined with other stressors, are capable of reducing the populations of native pollinators [1]. The pathogens associated with honey bees have been found in arthropods other than bees, likely transmitted at floral feeding sites or by predation on infected honey bees [3].

Compared with the provinces, Yukon Territory has very little legislative infrastructure to protect native pollinators from pathogen spillover imposed by honey bees and other managed bees. This makes Yukon vulnerable to the importation of infected honey bees from other Canadian jurisdictions. The territorial legislation is limited to the mandatory reporting of two "transmissible hazards": American foulbrood and the small hive beetle (Table 4). Dawson City is the only region in Yukon that reinforces this mandatory reporting.

Mandatory reporting is difficult to enforce without a system to register beekeepers and the location of their hives and a system to inspect and track imported honey bees. Unlike all the provinces, Yukon does not monitor honey bees or beekeeping equipment being moved into the territory and does not systematically regulate from where honey bees may be imported. Though, like the rest of Canada, Yukon beekeepers should be complying with the federal Health of Animals Regulation (C.R.C., c. 296), which includes restrictions on which countries honey bees can be imported from. For many years, a long-time beekeeper in Yukon was a proponent of sourcing honey bees from mite-free colonies to keep Yukon free of varroa mites [16], but there was no legislation to support this initiative. Today, honey bees in Yukon are commonly infected by varroa mite [16] and this could be widespread throughout Yukon. Beekeeping is widespread in Yukon, ranging north to Dawson City, west to Haines Junction, east to Watson Lake, and south to Teslin [34]. However, the actual number of beekeepers and number of hives remains unknown because the territory lacks a formal registry.

The pathogens associated with honey bees that could enter Yukon are not reflected in the territory's current legislation. The list of two pathogens identified for honey bees by Yukon's Animal Health Act (Statutes of Yukon 2013, c.10) is shorter than any province, and notably far less than the bordering provinces of British Columbia, with 23 listed pathogens, and Alberta with 18 (Table 4). It could be argued that the mandatory reporting of select pathogens in British Columbia offers some protection to Yukon; mandatory reporting in British Columbia facilitates the isolation and disinfection of affected hives. The same cannot be said of Alberta, where reporting is recommended but not legislated [28].

Across Canada, legislation abating the risk of pathogen spillover from managed bumble bees is less developed than that from honey bees. The necessary management approaches to mitigate the risks of managed bumble bees to native bees are as follows: the monitoring of pathogens in commercial bumble bee stock, preventing escape from greenhouses, and regulating the movement of commercial bumble bees [33]. Newfoundland's prohibition of any species other than those listed in their Wild Life Regulations (Consolidated Newfoundland and Labrador Regulation 1156/96 s.83) is the strictest provision pertaining to the movement of managed bumble bees. It prohibits any importation of bumble bees and is, in part, enforceable because it defines "wild animal" as "any live animal, including without limitation, any amphibian, arthropod, bird, coelenterate, crustacean, fish, other invertebrate, mammal, mollusk or reptile, whether or not bred, hatched or born in captivity and including any egg or offspring of them". By contrast, the Animal Protection and Control Act of Yukon (Statutes of Yukon, 2022, c.13) currently does not enable the control of any invertebrates. Furthermore, the Wildlife Act (Revised Statutes of Yukon 2002, C.229) for Yukon only includes vertebrate species in its definition of "wildlife", so is unable to offer any protection to invertebrates, such as insect pollinators.

Another avenue for protecting native pollinators from pathogen spillover is possible, specifically for pollinator species listed in Canada's Species at Risk Act (SARA). When species are listed under the SARA, the federal government is obligated to prepare Recovery Strategies for "endangered" and "threatened" species, and Management Plans for "special concern" species. Such action plans identify approaches to reducing threats to the species, and it is mandatory to track the progress of these strategies and plans (Species at Risk Act, Statutes of Canada 2002, c.29). An example of an action within a Recovery Strategy to protect a native pollinator against pathogen spillover is the promotion of the Bumblebee

Sector Guide to the National Bee Farm-level Biosecurity Standard [35]. This standard provides details on how to confine managed bumble bees to greenhouses and to prevent the spread of pathogens [35].

Action plans enabled by the federal Species at Risk Act (SARA) are limited in their ability to achieve the goal of arresting or reversing species decline. One main limitation is the fact that for terrestrial species, the SARA only applies to federal lands (those "that belong to His Majesty in right of Canada"). In the provinces, this is only 4% of the land [27]. The vast majority of land in Yukon is public land, still vested in the federal Crown but administered under territorial jurisdiction (Yukon Act S.C. 2002, c. 7) [26]. In practice, the application of the SARA in Yukon has come to refer only to federally controlled or administered lands, as is conducted in the provinces. Federally controlled lands in Yukon (mostly conservation lands) comprise c. 8% of the land base [36]. Unlike the Northwest Territories and seven provinces, Yukon does not have legislation that complements the federal SARA (Table 2) [27]. Provincial and territorial equivalents of the federal SARA usually adopt components of federal Recovery Strategies and Management Plans for use on public lands under provincial and territorial administration.

4.3. Competition

Unlike native bees, worker honey bees can recruit each other to harvest the best sources of nectar and pollen, thereby monopolizing the food source [10,37]. The level of numerical dominance by honey bees increases with flower abundance [38]. The amount of pollen a honey bee colony collects from June to August is equivalent to the amount needed to produce 100,000 progeny of an average solitary bee [39]. Under such competition from honey bees, the most affected native species would have difficulty maintaining the local population levels into the future.

Legislation that could reduce the risk of resource competition from honey bees and other managed bees is limited, with most examples coming from municipal bylaws. Urban beekeeping is a growing trend in Canada, with many municipalities moving away from the prohibition of beekeeping, not enforcing the prohibition, or developing regulations around beekeeping. Municipal regulations have the potential to control the amount of competition with native pollinators and the risk of pathogen spillover, by either prohibiting beekeeping or by limiting the number of honey bee hives. However, because public safety is the main rationale for restricting the number of bee hives within municipal boundaries, the overall density of bee hives and their risk to native pollinators has yet to be factored in. The risk of competition was highlighted by the finding that the high honey bee density in Montreal was negatively impacting wild bee species richness [40]. Similar findings have been documented from elsewhere in Canada and Europe [41,42]. Beyond municipal boundaries, most provinces track the location and number of honey bee hives by using a registry system, where such registries have the potential to track the density of bee hives and competition risk to native pollinators.

Legislation specifically restricting the placement of apiaries on public lands appears lacking for most provinces. The lack of regulation would indicate it is a prohibited activity if the particular provincial Lands Act is prescriptive in nature, meaning that by default, activities are prohibited unless authorized. British Columbia appears to be the only province that has explicitly authorized regulations pertaining to apiaries on public lands, and these were enabled by its provincial Lands Act (C.245, R.S.B.C., 1996, s.60.). Unlike British Columbia, which has a prescriptive Lands Act, the existing Yukon Lands Acts (Territorial Lands (Yukon) Act, Revised Statutes of Canada, 1985, c. T-7; Lands Act Revised Statutes of Yukon 2002, c. 132) are permissive, allowing any activities to occur without restrictions unless those activities are specifically addressed in the Act or by its enabled legislation. At present, this means that beekeeping on public lands administered by the Yukon Government is permitted without restriction. This may change as the Yukon's Lands Acts are undergoing major changes [26].

Providing sufficient forage is a general concern and recommendation for beekeepers [43]. Honey bees are essentially free-ranging livestock, travelling beyond property lines to collect nectar and pollen. Unrestricted access to public lands by beekeepers with apiaries potentially creates problems with competition between beekeepers and their honey bees, and competition between honey bees and native pollinators. For other livestock in Yukon, such as cattle, sheep, and horses, competition with wildlife is a consideration in deciding whether a land parcel can be a grazing lease. Adopting the approach of the Yukon Grazing Policy [44] for identifying public lands for grazing livestock could be a starting point for developing an assessment of where suitable sites would be for apiaries on public lands. Specifically, the Yukon Grazing Policy stipulates a minimum amount of available forage within a defined area, and how the period of time that the land can be used for grazing livestock is determined. The policy also summarizes a process to identify conflicting land uses and states that applications for grazing are subjected to review by the Yukon Government or the Yukon Environmental and Socio-economic Assessment Act.

We recommend the establishment of foraging leases as a necessary condition for private honey bee keepers to place apiaries on public lands. By using an approach like the Yukon Grazing Policy, regulators would limit the number of hives to the capacity of the forage (nectar and pollen) in defined areas (foraging leases), and such leases would be limited in time and space. These limitations could help mitigate the competition with native insects reliant on similar nectar and pollen sources.

Management practices need to be in place to reduce the number of apiaries on public lands during time periods outside the period of mass flowering [32]. Mass flowering refers to events during which a set of plant species produce many blooms within a short period of time. Competition with native pollinators would be the most intense before and after such mass-flowering periods when fewer food sources are available [32]. The prohibition of apiaries in protected areas is also recommended to avoid disruption to native pollinators [32]. This is supported by a study that found the carrying capacity of floral resources in protected areas was exceeded by the introduction of honey bees [45]. As is conducted in British Columbia, restricting the window of time during which apiaries can be on public lands and regulating cleanup would help ensure that the public lands will return to their original state as much as possible.

Precedents for limiting the density of honey bees and assessing risk to native pollinators on public lands are generally lacking in Canada. However, there are initiatives in the United States that demand environmental assessments prior to deciding whether to issue permits for apiaries on public land. Such assessments would consider the risk of displacing native pollinators from their essential food sources, and also the risk of potential pathogen spillover from honey bees to native bees [46,47].

4.4. Recommendations for Legislative Change in Yukon

Through a combination of amending the existing Yukon legislation and introducing new legislation, it would be possible to reduce the risk of pathogen spillover to native pollinators and mitigate the competition between introduced bees and native pollinators.

4.4.1. Pathogen Spillover

For Yukon, existing legislation can be used to reduce the risk of pathogen spillover from honey bees. The Animal Health Act has the ability to create regulations to prevent the entry and spread of disease to domestic and wild animals. We recommend creating a set of regulations enabled by the Animal Health Act specific to beekeeping. In the regulations, we suggest the inclusion of the following stipulations:

- Registration of apiaries, including the beekeeper's name, number of hives, and specific location(s) of hives;
- Onsite identification of apiaries to facilitate contact with the beekeeper if needed;
- Reporting of any movement of apiaries and equipment so as to track the potential spread of pathogens;
- Inspection of honey bees and equipment prior to import into Yukon, and certification to ensure that the honey bees and equipment are free of pathogens;
- Updated list of pathogens to include all pathogens at risk to honey bees in the provinces from which the bees are being imported;
- Continued mandatory reporting of "transmissible diseases" as in the current Animal Health Regulations.

To reduce the risk of pathogen spillover from managed bumble bees, there are two options: one that would prohibit the use of managed bumble bees altogether, and another that would control the use of managed bumble bees. We prefer the higher effectiveness of the first option but discuss how to implement either through legislative changes to existing Yukon Acts or Regulations.

Under option 1, prohibiting the use of managed bumble bees in Yukon would either need amendments to the Animals and Species Regulation (Ministerial Order 2024/12) or amendments to Yukon's Wildlife Act.

Currently, the Animals and Species Regulation enabled by the Animal Protection and Control Act prohibits "an animal that is not livestock and does not belong to an allowed species or a restricted species" (s 6(a)). This wording offers the possibility of preventing the import of non-native bumble bees if the definition of "animal" in the Animal Protection and Control Act were to include non-native bumble bees. That definition currently only includes vertebrates and "prescribed" species, and the "prescribed species" in the Animals and Species Regulation are limited to cephalopods, tarantulas, and scorpions. Adding bumble bees as "prescribed" species to Schedule 1 of the Animals and Species Regulation would enact a prohibition on the import of non-native bumble bees.

For the Wildlife Act to prohibit managed bumble bees, the definition of "wildlife" would need better clarity to include native insects so that stipulations to protect wildlife through the Wildlife Act are applicable to native pollinators. The current definition of "wildlife" as "wild by nature" may or may not apply to native insects. Then, a new section would be needed in the Act to prohibit the importation of any species unless authorized, akin to Newfoundland and Labrador's Wild Life Regulations (Consolidated Newfoundland and Labrador Regulation 1156/96 s.83). This would enable the prohibition of insects considered a risk to Yukon's wildlife and insects that are considered invasive.

Under option 2, controlling the use of managed bumble bees to minimize pathogen spillover to native pollinators would require amendments to the Animal Protection and Control Act and its regulations. For the purposes of the Animal Protection and Control Act and the Animal Protection and Control Regulations (O.I.C. 2024/62), livestock is defined by the species in Schedule 2 of the Animals Species Regulation and Schedule A of the Game Farm Regulations (O.I.C. 1995/015). Adding non-native bumble bees as livestock to Schedule 2 of the Animals and Species Regulation would enable the Animal Protection and Control Act to prescribe conditions for the use of managed bumble bees in Yukon. The conditions should include assurances that the bees are only permitted for use in winter (November to March) and that they are confined to greenhouses. This would prevent contact between the managed bumble bees and native bees; native bees are dormant in winter. We also recommend the destruction of the managed bumble bee colonies at the end of winter to prevent any individuals from escaping out of the greenhouses. If the option of a standardized disease-free stock of bumble bees becomes available [48], we recommend the mandatory sourcing of commercial bumble bees from facilities that produce such stock.

4.4.2. Competition

The changes to legislation that we recommend in the Pathogen Spillover Section (Section 4.4.1) are applicable to mitigating the competition between managed bees and native pollinators. Regarding honey bees, the establishment of a registry that can track the location and density of apiaries throughout Yukon is not only useful for understanding where the competition risk to native pollinators is the highest, but also useful for beekeepers wanting to ensure adequate food for their honey bees. Regarding managed bumble bees, by prohibiting the use of managed bumble bees or confining their use to greenhouses in winter, competition with native pollinators would be avoided.

Beyond the recommended changes discussed in the Pathogen Spillover Section, there are additional amendments to Yukon's territorial legislation and municipal bylaws that could reduce the competition with native pollinators by honey bees. Apiaries are kept on both public and private lands in Yukon, though in both circumstances, the honey bees have access to food on public lands and compete with native pollinators. This competition cannot be avoided altogether while beekeeping continues in Yukon because, unlike livestock that can be confined by fences, honey bees are free-ranging. However, restrictions on the number of hives and the placement of apiaries can reduce the competition with native pollinators from honey bees.

At the municipal level, having municipalities, such as Whitehorse, adopt wording similar to the Dawson City beekeeping regulations into its Zoning Bylaw or Animal Control Bylaw would potentially limit the placement and density of honey bees within the municipality. The Dawson City regulations include a maximum of two hives for properties smaller than 0.41 ha, and a maximum of four hives for properties 0.41 ha or larger. Additionally, imposing a total quota of apiaries within city limits could reduce competition with native pollinators.

At the territorial level, the new Public Lands Act that the Yukon Government is now developing could direct the conditions by which apiaries would be permitted on public lands. At one extreme, apiaries on public lands could be prohibited altogether, and this would produce the least risk to native pollinators. At the other extreme, in what currently exists, apiaries would be permitted without restriction because, by default, activities are allowed unless prohibited by the current Lands Acts and enabled regulations. Somewhere between these extremes would require a regulated approach that mitigates the risk of competition with native species while still allowing some use of public lands. We recommend that the new Public Lands Act be prescriptive, rather than permissive, so that it includes clause(s) that prohibit apiaries on public land except under specific conditions and through a permitting process similar to how grazing agreements are devised in Yukon.

To mitigate the negative consequences of competition with native species, the permitting process would include the following considerations:

- Forage: Identify sites that have a high abundance of floral resources, such as recent burns in forested lands, or repeatedly disturbed sites (rights of way and gravel pits). Estimate the longevity of the abundance, as some sites will lose their abundant floral resources as part of forest succession or changes to land use.
- Timing: Restrict the window of time during which honey bees can be at a site. This could reduce competition against native pollinators when food is most limiting (e.g., early spring), and permit apiaries when flowers are most abundant, and thus, least likely to be limiting to native pollinators.

- Quota: limit the total number of apiaries so that the total competition with native pollinators is reduced.
- Placement A: Avoid remote sites to reduce the chances of pathogen spillover to native pollinator populations that are currently less affected by introduced pathogens and to maintain native plant–pollinator networks without intrusion. Instead, only use sites with easy road access. For native species, road access often indicates that the area is already disturbed compared with a remote site. For beekeepers, road access to sites facilitates the easier placement of hives and removal at the end of the growing season.
- Placement B: Prohibit apiaries in rare ecosystems and protected areas. Disrupting the native plant–pollinator network in ecosystems that are already spatially limited compromises their future. In particular, native grassland ecosystems in Yukon often support rare and/or range-restricted plant species [49], and their pollinator networks are incompletely understood and may contain specialist species.

4.4.3. Species at Risk

Creating species at risk legislation in Yukon would help to implement Management Plans and Recovery Strategies for safeguarding our native pollinators that are federally listed species at risk. Yukon has previously signed on to the 1996 Accord for the Protection of Species at Risk, and the 2018 Pan Canadian Approach to Transforming Species at Risk Conservation in Canada. The Canada–Yukon Nature Agreement [50] states that its purpose "is to establish the framework for cooperation between the Participants for measures and plans by the Participants for the protection, conservation and recovery of biodiversity, habitat, and species at risk in the Yukon". These initiatives all point to the need and existing mandate for Yukon to enact species-at-risk legislation.

5. Conclusions

Given that imported non-native honey bees and bumble bees can threaten the viability of native bee pollinator species, and the native bees' service of pollinating northern boreal food crops, our review of the Canadian governance regime indicates that legislation needs to (i) track the number and location of honey bee hives; (ii) control the spread of pathogens; (iii) control competition with native pollinators; and (iv) make regulations applicable to all domesticated bees, in addition to honey bees. These four requirements or functions of governance need to occur in provincial or territorial legislation; currently, they are very unevenly satisfied across these jurisdictions within Canada. In particular, policies and regulations aimed at controlling honey bee competition with native pollinators are generally lacking. In response, we propose the concept of "foraging leases" to manage any proposed location of honey bee (i.e., domestic insect livestock) apiaries on public lands because these lands often support a full complement of native pollinators. Foraging leases would be modelled on "grazing leases", a management tool already well established for the management of domestic mammalian livestock on public lands.

Specific to Yukon Territory, we recommend new regulations under the Animal Health Act to satisfy the need to track the number and location of honey bee hives, and to control the potential spread of pathogens into and within the territory. We explain alternative options for managing domestic bumble bees under the Animal Protection and Control Act or the Wildlife Act, each of which would require the explicit listing of domestic bumble bees as species requiring management. To control the competition for floral resources between introduced bees and native bees, we recommend a new Public Lands Act (currently under development) that is prescriptive and controls all placement of apiaries on public lands within a set of conditions defined by the abundance of forage (flowers), the timing window within the growing season, a quota (density of hives), and the placement with respect to permitted and non-permitted land covers. The resulting spaces and times when apiaries are allowed on public lands could be called foraging leases. We also recommend the establishment of a Yukon Territory Species at Risk Act to provide habitat protection for pollinators already at risk.

Supplementary Materials: The following supporting information can be downloaded from https://www.mdpi.com/article/10.3390/biology14030282/s1: Table S1. Requirements within provincial legislation pertaining to beekeeping that potentially reduce risk to native pollinators. Requirements are organized first by general functions, and then by actions (identified in italics). Each action is followed by example wording directly quoted from existing legislation. Table S2. Sample list of bylaws pertaining to urban beekeeping.

Author Contributions: Conceptualization, M.L.; methodology, M.L.; validation, M.L.; investigation, M.L.; resources, M.L. and D.R.; writing—original draft preparation, M.L.; writing—review and editing, M.L. and D.R.; visualization, M.L.; project administration, D.R. and M.L.; funding acquisition, M.L. and D.R. All authors have read and agreed to the published version of the manuscript.

Funding: This research was funded by the Yukon Fish and Wildlife Enhancement Trust, Contribution Agreement 2024-25-11. The APC was funded by the authors.

Institutional Review Board Statement: Not applicable.

Informed Consent Statement: Not applicable.

Data Availability Statement: No new data were created.

Acknowledgments: We thank the reviewers for their time and comments, which have improved our manuscript.

Conflicts of Interest: The authors declare no conflicts of interest. The funders had no role in the design of the study; in the collection, analyses, or interpretation of data; in the writing of the manuscript; or in the decision to publish the results.

Abbreviations

The following abbreviations are used in this manuscript:

CFIA	Canadian Food Inspection Agency
COSEWIC	Committee on the Status of Endangered Wildlife in Canada
NWT	Northwest Territories
OIC	Order in Council
SARA	Species at Risk Act
SARC	Species at Risk Committee

References

1. Nalepa, R.; Colla, S.R. Toward a wild pollinator strategy for Canada: Expert-recommended solutions and policy levers. *Facets* **2023**, *8*, 1–18. [CrossRef]
2. Sánchez-Bayo, F.; Wyckhuys, K.A.G. Worldwide decline of the entomofauna: A review of its drivers. *Biol. Conserv.* **2019**, *232*, 8–27. [CrossRef]
3. Nanetti, A.; Bortolotti, L.; Cilia, G. Pathogens spillover from honey bees to other arthropods. *Pathogens* **2021**, *10*, 1044. [CrossRef]
4. Figueroa, L.L.; Sadd, B.M.; Tripodi, A.D.; Strange, J.P.; Colla, S.R.; Adams, L.D.; Duennes, M.A.; Evans, E.C.; Lehmann, D.M.; Moylett, H.; et al. Endosymbionts that threaten commercially raised and wild bumble bees (*Bombus* spp.). *J. Pollinat. Ecol.* **2023**, *32*, 14–36. [CrossRef]
5. Evans, E.C.; Strange, J.P.; Sadd, B.M.; Tripodi, A.D.; Figueroa, L.L.; Adams, L.D.; Colla, S.R.; Duennes, M.A.; Lehmann, D.M.; Moylett, H.; et al. Ectoparasites, parasitoids, and hive products that are potentially deleterious to wild and commercially raised bumble bees (*Bombus* spp.) in North America. *J. Pollinat. Ecol.* **2023**, *32*, 37–53. [CrossRef] [PubMed]
6. Cameron, S.A.; Lim, H.C.; Lozier, J.D.; Duennes, M.A.; Thorp, R. Test of the invasive pathogen hypothesis of bumble bee decline in North America. *Proc. Natl. Acad. Sci. USA* **2016**, *113*, 4386–4391. [CrossRef]

7. Graystock, P.; Goulson, D.; Hughes, W.O.H. The relationship between managed bees and the prevalence of parasites in bumblebees. *PeerJ* **2014**, *2*, e522. [CrossRef]
8. Iwasaki, J.M.; Hogendoorn, K. Mounting evidence that managed and introduced bees have negative impacts on wild bees: An updated review. *Curr. Res. Insect Sci.* **2022**, *2*, 100043. [CrossRef] [PubMed]
9. Hatfield, R.G.; Jepson, S.; Vaughan, M.; Black, S.; Lee-Mäader, E. *An Overview of the Potential Impacts of Honey Bees to Native Bees, Plant Communities, and Ecosystems in Wild Landscapes: Recommendations for Land Managers*; The Xerces Society for Invertebrate Conservation: Portland, OR, USA, 2018; pp. 1–12. Available online: https://xerces.org/publications/guidelines/overview-of-potential-impacts-of-honey-bees-to-native-bees-plant (accessed on 16 December 2024).
10. Valido, A.; Rodríguez-Rodríguez, M.C.; Jordano, P. Honeybees disrupt the structure and functionality of plant-pollinator networks. *Sci. Rep.* **2019**, *9*, 4711. [CrossRef]
11. Looney, C.; Strange, J.P.; Freeman, M.; Jennings, D. The expanding Pacific Northwest range of *Bombus impatiens* Cresson and its establishment in Washington State. *Biol. Invasions* **2019**, *21*, 1879–1885. [CrossRef]
12. iNaturalist Community. Observations of *Bombus impatiens* from Greater Vancouver Area, Canada Observed Before 1 October 2024. Available online: https://www.inaturalist.org (accessed on 1 October 2024).
13. Brown, A.O.; McNeil, J.N. Pollination ecology of the high latitude, dioecious cloudberry (*Rubus chamaemorus*; Rosaceae). *Am. J. Bot.* **2009**, *96*, 1096–1107. [CrossRef] [PubMed]
14. Hoshide, A.K.; Drummond, F.A.; Stevens, T.H.; Venturini, E.M.; Hanes, S.P.; Sylvia, M.M.; Loftin, C.S.; Yarborough, D.E.; Averill, A.L. What is the value of wild bee pollination for wild blueberries and cranberries, and who values It? *Environments* **2018**, *5*, 98. [CrossRef]
15. Leung, M.C.-Y.; Forrest, J.R.K. Insect pollinators of haskap (*Lonicera caerulea* L.: Caprifoliaceae) in subarctic Canada. *Open Agric.* **2019**, *4*, 676–683. [CrossRef]
16. Yukon Agriculture Branch. Pete McPeake, Takhini River honey and pollen. *Infarmation* **2016**, *29*, 7. Available online: https://archives-ftp.gov.yk.ca/library/normal/Infarmation_2016_Spring.pdf (accessed on 17 December 2024).
17. COSEWIC. *COSEWIC Assessment and Status Report on the Suckley's Cuckoo Bumble Bee Bombus suckleyi in Canada*; Committee on the Status of Endangered Wildlife in Canada: Ottawa, ON, Canada, 2019.
18. Environment and Climate Change Canada. *Management Plan for the Yellow Banded Bumble Bee (Bombus terricola) in Canada [Proposed]*; Species at Risk Act Management Plan Series; Environment and Climate Change Canada: Ottawa, ON, Canada, 2022.
19. Environment and Climate Change Canada. *Recovery Strategy for the Gypsy Cuckoo Bumble Bee (Bombus bohemicus) in Canada [Proposed]*; Species at Risk Act Recovery Strategy Series; Environment and Climate Change Canada: Ottawa, ON, Canada, 2022.
20. Environment and Climate Change Canada. *Management Plan for the Western Bumble Bee mckayi Subspecies (Bombus occidentalis mckayi) in Canada [In Review]*; Species at Risk Act Management Plan Series; Environment and Climate Change Canada: Ottawa, ON, Canada, 2024.
21. Convention on Biological Diversity. *Kunming-Montreal Global Biodiversity Framework (GBF)*; Conference of the Parties 15; Convention on Biological Diversity: Montreal, QC, Canada, 2022. Available online: https://www.cbd.int/gbf (accessed on 8 December 2024).
22. Government of Canada. *Canada's 2030 Nature Strategy: Halting and Reversing Biodiversity loss in Canada*; Environment and Climate Change Canada: Ottawa, ON, Canada, 2024. Available online: https://www.canada.ca/en/environment-climate-change/services/biodiversity/canada-2030-nature-strategy.html (accessed on 8 December 2024).
23. Canadian Food Inspection Agency (CFIA). *Bee Products*; CFIA Directive TAHD-DSAT-IE-2001-3-6, Updated January 5, 2011; Canadian Food Inspection Agency: Ottawa, ON, Canada, 2011. Available online: https://epe.lac-bac.gc.ca/100/206/301/cfia-acia/2011-09-21/inspection.gc.ca/english/anima/heasan/pol/ie-2001-3e.shtml (accessed on 12 December 2024).
24. Canadian Food Inspection Agency (CFIA). *Import Requirements for Potentially Injurious Organisms (Other Than Plants) to Prevent the Importation of Plant Pests in Canada)*; CFIA Directive-12-02, dated 24 October 2023; Canadian Food Inspection Agency: Ottawa, ON, Canada, 2023. Available online: https://inspection.canada.ca/en/plant-health/invasive-species/directives/imports/12-02 (accessed on 12 December 2024).
25. Canadian Food Inspection Agency (CFIA). *Alfalfa leafcutting Bee Producer Guide to the National Bee Farm-Level Biosecurity Standard*; Canadian Food Inspection Agency: Ottawa, ON, Canada, 2013. Available online: https://inspection.canada.ca/en/animal-health/terrestrial-animals/biosecurity/standards-and-principles/alfalfa-leaf-cutting-bee (accessed on 16 December 2024).
26. Government of Yukon. *Discussion Paper: A New Yukon Public Lands Act*; Government of Yukon: Whitehorse, YT, Canada, 2022; pp. 1–28. Available online: https://yukon.ca/sites/yukon.ca/files/emr/emr-yukon-public-lands-act-discussion-paper.pdf (accessed on 16 November 2024).
27. Gordon, S.C.; Duchesne, A.G.; Dusevic, M.R.; Galán-Acedo, C.; Haddaway, L.; Meister, S.; Olive, A.; Warren, M.; Vincent, J.G.; Cooke, S.J.; et al. Assessing species at risk legislation across Canadian provinces and territories. *Facets* **2024**, *9*, 1–18. [CrossRef]

28. Nasr, M.; Muirhead, S. *Recommendations for Management of Honey Bee Pests in Alberta 2018–2019*; Pest Surveillance Section, Crop Research and Extension Branch, Alberta Agriculture and Forestry: Edmonton, AB, Canada, 2019; pp. 1–22. Available online: https://www1.agric.gov.ab.ca/$Department/deptdocs.nsf/ba3468a2a8681f69872569d60073fde1/e769c0b81e297276872580bc0079cf81/$FILE/Recommendations-for-management-of-honey-bee-pests-in-alberta-2018-2019.pdf (accessed on 16 December 2024).
29. Government of Manitoba. *The Municipal Planning Guide to Zoning Bylaws in Manitoba. Component C: Plug-In Sections of Zoning Tools*; Government of Manitoba: Winnipeg, MB, Canada, 2015; pp. 1–73. Available online: https://www.gov.mb.ca/mr/land_use_dev/pubs/zoning_plugins.pdf (accessed on 16 December 2024).
30. NatureServe. *NatureServe Explorer [Web Application]*; NatureServe: Arlington, VA, USA, 2024. Available online: https://explorer.natureserve.org/ (accessed on 15 October 2024).
31. Tang, J.; Wice, J.; Thomas, V.G.; Kevan, P.G. Assessment of Canadian federal and provincial legislation to conserve native and managed pollinators. *Int. J. Biodivers. Sci. Manag.* **2007**, *3*, 46–55. [CrossRef]
32. Geldmann, J.; González-Varo, J.P. High densities of managed honey bees can harm populations of wild pollinators. *Science* **2018**, *359*, 392–393. [CrossRef] [PubMed]
33. Evans, E. From humble bee to greenhouse pollination workhorse: Can we mitigate risks for bumble bees? *Bee World* **2017**, *94*, 34–41. [CrossRef]
34. Viger, A. Produire du miel au Yukon. *L'aurore Boreale* **2022**, *39*, 37–39. Available online: https://auroreboreale.ca/actualites/sciences-et-environnement/2022/06/30/produire-du-miel-au-yukon-un-defi-de-taille/ (accessed on 16 December 2024).
35. Canadian Food Inspection Agency (CFIA). *Bumblebee Sector Guide to the National Bee Farm-Level Biosecurity Standard*; Canadian Food Inspection Agency: Ottawa, ON, Canada, 2013. Available online: https://inspection.canada.ca/en/animal-health/terrestrial-animals/biosecurity/standards-and-principles/bumblebee-sector-guide (accessed on 16 December 2024).
36. Cooke, H.A. *Securing a Wild Future: Planning for Landscape-Scale Conservation of Yukon's Boreal Mountains*; Conservation Report #9; Wildlife Conservation Society Canada: Toronto, ON, Canada, 2017; pp. 1–114. Available online: https://wcscanada.org/resources/securing-a-wild-future-planning-for-landscape-scale-conservation-of-yukons-boreal-mountains-wcs-canada-conservation-report-n/ (accessed on 16 December 2024).
37. Torné-Noguera, A.; Rodrigo, A.; Osorio, S.; Bosch, J. Collateral effects of beekeeping: Impacts on pollen-nectar resources and wild bee communities. *Basic Appl. Ecol.* **2016**, *17*, 199–209. [CrossRef]
38. Hung, K.J.; Kingston, J.M.; Lee, A.; Holway, D.A.; Kohn, J.R. Non-native honey bees disproportionately dominate the most abundant floral resources in a biodiversity hotspot. *Proc. R. Soc. B* **2018**, *286*, 20182901. [CrossRef]
39. Cane, J.H.; Tepedino, V.J. Gauging the effect of honey bee pollen collection on native bee communities. *Conserv. Lett.* **2017**, *10*, 205–210. [CrossRef]
40. MacInnis, G.; Normandin, E.; Ziter, C.D. Decline in wild bee species richness associated with honey bee (*Apis mellifera* L.) abundance in an urban ecosystem. *PeerJ* **2023**, *11*, e14699. [CrossRef] [PubMed]
41. MacKell, S.; Elsayed, H.; Colla, S.R. Assessing the impacts of urban beehives on wild bees using individual, community, and population-level metrics. *Urban Ecosyst.* **2023**, *26*, 1209–1223. [CrossRef]
42. Renner, S.S.; Graf, M.S.; Hentschel, Z.; Krause, H.; Fleischmann, A. High honeybee abundances reduce wild bee abundances on flowers in the city of Munich. *Oecologia* **2021**, *195*, 825–831. [CrossRef]
43. Eccles, L.; Kempers, M.; Gonzalez, R.M.; Thurston, D.; Borges, D. *Canadian Best Management Practices for Honey Bee Health: Industry Analysis and Harmonization*; Bee Health Roundtable and Agriculture and Agri-Food Canada: Ottawa, ON, Canada, 2017; pp. 1–61. Available online: https://honeycouncil.ca/wp-content/uploads/2016/12/BMP-manual-for-honey-bee-health-Feb-2017-English.pdf (accessed on 16 December 2024).
44. Government of Yukon. *Yukon Grazing Policy*; Agriculture Branch, Yukon Energy, Mines and Resources: Whitehorse, YT, Canada, 2015; pp. 1–14. Available online: https://yukon.ca/sites/default/files/emr/emr-2015-yukon-grazing-policy.pdf (accessed on 17 December 2024).
45. Henry, M.; Rodet, G. Controlling the impact of the managed honeybee on wild bees in protected areas. *Sci. Rep.* **2018**, *8*, 9308. [CrossRef]
46. Bureau of Land Management. *Strategic Plan for Pollinator Conservation*; U.S. Department of the Interior, Bureau of Land Management, Division of Wildlife Conservation, Aquatics, and Environmental Protection: Lakewood, CO, USA, 2022; pp. 1–38. Available online: https://www.blm.gov/sites/default/files/docs/2022-06/StrategicPlanPollinatorConservation_062022.pdf (accessed on 16 December 2024).
47. O'Brien, M.; Zafar, K.; Burd, L.A.; Cornelisse, T.; Frates, T.; Black, S.H. *Petition to the United States Department of Agriculture's Forest Service Regarding Apiaries on Public Lands*; Grand Canyon Trust, Center for Biological Diversity, and The Xerces Society for Invertebrate Conservation: Portland, OR, USA, 2020. Available online: https://www.biologicaldiversity.org/programs/environmental_health/pdfs/APA-Petition-to-Forest-Service-on-Apiary-CEs-7-29-20.pdf (accessed on 16 December 2024).

48. Strange, J.P.; Colla, S.R.; Duennes, M.A.; Evans, E.C.; Figueroa, L.L.; Lehmann, D.M.; Moylett, H.; Richardson, L.; Sadd, B.M.; Smith, J.W.; et al. An evidence-based rationale for a North American commercial bumble bee clean stock certification program. *J. Pollinat. Ecol.* **2023**, *32*, 1. [CrossRef]
49. Vetter, M.A. Grasslands of the Aishihik-Sekelmun Lakes area, Yukon Territory, Canada. *Arctic* **2000**, *53*, 165–173. [CrossRef]
50. Government of Canada. *Canada-Yukon Nature Agreement*; Environment and Climate Change Canada: Ottawa, ON, Canada, 2022. Available online: https://www.canada.ca/en/environment-climate-change/services/integrated-nature-initiatives/nature-agreements/canada-yukon-nature-agreement.html (accessed on 8 December 2024).

biology

MDPI

Article

Space and Time Dynamics of Honeybee (*Apis mellifera* L.)-Melliferous Resource Interactions Within a Foraging Area: A Case Study in the Banja Luka Region (Bosnia & Herzegovina)

Samuel Laboisse [1,*], Michel Vaillant [1], Clovis Cazenave [1], Biljana Kelečević [2], Iris Chevalier [1] and Ludovic Andres [1]

[1] Agricultural Development and Innovations for the Souths Research Unit (UR ADI-Suds), Engineering School in International Agro-Development (ISTOM), 49000 Angers, France; m.vaillant@istom.fr (M.V.); c.cazenave@istom.fr (C.C.); i.chevalier@istom.fr (I.C.); l.andres@istom.fr (L.A.)

[2] Faculty of Agriculture, University of Banja Luka, 78000 Banja Luka, Bosnia and Herzegovina; biljana.kelecevic@agro.unibl.org

* Correspondence: s.laboisse@istom.fr

Simple Summary: A better understanding of resource exploitation by honeybees is fundamental for beekeeping production. More specifically, the spatial and temporal variability of the exploration area around the hive helps beekeepers to anticipate production levels and quality. The present study aimed at understanding the fine variability of these resources in space and time in northern Bosnia & Herzegovina. The combined use of geographical information system tools, field observations, and modelling contributed to elaborate monitoring maps of potential honey production levels over time, and to assess the impact of different land uses on the area, particularly on the plant species that contributed most to this potential. Forest and grassland had better contribution on potential, respectively in May–June and in July–August. Nevertheless, some specific land uses contributed to increasing the honey production potential, such as orchards in late March to the beginning of April, and wasteland for some plant species.

Abstract: Interactions between honeybees and the environment are often difficult to achieve, particularly when the purpose is to optimize beekeeping production. The present study proposed to monitor the space-time variations of melliferous resources potentially exploited by colonies within a foraging area in Bosnia & Herzegovina, characterized by contrasting landscapes. The combination of methods involving Geographical Information Systems, floristic monitoring, and modelling enabled honey production potential to be calculated for the entire foraging area. In particular, the location of taxa, their abundance, diversity, and phenology enabled us to determine the spatial distribution and temporal variation of production potential. *Robinia pseudoacacia* and *Rubus* sp. made a major contribution. This potential was highly contrasted, with distant areas from the apiary more attractive than closer ones, depending on the moment. Specific periods, such as June were particularly conducive to establishing a high potential. Forest and grassland played a major role in the temporal succession, mainly because of the area covered, but moments with lower potential were supported by specific land uses (orchards). Land uses with a small surface area, such as orchards, wasteland, and riparian zones had a high potential per unit area, and improving the production potential within a foraging area could involve increasing these specific surfaces.

https://doi.org/10.3390/biology14040422

Keywords: *Apis mellifera* L.; honeybee; honey production potential; landscape; land use; melliferous resources; pollen and nectar; seasonality; space and time variations

1. Introduction

Animal pollination, at the forefront of which is entomophilic pollination, is of vital importance in the production of many fruits and seeds used in food and feed [1]. In this context, the foraging activity of honeybees *Apis mellifera* L. represents an interesting opportunity to optimize pollination of many crops, but also to ensure the reproductive function of plant biodiversity in natural environments. Among the diversity of pollinators that play a major role in pollination of ecosystems [2], the honeybee is singular because it is dependent not only on the beekeeper's practices but also on its physiological and behavioral specificities [3]. Many publications deal with the loss of pollinators [4–7], but it seems crucial to understand the involvement of bees in the current landscape, the one that still allows us to survive [8].

Bees are eusocial insects. Each individual interacts with others within a colony, which is based in a hive in the case of beekeeping activity. This colony is spatially positioned within a territory, and interactions with landscape elements will determine its survival. Therefore, it is interesting to know in detail the spatial distribution of resources available among the plant diversity with the foraging area, which is defined as the area of interactions favoured by foragers to gather the various resources the colony needs to survive. In order to collect these resources (mainly nectar and pollen), they can explore the foraging area over quite long distances, in excess of ten kilometers [9]. Most often, however, these distances remain between 1 and 3 km in radius around the hive [10]. The variability of this distance depends on the type of habitat considered [9,11–15]. The structure of the landscape, particularly in terms of quality, quantity, density, homogeneity, and seasonal variations in availability, also has an influence on the definition of the foraging area for the colony in question [14]. Each type of landscape has a different performance, and even forests that are not considered a good type could have some interest according to the period or the specific honey that beekeeper wants to obtain [16].

The match between melliferous resources and bee foragers is at the heart of beekeepers' strategies for optimizing the resources used by the colony and guiding bee management, particularly in the choice of location and/or productive activities [17,18]. Several studies have evaluated this spatial distribution, notably [19], which have modeled the dynamics of nectar supply within the landscape, and on several temporal scales, from very fine (3–4 days) to coarser (months). Therefore, the foraging area needs to be known in detail in order to give beekeepers the ability to evaluate a territory in terms of production potential. Nevertheless, the fine monitoring of space and time variations of melliferous resources available within a precise foraging area remains poorly evaluated in this context.

The melliferous resource can be considered in first approximation as a surface, for which the nectar and/or pollen supply is provided by a plant species [3]. However, except in special cases (intensive monocultures), several plant species coexist on a given surface, which can give rise to competition between different melliferous plants, as several plant species compete to provide the best possible resource for the bee, and hence to be visited. In such cases, we need to be able to reason about the bee's choice of the most interesting melliferous resource.

A number of ethological characteristics influence the interest of a geographical position in terms of melliferous resources. Floral constancy is defined as the preference of foragers to visit only one resource at a time during a trip [20]. Many foragers are conditioned during their flight by the recruitment of scouts within the hive, with dancing and trophallaxing behaviors [21]. Numerous studies chose dance activity as a tool for assessing interaction with melliferous resources [14,22,23]. This choice allows the clear identification of resources actually foraged by the recruiters, even if some variability can exist between previous resources exploited by recruiters and the next resources visited by recruits [22]. On the other hand, it does not give access to the potential of the foraging area and hence to the possible anticipation of impacts of weather conditions or agricultural choices in cultivated areas.

Numerous studies have modelled the functioning of bee colonies, proposing dynamics in terms of population and foraging behavior (for example [24] for a review). The presence and attractiveness of sectors to pollinators as a function of the geographical characterization of the landscape have been investigated [25,26], but few of these studies also consider the ecological variation that may exist within these environments [27]. Geographic Information System (GIS) approaches are often used to characterize the landscape as part of a model of bee colony functioning. However, they are very often used for regional or national scales [3] which makes them difficult to apply to the assessment of melliferous resources around a hive within the foraging area. In order to model the space and time variations of honey production potential in fine detail within the restricted territory of the foraging area, the present work proposes an approach combining a geographical approach using GIS tools, botanical inventories, phenological monitoring of flowering, and mathematical modelling. By taking into account not only the landscape but also plant development, we can gain a more accurate picture of variations in the nutritional resources available to colonies.

2. Materials and Methods

2.1. Study Site

Bosnia and Herzegovina (B&H) is located in the western part of the Balkans Peninsula. The north comprises the southern edge of the Pannonian plain, belonging to the peripannonian region. The study area was located near the village of Bistrica (44.81° N, 17.05° E), on the territory of the region of Banja Luka, The Republic of Srpska (B&H). As suggested by [14], a radius of 1.5 km around the apiary location was used to define the foraging area. This territory is characterized by a heterogeneous landscape and a relief structuring the landscape into valleys and hills, forests, field crops, orchards, and residential areas that follow one another without really dominating the landscape (Figure 1). Landscape observation and preliminary GIS assessment were carried out at the start of the study to confirm the heterogeneous nature of the area.

2.2. Environmental Data

Field monitoring was carried out over two consecutive years (2022 and 2023) to capture inter-annual variability. These years were consistent in terms of average, minimum, and maximum temperatures, with the typical trend for this area (Table 1).

Figure 1. Location of the area and examples of landscapes. The study area is located in northern B&H, North-West of Banja Luka.

Table 1. Mean monthly, maximum, and minimum temperatures for 2022 [28], 2023 [29], and 1991–2020 period [30].

	Month	Jan.	Feb.	Mar.	Apr.	May	June	Jul.	Aug.	Sept.	Oct.	Nov.	Dec.
Temperature (°C)	Mean monthly in 2022	0.60	5.00	4.90	10.3	17.8	22.7	22.4	21.9	16.3	13.5	7.70	5.30
	Mean monthly in 2023	3.80	3.20	8.70	10.3	15.6	20.2	22.7	21.6	18.8	15.6	8.40	4.90
	1991–2020 mean	0.64	1.95	5.73	10.2	14.7	18.4	20.4	20.6	15.9	11.0	6.03	1.09
	Mean monthly minimum in 2022	−4.00	−0.60	−2.00	3.70	10.8	15.2	14.6	15.8	11.0	8.50	4.20	2.10
	Mean monthly minimum in 2023	0.30	−2.30	2.40	5.00	11.4	14.4	16.4	15.4	12.7	9.80	3.30	0.40
	Mean 1991–2020 minimum	−2.70	−1.97	0.94	5.10	9.23	12.8	14.3	14.4	10.1	6.09	2.26	−2.03
	Mean monthly maximum in 2022	6.30	12.6	13.1	17.6	25.2	29.8	30.6	28.9	23.8	21.4	12.3	9.20
	Mean monthly maximum in 2023	8.40	10.0	16.1	16.2	20.8	26.6	30.4	29.3	26.8	23.4	14.0	10.9
	Mean 1991–2020 maximum	3.98	5.89	10.6	15.4	20.2	23.9	26.6	26.9	21.8	16.0	9.84	4.21

Data for 2022 and 2023 were measured at the Sanski Most station, 30 km from the study site, and data for the period 1991–2020 were assessed for the Republic of Srpska using data compiled by the Climatic Research Unit of the University of East Anglia.

2.3. Characterisation of Land Use Using a Combined Field and GIS Approach

All geospatial analyses were represented with QGIS software, version 3.22.15 [31]. The aim was to create a graphical representation that shows the Land use Station (LUS) [32]. This is a spatial entity considered homogeneous, with specific biophysical characteristics that it covers (without taking into account its socio-economic use, although the two are linked). Therefore, it corresponds to the whole surface with the same land use, although it is not necessarily continuous. In order to refer more specifically to a continuous area with the same land use, we defined the term Land Use Unit (LUU). This is the spatial structuring unit of the foraging area in question. Two databases were used in the foraging area. The first was the "WorldCover 10 m 2020" developed by [33]. With a resolution of 10 m, this image is derived from remote sensing data from the Sentinel-1 and Sentinel-2 satellites. The second used the NDVI (Normalized Difference Vegetation Index) to highlight differences in vegetation density over the foraging area under consideration [34]. It was calculated using the 4 Sentinel-2 L2A red and 8 Sentinel-2 LSA near infrared spectral bands [33].

In addition to these GIS approaches, further field observations were carried out in order to refine the initial representations of land use. For all of the forest sectors, main tree species were evaluated along transects, and LUU was divided up if changed. All the crop fields were observed on site to identify orchards and plots of arable land. For this land use, only wheat, barley, and maize plots were identified. Maize was isolated because can play a role in bee nutrition as a pollen source. For most field crops, crop rotation was very low or non-existent in the study area. Therefore, we decided to fix the land use of field crops for the year 2023. For wasteland, the representation based on satellite images was not very accurate. It was identified by analysing aerial photographs using Google Earth software version 10.78.0.2 [35] and verifying them in the field (geolocated field surveys using the Field Maps application developed by ArcGIS Pro [36]). Riparian vegetation was very difficult to characterize by using GIS tools alone, and therefore it is very rarely mentioned in similar research [32]. Streams were exhaustively mapped in the foraging area during the month of April, after snowmelt and heavy rainfall, to increase their visibility, including temporary ones [37]. Each stream was followed from downstream to upstream and all the bends and branches encountered were recorded using Field Maps, in order to map the structure of the associated riparian forest in detail. In this way, eight LUSs were selected and described: forest, grassland, wasteland, riparian forest, arable crops (excluding maize), maize crops, orchards, habitations and others.

2.4. Analysis of Landscape Structure

Two levels of scale are used to assess the overall organization landscape within the foraging area under consideration: the LUU, which characterises the degree of fragmentation within the area, and the general landscape, which characterises its diversity [38]. For a quantitative approach, various indicators from the Fragstats software version 4.2 [39] were used, as detailed below.

S_i is the average surface area of LUU for a given LUS i (ha), which gives an idea of the average size of LUU within an area:

$$S_i = \frac{\sum_{j=1}^{n_i} a_{ij}}{n_i} \quad (1)$$

With a_{ij} the surface area (m^2) of the LUU ij and n_i the total number of LUU within a LUS i.

Dt_i is the density of LUU per 100 ha for a LUS i. This gives an idea of the fragmentation of LUU and hence of the landscape:

$$Dt_i = \frac{n_i}{A}(10,000)(100) \quad (2)$$

where A is the total area of the landscape (m^2).

ENN_i is the average distance between two LUU of the same LUS i (m), which makes it possible to estimate the connectivity and dispersion of land use types:

$$ENN_i = \frac{\sum_{j=1}^{n_i} d_{ij}}{n_i} \quad (3)$$

where d_{ij} is the distance between two nearest neighboring spots of the same LUS i (m).

2.5. Characterisation of Melliferous Flora

2.5.1. Sampling Method

In 2022, systematic sampling was set up to cover the entire foraging area. 150 m stretch separated each observation point from its neighbors in all directions. Only points that were inaccessible in the field or were not in exploitable areas were not determined. A total of 142 points were studied. The year 2023 was used to take account of inter-annual changes and to provide a more detailed description of certain areas/landscapes that are often given little consideration in this type of study. The observation points were placed according to a stratified sampling strategy. Each type of land use was assessed by setting up botanical observation points, with the exception of inhabited areas which were too difficult to access and fields cultivated with grasses, which are all characterized by the very low percentage of weeds present, the absence of rotations involving melliferous crops and the low melliferous interest of the main crops [40]. Within each type, the number of observation points was determined based on the share of the surface area allocated to the foraging area and the botanical diversity characterized in 2022. Each observation point was positioned at random within the LUUs of the same type.

2.5.2. Botanical Parameter

At each observation point, a circular plot was positioned, the area of which depended on the stratum considered [41]. For an herbaceous layer (height less than 0.5 m) the radius was 0.56 m. For a shrub layer (height between 0.5 and 2 m) the radius was 4 m. Finally, for a tree layer (height greater than 2 m), the radius was 8 m. Within this plot, all the species were taken into account. Only grasses were not considered. The nomenclature and determination of species were conducted using modern systematic principles according to [42,43], and specific references linked to the region [44–48].

2.5.3. Phenological Parameter

The floral development of the various melliferous species was monitored. Three specific phenophases, which can have an impact on nectar availability for bees, were considered on the basis of [49]:

- the early flowering phase, where less than 50% of the flowers were open (the rest were in the form of flower buds),
- the full flowering phase, where more than 50% of the flowers were open,
- the late flowering phase, where less than 50% of the flowers were open and the rest were wilted.

A qualitative assessment by direct observation for each melliferous species present at the observation points was carried out over weeks 10 to 34, from March to August to describe the evolution of phenology.

2.5.4. Ecological Parameter

For each taxon present within an LUU, the rate of coverage was assessed using the Braun-Blanquet abundance-dominance coefficient methodology [50,51]. The total plant species, which is the total number of species present on a LUS, and the average plant species number per plot for each LUS, defined as the quotient of the sum of the species and the number of LUU for the LUS in question were evaluated.

2.6. Modelling

2.6.1. Estimation of Honey Production Potential

Initially developed by [52] and taken up by [32], the honey production potential (HPP), which includes the main parameters that affect the attractiveness of a melliferous resource [53], was calculated by integrating the various parameters developed above, in particular phenology, in order to accurately monitor the temporal variability of the resources. This landscape element represents the sum of the HPPs of each melliferous species present in the LUU under consideration. It varied on a weekly basis, in contrast to previous studies:

$$\text{HPP} = \left(1 - \frac{\text{D}}{10,000}\right) * \text{S} * \sum_{\text{t}} (\text{R}_\text{t} * \text{pnh}_\text{t}) \tag{4}$$

With:
HPP the honey production potential per LUU and per unit of time (kg/week),
t the number of considered nectariferous taxons,
D the distance between the center of gravity of the LUU and the apiary (m),
S the area covered by the LUU (ha),
R_t the coverage rate of the taxon on the LUU worked (%),
pnh_t the weekly nectariferous potential of the taxon t (kg/ha), calculated as follows:

$$\text{pnh}_\text{t} = \frac{\text{pn}_t}{\text{Sf}} \text{I}_\text{f} \tag{5}$$

With:

pn_t the annual nectariferous potential of a given taxon t. This value was taken from the literature, in particular [46–48,52,54,55]. For species not listed in the literature, either values for the generic genus were still used if available (for *Campanula* sp., *Knautia* sp., *Ranunculus* sp., *Rubus* sp., *Salix* sp. and *Stachys* sp.), or they were not included due to lack of data,

Sf the number of weeks of flowering,

I_f the flowering intensity, value 0.5 for early and late flowering, 1 for full flowering, 0 otherwise.

2.6.2. Mapping the Space and Time Evolution of Melliferous Resources

For each LUU considered, the botanical, phenological, and ecological parameters used to calculate an HPP were associated with at least one observation made within the LUU. For the others, the nearest observation point located on another LUU was used to feed the equation. In this way, each pixel making up the foraging area was characterized by an LUU identification and a HPP value per week. This information was then vectorized on the mapping of the area using QGIS software in order to visualise the space and time variability of the HPP over the entire foraging area.

2.7. Statistical Analysis

A non-parametric test (Kruskal-Wallis test) and a two-by-two multiple comparison test were carried out to assess the significance of the effect of the LUS on the mean LUU surface area, the mean distance between the nearest LUUs and the average plant species number per observation plot, using R version 4.0.4 [56].

3. Results

3.1. Landscape Characterisation

The total surface area of the foraging area was 707 ha. 36.3% was covered by forest, 26.1% by grassland, 12.5% by housing, 9.06% by wasteland, 6.11% by orchards, 4.17% by arable crops excluding maize, 4.04% by maize and 1.71% by riparian vegetation (Figure 2). The average surface area of the LUU was fairly small (1.75 ha), with cultivated LUU being even smaller on average, at around 0.5 ha (Table 2). Only the forested LUUs were larger, although they vary greatly in size. The landscape mosaic was particularly represented by the grassland and orchard LUU, which have a higher Dt_i and a lower ENN_i than the other LUSs. On the other hand, riparian vegetation was very little represented in the landscape and was exclusively linked to water (therefore very punctual in the landscape and linked to the watercourse and the small lakes present in the foraging area). The arable LUUs were particularly remote from one another. It should be noted that the average distance between two LUUs of the same type varies greatly from one LUS to another.

Table 2. Landscape characterization of the type of land use.

Type of Land Use	Number of LUUs	S_i (ha)	Dt_i (LUU/100 ha)	ENN_i (m)
Cropfield excluding corn	43	0.685 ± 0.52^{ab}	6.08	118 ± 110^{a}
Forest	40	6.41 ± 11.0^{a}	5.66	51.4 ± 51.7^{bc}
Wasteland	38	1.68 ± 2.95^{a}	5.38	81.9 ± 112^{abcd}
Maize	49	0.583 ± 0.50^{b}	6.93	99.7 ± 106^{ab}
Grassland	114	1.62 ± 3.44^{ab}	16.1	24.7 ± 28.7^{d}
Riparian forest	13	0.929 ± 1.28^{abc}	1.84	72.3 ± 142^{abcd}
Orchard	132	0.327 ± 0.545^{c}	18.7	29.5 ± 45.5^{cd}
Statistical analysis	-	$X^2 = 116.84$, df = 6, $p < 0.001$	-	$X^2 = 71.35$, df = 6, $p < 0.001$

Values are given in mean ± standard deviation. Different superscript letters within a column indicate significant differences (two-by-two multiple comparison test).

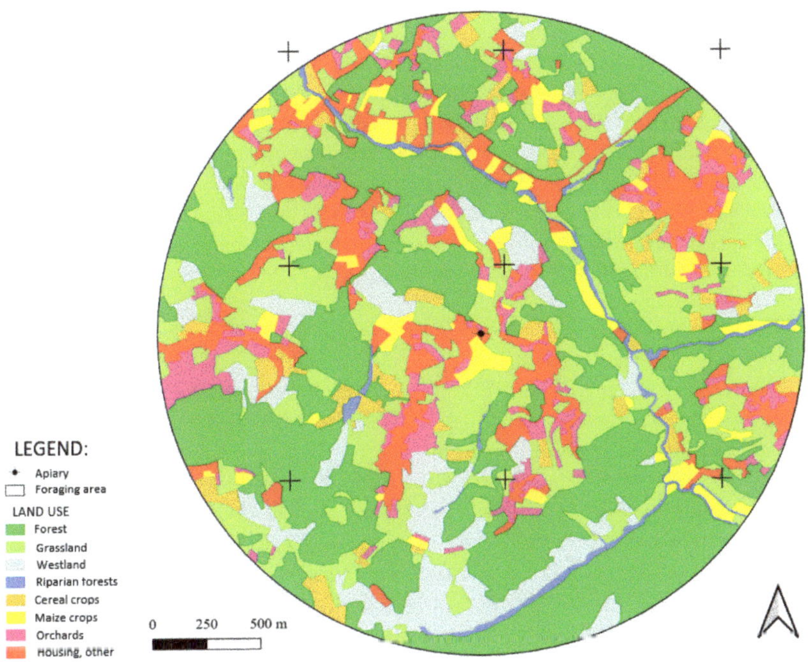

Figure 2. Distribution of land use types within the foraging area.

3.2. *Link Between Land Use and Botanical Diversity*

Total plant species were highest in grassland throughout the foraging area (Table 3). Orchards and wasteland had the lowest one. For this first LUS, non-cultivated flora was considered a weed and the management contributed to limiting the associated diversity. In the case of wasteland, these areas were in transition and were mainly colonized by plants considered to be pioneers or even ruderals, with fairly strong competition, particularly from *Rubus* sp. On the other hand, if we considered the number of species present on average within a plot, the diversity was greater for grassland and riparian zones, unlike for woodland and wasteland. Thus, the use of a restricted area (a LUU) by bees would allow a greater potential number of species to be used. The average plant species number for the orchard plots had not been evaluated, as the measurement process was different for this LUU, as most of the plots were inaccessible. An average value based on the only accessible plots was therefore generalized to the entire LUS.

Table 3. Total and average plant richness of the research area.

Land Use Type	Total Plant Species	Average Plant Species Number per Observation Plot
Forest	66	5.39 ± 3.07 [a]
Wasteland	50	5.78 ± 2.65 [a]
Grassland	84	14.4 ± 6.05 [b]
Riparian forest	69	16.7 ± 5.88 [b]
Orchard	45	-
Statistical analysis	-	$X^2 = 59.12$, df = 3, $p < 0.001$

For the statistical analysis, values are given in mean ± standard deviation. Different superscript letters within a column indicate significant differences (two-by-two multiple comparison test).

3.3. Spatial and Temporal Distribution of HPP

The detailed space and time characterization of the foraging area (Figure 3) enabled the HPP to be monitored precisely over the weeks making up the beekeeping season. From a very low potential in the first few weeks (W10 and W11), concentrated on a few LUUs giving access to HPP of a few dozen kg but relatively far from the apiary, to a more substantial potential in W12 and W13, this led to a general increase in HPP in the area (W14 to W20), where less than half of the LUUs exceed an HPP of 10 kg and some were already at 100 kg or more, culminating in the most generous periods between W21 and W27, where almost all the LUUs were likely to provide an HPP of more than 50 kg. From W28 onwards, there were specific areas that were very attractive among others less generous. These areas of interest remained relatively close to the apiary.

3.4. Temporal Variations in Cumulative Weekly HPP in the Foraging Area

Figure 4 provided a clear picture of the temporal variability and the general impact of each type of land cover considered. Four periods followed one another in the honey production potential between weeks 10 and 34 (Figure 4a). The first concerned weeks were 10 to 15. It showed rather low levels of HPP with a major contribution from the forests. However, in the downward phase, the orchards made a significant contribution to the weekly HPP in weeks 12 to 15, with, for example, up to 45% of the HPP in week 14 coming from the orchards. Weeks 14 and 15 showed the lowest weekly HPP and similar values were not seen again until August. The second period ran from weeks 16 to 19 and was characterized by a greater contribution from grassland, up to 21.5% in week 17. The contribution from orchards became negligible from week 16 onwards. The third period ran from week 20 to 27. It was characterized by an increase in the contribution from all the LUSs, which contributed to a very significant increase in the weekly HPP, with a maximum value in week 23 for the season. From weeks 20 to 26, the forest made the largest contribution then equaled the contribution from grassland in weeks 25, 26, and 27. Weeks 21 to 27 allowed a contribution from wasteland, up to 10% in week 25. In the fourth period, from week 28 and almost all in weeks 29 to 34, grassland accounted for most of the HPP. Although forests represented the largest share of potential honey production, meadows provided a steady and substantial foraging resource throughout June, July, and August.

This indicated the importance of grassland as a significant foraging area for bees. The contribution of riparian vegetation remained very low, with an increase between weeks 21 and 27.

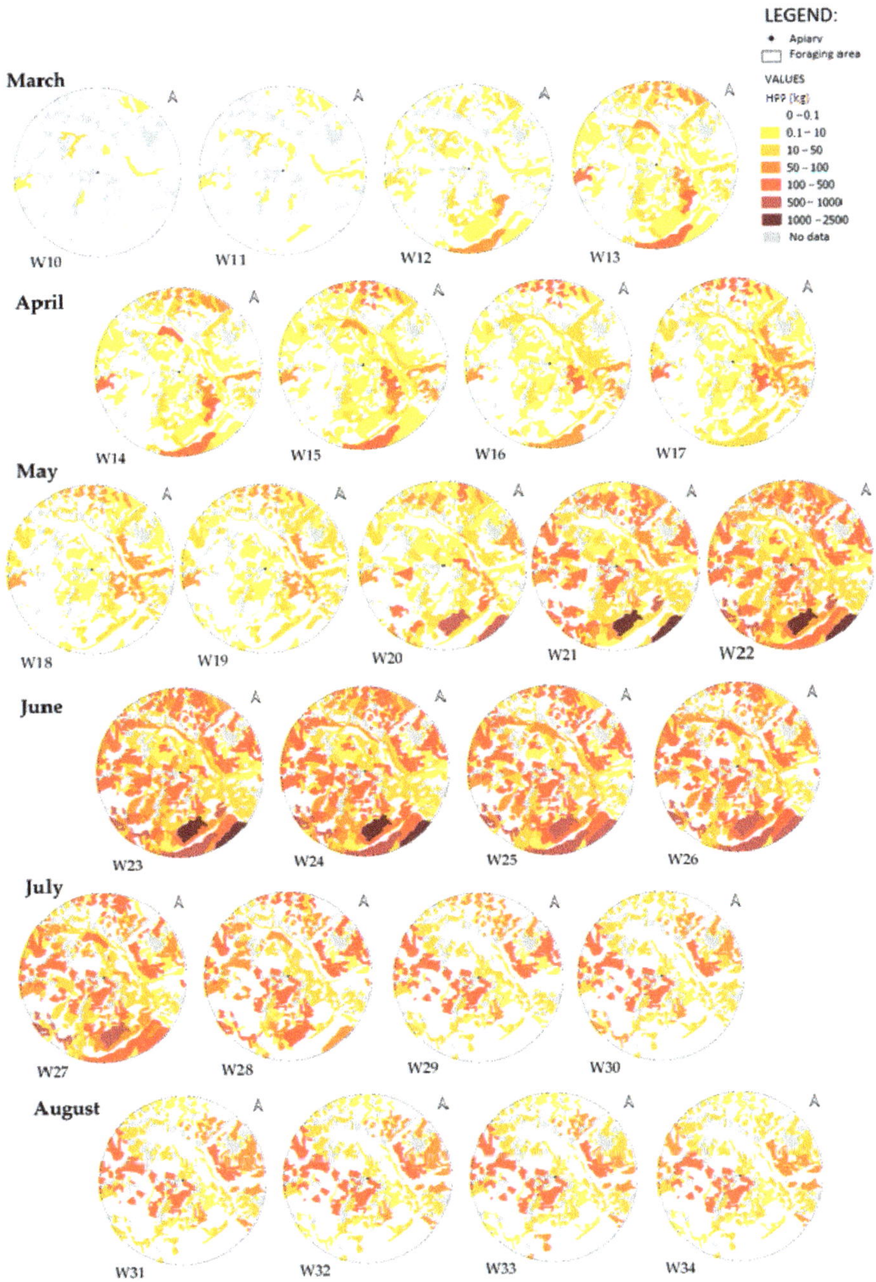

Figure 3. Space and time variation of HPP. The represented time step is the week, each card is linked to a week number during the year (for example W10 for week 10). Each circle has a diameter equivalent to 3 km.

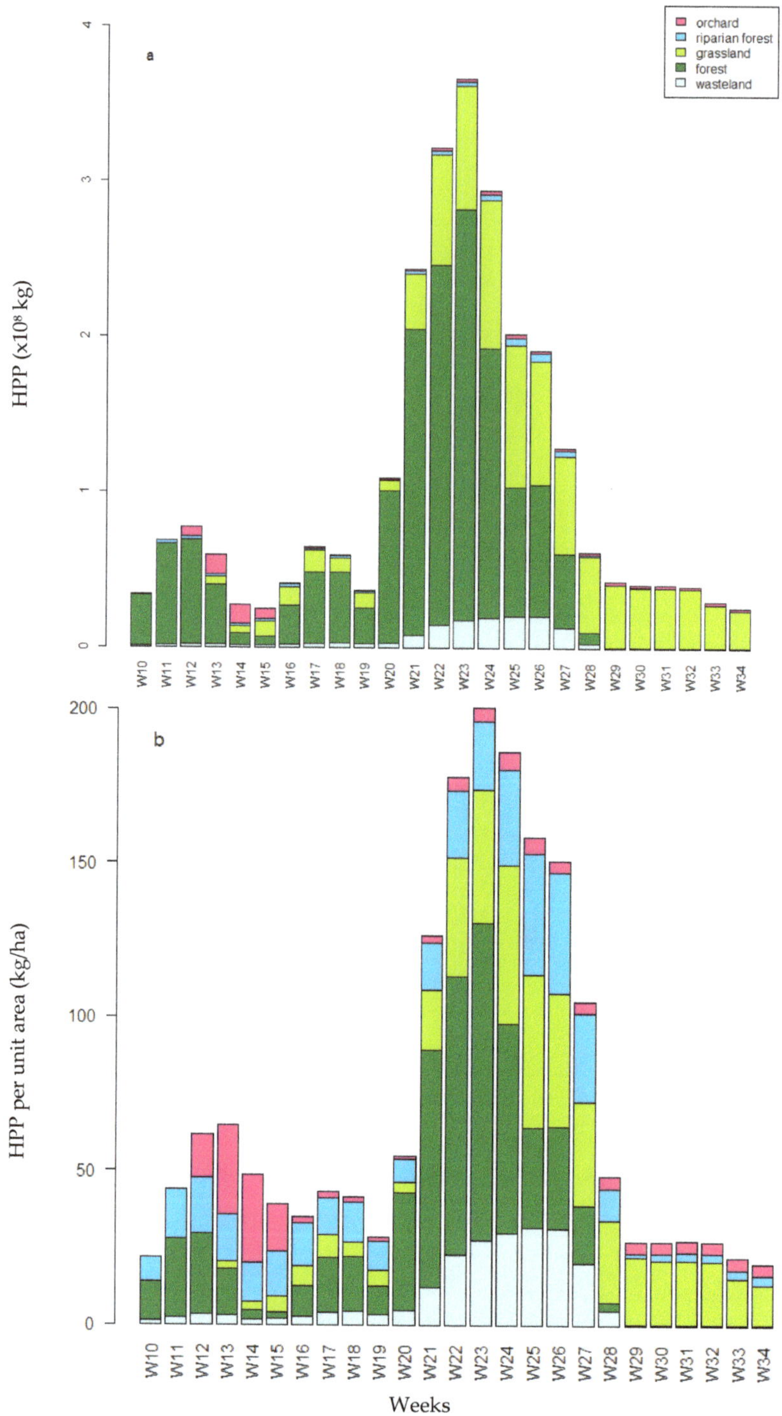

Figure 4. Temporal variation in cumulative HPP per week as a function of land use type: (**a**) total HPP ($\times 10^8$ kg); (**b**) HPP per unit area (kg/ha).

The importance of forest and grassland in the establishment of total HPP was clearly visible. Nonetheless, rarer land used such as orchards, riparian forests, and wasteland were useful insofar as they represented a significant contribution to the HPP in relation to their surface area (Figure 4b). Using the periods described above, their relative involvement varied greatly. In the first phase, the involvement of orchards and riparian vegetation was very high, reaching 59.3% and 25.7% respectively in week 14, while grassland and forest were at 5.31% and 6.20%, respectively. In the second phase, forest and riparian forest shared the bulk of the contributions, as in week 18, when they accounted for 42.6% and 31.2% respectively. The third phase saw the most significant implications for all land uses except orchards. Initially dominated by forest, the implications were equivalent in weeks 25, 26, and 27. Finally, in the last phase, grassland dominated the implications in relation to the other land uses.

3.5. Contribution of the Main Taxa to Overall HPP

Identifying the contribution of each taxon to the overall HPP by considering all the observation points enabled to characterize the most abundant taxa (Figure 5). These were mainly found in forests, with trees such as *Carpinus betulus* or *Robinia pseudoacacia*, but also *Rubus*. Grasslands also played an important role, as ten of these major taxa were mainly found in this LUS. Forests and grasslands were the main hosts for these taxa, but land-use types less important in terms of surface area were heavily involved for certain abundant taxa, where, for example, 24.7% of the surface area covered by *Rubus caesius*, 48.8% for *Rubus plicatus* and up to 55.8% for *Cornus sanguinea* were attributed to wasteland and riparian vegetation. Similarly, in the case of orchards, *Filipendula vulgaris* accounted for 6.02%, *Centaurea jacea* for 23.3%, and up to 96.9% for *Prunus domestica*, the main species cultivated in these orchards.

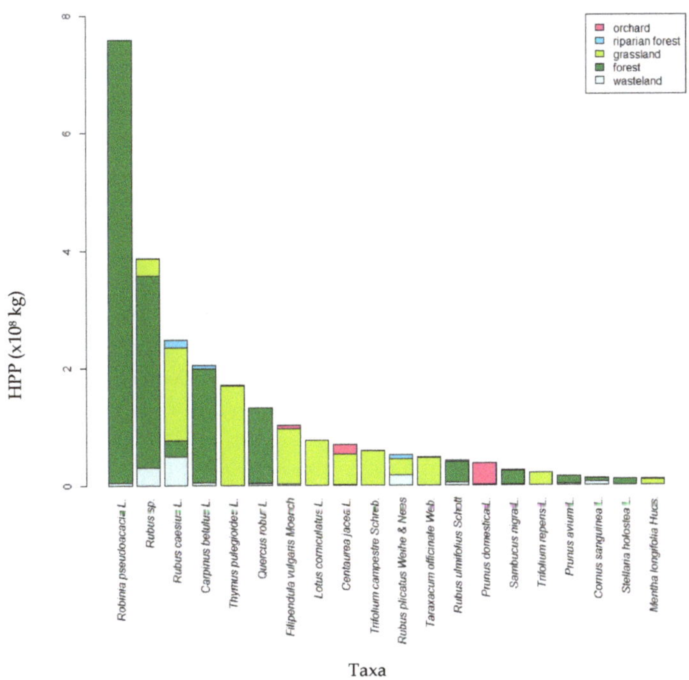

Figure 5. Cumulative HPP over 25 weeks of the 20 taxa contributed most to the total HPP of the foraging area.

4. Discussion

The diversity of melliferous resources within a landscape makes it possible to supply the colony with complementary nutritional inputs that maximize its fitness [57]. A variety of melliferous resources which will follow one another in time due to their specific phenology, and in space due to the diversity of communities associated with land use, will make it possible to offer colonies resources in sufficient quantity and quality to meet the diverse of their physiological needs. As *Apis mellifera* is a polylectic pollinator [58] and the colony is structured as a complex system [59], the organization of foraging makes it possible to take advantage of the different spatial and temporal opportunities for access to melliferous resources [60–62]. Wasteland and riparian zones are particularly interesting because they combine several vegetation strata, often in a small area, making available a diversified melliferous resource [63] and offering staggered flowering throughout the season [54]. On the other hand, landscapes characterized by a large number of arable plots have a strong impact on bee colony dynamics [64]. In our example, the presence of arable plots did not provide colonies with melliferous resources, apart from maize, which provides a potential pollen resource. Nevertheless, the small surface area allocated (less than 10% of the foraging area considered) remained small compared with other types of LUS and the impact on colony dynamics hence remained small. Similarly, the HPP responded to a question about honey production from nectar resources, which excluded maize. Studies carried out in landscapes with a high proportion of arable crops [62,65] had identified a harvest peak in April followed by a trough in June. In the case of our study, April was a low period, partly offset by the flowering of orchard plots, and June corresponded to the highest potential with the transition in exploitable melliferous resources from forests to meadows. Several studies [10,16,66] have shown an increase in flight distances due to the scarcity of melliferous resources during the summer months, particularly in July. Our study is in line with this trend, showing a decrease in HPP in summer, with resources almost exclusively provided by meadows. Similarly, [16] showed that the 'summer blues' are more pronounced in forested landscapes than in more open environments. Our study, which shows a forest landscape at less than 40%, shows a 5.7-fold decrease in overall HPP between periods 3, which relied mainly on forest resources, and 4, which relied on grassland resources.

The number of species potentially exploitable by bees within the foraging area (162 species) is consistent with other results of pollen analysis studies (between 149 and 168 species, in particular for [67] which worked on the subspecies *Apis mellifera carnica* present in the Balkans). Four of the twenty most important plant species contributing to the overall HPP for our study are found in [55], nine if we generalize to genera, for a study area located around 200 km away as the crow flies. Most of the HPP identified in the foraging area is borne by a small number of plant species. However, it has been shown that the diversity of the resources visited, as well as their sufficient availability throughout the colony's activity, has a strong impact on the colony's long-term survival [11]. This diversity will make it possible to meet the colony's nutritional needs and improve the general health of the individuals making up the colony [68]. Moreover, landscape diversity does not influence the abundance and diversity of the resources collected. Low values in the nearby landscape are compensated for by the exploitation of a larger foraging area [11]. This leads to a greater flight distance per forager in simple landscapes compared with more complex ones [14]. Wasteland could be interesting to increase the diversity of potential melliferous resources. This LUS was characterized by several strata (shrub, bush, sometimes with a herbaceous stratum) made up of spontaneous vegetation, formerly maintained and/or cultivated then abandoned and evolving towards a forest formation [69]. Several species are abundant there, especially *Rubus* sp., even if, in our study, the total plant species was lower than other LUSs, and weeks with the highest values of HPP contribution coincide with forest and grassland ones.

This work enabled us to model the theoretical availability of the melliferous resource for bees by identifying the most attractive sectors in space and time. However, this upstream availability does not condition the actual ability of bees to visit flowers and bring nectar and pollen back to the hive. In [70], only half of the available melliferous resources were found in honey and beebread. In natural habitats, the proportion of flowering species not utilized in colony products can reach up to 60%, depending on the time of year. Estimating the nectariferous potential of each taxon poses a challenge. Nectar production depends on the plant's physiology. The quantity of nectar produced and available fluctuates during the day, via hormonal control, particularly by jasmonate [71], but also by competition for food with other pollinators [72]. Nectar produced but not used by a pollinator may even be reabsorbed by the plant [73]. Experimental methods can be used to monitor fluctuations in production and availability, in particular by collecting nectar inside the flowers [74,75], but they are difficult to set up and monitor. Modelling can also be used to monitor nectar potential [74,76]. Finally, remote sensing approaches have been proposed to anticipate and assess flowering, some of them even proposing an evaluation of honey production [77–80]. However, although these approaches are useful for obtaining a general view of the production potential over a large area with homogeneous resources, such as in the context of large plots, large orchards, or little diversified forests, it is difficult to use in the context of specific work on the capacity of an apiary to exploit a restricted area such as a foraging area, in a heterogeneous landscape such as northern B&H.

The ability of a colony to exploit a foraging area and its resources depends on a variety of factors: organoleptic and chemical characteristics of the products harvested from the resource [81–83], ecological characteristics of the populations and communities of the plants of interest, and structure of the landscape [84], phenology and capacity to supply nectar and/or pollen [85], individual and collective ethological characteristics of bees (distance to the hive, resource encountered by a scout, recruitment within the hive) [22,86], weather conditions which will influence the phenology of the plant and the capacity of the foragers to leave the hive and ensure the harvest [87–89]. Our model focused on the capacity of the resource to produce and the interaction with a colony but did not take into account the dynamics of the colony and its capacity to forage effectively. Models of interaction with the climate and colony dynamics [76,90–92] could be coupled to our approach in order to model more finely the effective interaction between a bee colony and a melliferous resource in order to come closer to a coherent prediction of the productivity of an apiary. Similarly, pollinator communities within ecosystems coexist and interact to exploit resources. A detailed understanding of an area's capacity to provide effective honey production therefore requires the integration of the effect of other pollinators. Pollinator-resource networks are complex and specific [72,93] and contribute to the ecosystem. Honeybees integrate these networks through direct and indirect competition [94]. Competition between different apiaries also remains problematic insofar as the functioning of colonies remains similar. Several studies have attempted to optimize the positioning of apiaries according to the potential productivity of a zone [77,95,96]. One of the possible perspectives of our study could be the generalization of the understanding of the space-time variation of HPP over a zone larger than the foraging area in order to optimize the positioning of apiaries and minimize mutual competition.

5. Conclusions

This study is part of the drive to better understand the interaction between bees and the environment. In particular, it enabled us to map the spatial and temporal variations in

HPP in an area close to an apiary, which is most used by the colonies. It made it possible to associate the specific features of the landscape, particularly in terms of land use, with the production capacity of an apiary. Successions over time in terms of the contribution made by LUS, as well as the intensity of the production potential, could be highlighted in detail, down to the week. It was also possible to identify the plant species that make the greatest contribution to HPP. The results presented in this study are largely in line with the literature and reflect the specific features of a contrasting landscape environment where forest dominates in the establishment of HPP. This work is an example of a detailed understanding of the environment around the apiary, corresponding to the questions beekeepers are asking about their ability to anticipate the production potential of an area, both in terms of the reasoning behind honey flows and the triggering of transhumance.

Author Contributions: Conceptualization, S.L., M.V. and L.A.; methodology, S.L., C.C. and I.C.; software, S.L., C.C. and I.C.; validation, S.L.; formal analysis, S.L., C.C. and I.C.; investigation, C.C., B.K. and I.C.; data curation, S.L., C.C. and I.C.; writing—original draft preparation, S.L.; writing—review and editing, S.L., M.V., B.K. and L.A.; visualization, S.L., C.C. and I.C.; supervision, S.L., M.V. and L.A.; project administration, L.A.; funding acquisition, L.A. All authors have read and agreed to the published version of the manuscript.

Funding: This work was realized within the framework of PACAD project, which received fundings in 2022 by AUF-Co//ectif program, grant number DREO_6686 and from 2023 to 2025 by Erasmus+ program, grant number 2022-1-FR01-KA171-HED-000072876.

Institutional Review Board Statement: Not applicable.

Informed Consent Statement: Not applicable.

Data Availability Statement: Data are available upon request from the authors.

Acknowledgments: The authors thank the edition team and anonymous reviewers for their contributions, AUF and Erasmus+ programs for the financial support, the Faculty of agriculture of the University of Banja Luka for the hospitality and help during the data collection campaigns and Dejan Ilić for the access to the apiary and advice on the ground.

Conflicts of Interest: The authors declare no conflicts of interest.

Abbreviations

The following abbreviations are used in this manuscript:

B&H	Bosnia & Herzegovina
GIS	Geographical Information System
HPP	Honey Production Potential
LUS	Land Use Station
LUU	Land Use Unit
NDVI	Normalized Difference Vegetation Index

References

1. Eilers, E.J.; Kremen, C.; Smith Greenleaf, S.; Garber, A.K.; Klein, A.-M. Contribution of Pollinator-Mediated Crops to Nutrients in the Human Food Supply. *PLoS ONE* **2011**, *6*, e21363. [CrossRef] [PubMed]
2. Alignier, A.; Aviron, S.; Baude, M.; Jeavons, E.; Michelot-Antalik, A.; Porcher, E. Les Plantes de Services Favorisant La Pollinisation. In *Les Plantes de Services, vers de Nouveaux Agroécosystèmes*; Editions Quae: Versailles, France, 2024; p. 388.
3. Baveco, J.M.; Focks, A.; Belgers, D.; van der Steen, J.J.M.; Boesten, J.J.T.I.; Roessink, I. An Energetics-Based Honeybee Nectar-Foraging Model Used to Assess the Potential for Landscape-Level Pesticide Exposure Dilution. *PeerJ* **2016**, *4*, e2293. [CrossRef] [PubMed]

4. Dicks, L.V.; Breeze, T.D.; Ngo, H.T.; Senapathi, D.; An, J.; Aizen, M.A.; Basu, P.; Buchori, D.; Galetto, L.; Garibaldi, L.A.; et al. A Global-Scale Expert Assessment of Drivers and Risks Associated with Pollinator Decline. *Nat. Ecol. Evol.* **2021**, *5*, 1453–1461. [CrossRef] [PubMed]
5. Potts, S.G.; Imperatriz-Fonseca, V.; Ngo, H.T.; Aizen, M.A.; Biesmeijer, J.C.; Breeze, T.D.; Dicks, L.V.; Garibaldi, L.A.; Hill, R.; Settele, J.; et al. Safeguarding Pollinators and Their Values to Human Well-Being. *Nature* **2016**, *540*, 220–229. [CrossRef]
6. Potts, S.G.; Biesmeijer, J.C.; Kremen, C.; Neumann, P.; Schweiger, O.; Kunin, W.E. Global Pollinator Declines: Trends, Impacts and Drivers. *Trends Ecol. Evol.* **2010**, *25*, 345–353. [CrossRef]
7. Biesmeijer, J.C.; Roberts, S.P.M.; Reemer, M.; Ohlemüller, R.; Edwards, M.; Peeters, T.; Schaffers, A.P.; Potts, S.G.; Kleukers, R.; Thomas, C.D.; et al. Parallel Declines in Pollinators and Insect-Pollinated Plants in Britain and the Netherlands. *Science* **2006**, *313*, 351–354. [CrossRef]
8. Mugnier, R. Le travail interespèces de pollinisation dans les écologies perturbées. *Rev. D'anthropologie Connaiss.* **2023**, *17*, 1. [CrossRef]
9. Beekman, M.; Ratnieks, F.L.W. Long-Range Foraging by the Honey-Bee, *Apis mellifera* L. *Funct. Ecol.* **2000**, *14*, 490–496. [CrossRef]
10. Couvillon, M.J.; Riddell Pearce, F.C.; Accleton, C.; Fensome, K.A.; Quah, S.K.L.; Taylor, E.L.; Ratnieks, F.L.W. Honey Bee Foraging Distance Depends on Month and Forage Type. *Apidologie* **2015**, *46*, 61–70. [CrossRef]
11. Danner, N. Honey Bee Foraging in Agricultural Landscapes. Ph.D. Thesis, Julius-Maximilians Universität, Würzburg, Germany, 2016.
12. Greenleaf, S.S.; Williams, N.M.; Winfree, R.; Kremen, C. Bee Foraging Ranges and Their Relationship to Body Size. *Oecologia* **2007**, *153*, 589–596. [CrossRef]
13. Hagler, J.R.; Mueller, S.; Teuber, L.R.; Machtley, S.A.; Van Deynze, A. Foraging Range of Honey Bees, Apis Mellifera, in Alfalfa Seed Production Fields. *J. Insect Sci. Online* **2011**, *11*, 144. [CrossRef] [PubMed]
14. Steffan-Dewenter, I.; Kuhn, A. Honeybee Foraging in Differentially Structured Landscapes. *Proc. R. Soc. Lond. B Biol. Sci.* **2003**, *270*, 569–575. [CrossRef] [PubMed]
15. Visscher, P.K.; Seeley, T.D. Foraging Strategy of Honeybee Colonies in a Temperate Deciduous Forest. *Ecology* **1982**, *63*, 1790–1801. [CrossRef]
16. Rutschmann, B.; Kohl, P.L.; Steffan-Dewenter, I. Foraging Distances, Habitat Preferences and Seasonal Colony Performance of Honeybees in Central European Forest Landscapes. *J. Appl. Ecol.* **2023**, *60*, 1056–1066. [CrossRef]
17. Béguin, C. Contribution à la cartographie des potentialités mellifères de Haut-Jura: Exemples de variations spatio-temporelles autour d'un rucher à Chaumont/NE. *Geogr. Helvetica* **1994**, *49*, 115–124. [CrossRef]
18. Briane, G.; Cabrol, J.-L. L'abeille dans le géosystème: Essai de cartographie des ressources mellifères. *Rev. Géographique Pyrén. Sud-Ouest* **1986**, *57*, 363–373. [CrossRef]
19. Timberlake, T.P.; Vaughan, I.P.; Memmott, J. Phenology of Farmland Floral Resources Reveals Seasonal Gaps in Nectar Availability for Bumblebees. *J. Appl. Ecol.* **2019**, *56*, 1585–1596. [CrossRef]
20. Grüter, C.; Moore, H.; Firmin, N.; Helanterä, H.; Ratnieks, F.L.W. Flower Constancy in Honey Bee Workers (*Apis mellifera*) Depends on Ecologically Realistic Rewards. *J. Exp. Biol.* **2011**, *214*, 1397–1402. [CrossRef]
21. von Frisch, K. *The Dance Language and Orientation of Bees*; Harvard Univ. Press: Cambridge, UK, 1967.
22. Shackleton, K.; Balfour, N.J.; Al Toufailia, H.; James, E.; Ratnieks, F.L.W. Honey Bee Waggle Dances Facilitate Shorter Foraging Distances and Increased Foraging Aggregation. *Anim. Behav.* **2023**, *198*, 11–19. [CrossRef]
23. Sherman, G.; Visscher, P.K. Honeybee Colonies Achieve Fitness through Dancing. *Nature* **2002**, *419*, 920–922. [CrossRef]
24. Becher, M.A.; Osborne, J.L.; Thorbek, P.; Kennedy, P.J.; Grimm, V. REVIEW: Towards a Systems Approach for Understanding Honeybee Decline: A Stocktaking and Synthesis of Existing Models. *J. Appl. Ecol.* **2013**, *50*, 868–880. [CrossRef] [PubMed]
25. Polce, C.; Maes, J.; Rotllan-Puig, X.; Michez, D.; Castro, L.; Dvorak, L.; Fitzpatrick, Ú.; Francis, F.; Neumayer, J.; Manino, A.; et al. Distribution of Bumblebees across Europe. *One Ecosyst.* **2018**, *3*, e28143. [CrossRef]
26. Sárospataki, M.; Bakos, R.; Horváth, A.; Neidert, D.; Horváth, V.; Vaskor, D.; Szita, É.; Samu, F. The Role of Local and Landscape Level Factors in Determining Bumblebee Abundance and Richness. *Acta Zool. Acad. Sci. Hung.* **2016**, *62*, 387–407. [CrossRef]
27. Whitehorn, P.R.; Seo, B.; Comont, R.F.; Rounsevell, M.; Brown, C. The Effects of Climate and Land Use on British Bumblebees: Findings from a Decade of Citizen-science Observations. *J. Appl. Ecol.* **2022**, *59*, 1837–1851. [CrossRef]
28. Bijelčić, A.; Krajinović, B.; Hodžić, S.; Zulum, D.; Voljevica, N.; Tucaković, A. *Meteorološki Godišnjak 2022*; Federalni Hidrometeorološki Zavod: Sarajevo, Bosnia & Herzegovina, 2022; p. 62.
29. Bijedić, A.; Krajinović, B.; Hodžić, S.; Zulum, D.; Voljevica, N.; Tucaković, A. *Meteorološki Godišnjak 2023*; Federalni Hidrometeorološki Zavod: Sarajevo, Bosnia & Herzegovina, 2023; p. 56.
30. World Bank Climate Change Knowledge Portal. Available online: https://climateknowledgeportal.worldbank.org/ (accessed on 26 March 2025).
31. *QGIS Development Team QGIS Geographic Information System*; Open Source Geospatial Foundation: Beaverton, OR, USA, 2023.

32. Rhoné, F. L'abeille à Travers Champs: Quelles Interactions Entre *Apis mellifera* L. et le Paysage Agricole (Gers 32)?: Le rôle des Structures Paysagères Ligneuses Dans L'apport de Ressources Trophiques et Leurs Répercussions sur les Traits D'histoire De vie des Colonies. Ph.D. Thesis, Université Toulouse-Jean Jaurès, Toulouse, France, 2015.
33. ESA. Sentinel Copernicus. Available online: https://sentinels.copernicus.eu/web/sentinel/home (accessed on 4 May 2022).
34. Chuvieco, E. *Fundamentals of Satellite Remote Sensing: An Environmental Approach*, 3rd ed.; CRC Press: Boca Raton, FL, USA, 2020; ISBN 978-0-429-50648-2.
35. Google Earth. Available online: https://earth.google.com/web/ (accessed on 4 May 2022).
36. Fields Maps. Available online: https://www.esri.com/en-us/arcgis/products/arcgis-field-maps/overview (accessed on 4 May 2022).
37. Williams, D.D.; Hynes, H.B.N. The Ecology of Temporary Streams II. General Remarks on Temporary Streams. *Int. Rev. Gesamten Hydrobiol. Hydrogr.* **1977**, *62*, 53–61. [CrossRef]
38. Cushman, S.A.; McGarigal, K.; Neel, M.C. Parsimony in Landscape Metrics: Strength, Universality, and Consistency. *Ecol. Indic.* **2008**, *8*, 691–703. [CrossRef]
39. McGarigal, K.; Cushman, S.A.; Ene, E. *FRAGSTATS*, Version 4. Spatial Pattern Analysis Program for Categorical Maps. University of Massachusetts: Amherst, MA, USA, 2012.
40. Decourtye, A.; Lecompte, P.; Pierre, J.; Chauzat, M.-P.; Thiébeau, P. Introduction de jachères florales en zones de grandes cultures: Comment mieux concilier agriculture, biodiversité et apiculture? *Courr. Environ. INRA* **2007**, *16*, 3.
41. Ponce-Hernandez, R. *Assessing Carbon Stocks and Modelling Win-Win Scenarios of Carbon Sequestration Through Land-Use Changes*; Food and Agriculture Organization of the United Nations: Rome, Italy, 2004.
42. Beck, G.; Maly, K.; Bjelčić, Ž. *Flora Bosnae et Hercegovinae*; Sympetalae; Zemaljska Štamparija: Sarajevo, Bosnia & Herzegovina, 1983.
43. Euro+Med PlantBase the Information Ressource for Euro-Mediterranean Plant Diversity. 2006. Available online: https://www.emplantbase.org/home.html (accessed on 4 May 2022).
44. Janjić, V.; Kojić, M. *Atlas Travnih Korova*; Institut za Istraživanja u Poljoprivredi SRBIJA: Beograd, Serbia, 2003.
45. Kojić, M. *Livadske Biljke*; Naučna Knjiga: Baograd, Serbia, 1990.
46. Šumatić, N.; Topalić Trivunović, L.; Komljenović, I.; Todorović, J. *Najčešći Korovi Regije Banja Luka*; Grafomark: Laktaši, Bosnia & Herzegovina, 2006.
47. Umeljic, V. *In the World of Bees and Flowers. Atlas of Melliferous Plants*; part 2; Kolor Press: Kragujevac, Serbia, 2003.
48. Umeljic, V. *In the World of Bees and Flowers. Atlas of Melliferous Plants*; part 1; Kolor Press: Kragujevac, Serbia, 1999.
49. Frankl, R.; Wanning, S.; Braun, R. Quantitative Floral Phenology at the Landscape Scale: Is a Comparative Spatio-Temporal Description of "Flowering Landscapes" Possible? *J. Nat. Conserv.* **2005**, *13*, 219–229. [CrossRef]
50. Braun-Blanquet, J.; Roussine, N.; Negre, R. *Les Groupements Végétaux de La France Méditerranéenne*; CNRS: Paris, France, 1952.
51. Rodwell, J.S. *National Vegetation Classification: Users' Handbook*; Joint Nature Conservation Committee: Peterborough, UK, 2006; ISBN 978-1-86107-574.
52. Janssens, X.; Bruneau, É.; Lebrun, P. Prévision des potentialités de production de miel à l'échelle d'un rucher au moyen d'un système d'information géographique. *Apidologie* **2006**, *37*, 351–365. [CrossRef]
53. Timberlake, T. Mind the Gap: The Importance of FLowering Phenology in Pollinator Conservation. Ph.D. Thesis, University of Bristol, Bristol, UK, 2019.
54. Guerriat, H. Valeur Apicole Des Haies Dans l'Entre-Sambre et Meuse. *Abeille Cie* **1999**, *73*, 24–28.
55. Mačukanović-Jocić, M.; Jarić, S. The melliferous potential of Apiflora of south Western Vojvodina (Serbia). *Arch. Biol. Sci.* **2016**, *68*, 130. [CrossRef]
56. R Core Team. *R: A Language and Environement for Statistical Computing*; R Foundation for Statistical Computing: Vienna, Austria, 2021.
57. Hendriksma, H.P.; Shafir, S. Honey Bee Foragers Balance Colony Nutritional Deficiencies. *Behav. Ecol. Sociobiol.* **2016**, *70*, 509–517. [CrossRef]
58. Schmidt, J.O. Feeding Preferences of *Apis mellifera* L. (Hymenoptera: Apidae): Individual versus Mixed Pollen Species. *J. Kans. Entomol. Soc.* **1984**, *572*, 323–327.
59. Boccara, N. Modeling Complex Systems. In *Graduate Texts in Contemporary Physics*; Springer: New York, NY, USA, 2004; ISBN 978-0-387-40462-2.
60. Hanley, M.E.; Franco, M.; Pichon, S.; Darvill, B.; Goulson, D. Breeding System, Pollinator Choice and Variation in Pollen Quality in British Herbaceous Plants. *Funct. Ecol.* **2008**, *22*, 592–598. [CrossRef]
61. Avni, D.; Dag, A.; Shafir, S. The Effect of Surface Area of Pollen Patties Fed to Honey Bee (*Apis mellifera*) Colonies on Their Consumption, Brood Production and Honey Yields. *J. Apic. Res.* **2009**, *48*, 23–28. [CrossRef]
62. Requier, F.; Odoux, J.-F.; Tamic, T.; Moreau, N.; Henry, M.; Decourtye, A.; Bretagnolle, V. Honey Bee Diet in Intensive Farmland Habitats Reveals an Unexpectedly High Flower Richness and a Major Role of Weeds. *Ecol. Appl.* **2015**, *25*, 881–890. [CrossRef]

63. Bailey, S.; Requier, F.; Nusillard, B.; Roberts, S.P.M.; Potts, S.G.; Bouget, C. Distance from Forest Edge Affects Bee Pollinators in Oilseed Rape Fields. *Ecol. Evol.* **2014**, *4*, 370–380. [CrossRef]
64. Decourtye, A.A.; Alaux, C.; Odoux, J.F.; Henry, M.; Vaissière, B.; Conte, Y.L. Why Enhancement of Floral Resources in Agro-Ecosystems Benefit Honeybees and Beekeepers? *Ecosyst. Biodivers.* **2011**, *19*, 371–388.
65. Danner, N.; Keller, A.; Härtel, S.; Steffan-Dewenter, I. Honey Bee Foraging Ecology: Season but Not Landscape Diversity Shapes the Amount and Diversity of Collected Pollen. *PLoS ONE* **2017**, *12*, e0183716. [CrossRef]
66. Garbuzov, M.; Balfour, N.J.; Shackleton, K.; Al Toufailia, H.; Scandian, L.; Ratnieks, F.L.W. Multiple Methods of Assessing Nectar Foraging Conditions Indicate Peak Foraging Difficulty in Late Season. *Insect Conserv. Divers.* **2020**, *13*, 532–542. [CrossRef]
67. Köppler, K.; Vorwohl, G.; Koeniger, N. Comparison of Pollen Spectra Collected by Four Different Subspecies of the Honey Bee Apis Mellifera. *Apidologie* **2007**, *38*, 341–353. [CrossRef]
68. Alaux, C.; Ducloz, F.; Crauser, D.; Le Conte, Y. Diet Effects on Honeybee Immunocompetence. *Biol. Lett.* **2010**, *6*, 562–565. [CrossRef] [PubMed]
69. Derioz, P. Arrière-pays méditerranéen entre déprise et reprise: l'Exemple du Haut-Languedoc Occidental. *Econ. Rurale* **1994**, *223*, 32–38. [CrossRef]
70. Leponiemi, M.; Freitak, D.; Moreno-Torres, M.; Pferschy-Wenzig, E.-M.; Becker-Scarpitta, A.; Tiusanen, M.; Vesterinen, E.J.; Wirta, H. Honeybees' Foraging Choices for Nectar and Pollen Revealed by DNA Metabarcoding. *Sci. Rep.* **2023**, *13*, 14753. [CrossRef]
71. Radhika, V.; Kost, C.; Boland, W.; Heil, M. The Role of Jasmonates in Floral Nectar Secretion. *PLoS ONE* **2010**, *5*, e9265. [CrossRef]
72. Nielsen, A.; Reitan, T.; Rinvoll, A.W.; Brysting, A.K. Effects of Competition and Climate on a Crop Pollinator Community. *Agric. Ecosyst. Environ.* **2017**, *246*, 253–260. [CrossRef]
73. Pacini, E.; Nepi, M.; Vesprini, J.L. Nectar Biodiversity: A Short Review. *Plant Syst. Evol.* **2003**, *238*, 7–21. [CrossRef]
74. Baude, M.; Kunin, W.E.; Boatman, N.D.; Conyers, S.; Davies, N.; Gillespie, M.A.K.; Morton, R.D.; Smart, S.M.; Memmott, J. Historical Nectar Assessment Reveals the Fall and Rise of Britain in Bloom. *Nature* **2016**, *535*, 85–88. [CrossRef]
75. Power, E.F.; Stabler, D.; Borland, A.M.; Barnes, J.; Wright, G.A. Analysis of Nectar from Low-volume Flowers: A Comparison of Collection Methods for Free Amino Acids. *Methods Ecol. Evol.* **2018**, *9*, 734–743. [CrossRef]
76. Schmickl, T.; Crailsheim, K. HoPoMo: A Model of Honeybee Intracolonial Population Dynamics and Resource Management. *Ecol. Model.* **2007**, *204*, 219–245. [CrossRef]
77. Arundel, J.; Winter, S.; Gui, G.; Keatley, M. A Web-Based Application for Beekeepers to Visualise Patterns of Growth in Floral Resources Using MODIS Data. *Environ. Model. Softw.* **2016**, *83*, 116–125. [CrossRef]
78. Campbell, T.; Fearns, P. Honey Crop Estimation from Space: Detection of Large Flowering Events in Western Australian Forests. *Int. Arch. Photogramm. Remote Sens. Spat. Inf. Sci.* **2018**, *XLII-1*, 79–86. [CrossRef]
79. Makori, D.M.; Abdel-Rahman, E.M.; Landmann, T.; Mutanga, O.; Odindi, J.; Nguku, E.; Tonnang, H.E.; Raina, S. Suitability of Resampled Multispectral Datasets for Mapping Flowering Plants in the Kenyan Savannah. *PLoS ONE* **2020**, *15*, e0232313. [CrossRef] [PubMed]
80. Chen, B.; Jin, Y.; Brown, P. An Enhanced Bloom Index for Quantifying Floral Phenology Using Multi-Scale Remote Sensing Observations. *ISPRS J. Photogramm. Remote Sens.* **2019**, *156*, 108–120. [CrossRef]
81. Nicholls, E.; Hempel de Ibarra, N. Assessment of Pollen Rewards by Foraging Bees. *Funct. Ecol.* **2017**, *31*, 76–87. [CrossRef]
82. Nepi, M. Beyond Nectar Sweetness: The Hidden Ecological Role of Non-Protein Amino Acids in Nectar. *J. Ecol.* **2014**, *102*, 108–115. [CrossRef]
83. Nepi, M.; Grasso, D.A.; Mancuso, S. Nectar in Plant–Insect Mutualistic Relationships: From Food Reward to Partner Manipulation. *Front. Plant Sci.* **2018**, *9*, 01063. [CrossRef]
84. Nicholls, C.; Altieri, M. Plant Biodiversity Enhances Bees and Other Insect Pollinators in Agroecosystems. A Review. *Agron. Sustain. Dev.* **2013**, *33*, 257–274. [CrossRef]
85. Lonsdorf, E.; Kremen, C.; Ricketts, T.; Winfree, R.; Williams, N.; Greenleaf, S. Modelling Pollination Services across Agricultural Landscapes. *Ann. Bot.* **2009**, *103*, 1589–1600. [CrossRef]
86. Seeley, T.D.; Camazine, S.; Sneyd, J. Collective Decision-Making in Honey Bees: How Colonies Choose among Nectar Sources. *Behav. Ecol. Sociobiol.* **1991**, *28*, 277–290. [CrossRef]
87. Kjøhl, M.; Nielsen, A.; Stenseth, N.C. Potential Effects of Climate Change on Crop Pollination: Extension of Knowledge Base, Adaptive Management, Capacity Building, Mainstreaming. In *Pollitnation Services for Sustainable Agriculture*; Food and Agriculture Organization of the United Nations: Rome, Italy, 2011; ISBN 978-92-5-106878-6.
88. Harris, C.; Balfour, N.J.; Ratnieks, F.L.W. Seasonal Variation in the General Availability of Floral Resources for Pollinators in Northwest Europe: A Review of the Data. *Biol. Conserv.* **2024**, *298*, 110774. [CrossRef]
89. Karbassioon, A.; Yearlsey, J.; Dirilgen, T.; Hodge, S.; Stout, J.C.; Stanley, D.A. Responses in Honeybee and Bumblebee Activity to Changes in Weather Conditions. *Oecologia* **2023**, *201*, 689–701. [CrossRef] [PubMed]
90. Rivière, J.; Alves, T.; Alaux, C.; Conte, Y.L.; Layec, Y.; Lozac'h, A.; Singhoff, F.; Rodin, V. Modèle multi-agent d'auto-organisation pour le butinage au sein d'une colonie d'abeilles. *Rev. Ouverte Intell. Artif.* **2022**, *3*, 5–6. [CrossRef]

91. Russell, S.; Barron, A.B.; Harris, D. Dynamic Modelling of Honey Bee (*Apis mellifera*) Colony Growth and Failure. *Ecol. Model.* **2013**, *265*, 158–169. [CrossRef]
92. Betti, M.; LeClair, J.; Wahl, L.; Zamir, M. Bee++: An Object-Oriented, Agent-Based Simulator for Honey Bee Colonies. *Insects* **2017**, *8*, 31. [CrossRef]
93. Arroyo-Correa, B.; Bartomeus, I.; Jordano, P. Individual-based Plant–Pollinator Networks Are Structured by Phenotypic and Microsite Plant Traits. *J. Ecol.* **2021**, *109*, 2832–2844. [CrossRef]
94. Torné-Noguera, A.; Rodrigo, A.; Osorio, S.; Bosch, J. Collateral Effects of Beekeeping: Impacts on Pollen-Nectar Resources and Wild Bee Communities. *Basic Appl. Ecol.* **2016**, *17*, 199–209. [CrossRef]
95. Atanasov, A.Z.; Georgiev, I.R. A Multicriteria Model for Optimal Location of Honey Bee Colonies in Regions without Overpopulation. In Proceedings of the AIP Conference Proceedings, Sofia, Bulgaria, 8 March 2021; p. 090008.
96. Komasilova, O.; Komasilovs, V.; Kviesis, A.; Zacepins, A. Model for Finding the Number of Honey Bee Colonies Needed for the Optimal Foraging Process in a Specific Geographical Location. *PeerJ* **2021**, *9*, e12178. [CrossRef]

biology

MDPI

Article

Enhancing Alfalfa (*Medicago sativa*) Seed Yield: The Effect of Honey Bee (*Apis mellifera*) Supplementation and Efficiency of Other Pollinator

Kamran Ejaz [1], Mudssar Ali [1,*], Fawad Zafar Ahmad Khan [1,2] and Raimondas Mozūratis [3,4,*]

1 Institute of Plant Protection, Muhammad Nawaz Shareef University of Agriculture Multan, Multan 60000, Pakistan; kamran3194@gmail.com (K.E.); fawad.zafar@mnsuam.edu.pk (F.Z.A.K.)
2 Department of Outreach and Continuing Education, Muhammad Nawaz Shareef University of Agriculture Multan, Multan 60000, Pakistan
3 Department of Zoology, Stockholm University, SE-10691 Stockholm, Sweden
4 Laboratory of Chemical and Behavioural Ecology, The State Scientific Research Institute Nature Research Centre, LT-08412 Vilnius, Lithuania
* Correspondence: mudssar.ali@mnsuam.edu.pk (M.A.); raimondas.mozuraitis@su.se (R.M.)

Simple Summary: Alfalfa is a widely grown crop used as an animal feed because of its protein and fiber content. However, alfalfa plants cannot produce seeds without the assistance of flower-visiting insects, which transfer pollen between flowers. In this study, we tested whether placing honey bee colonies near alfalfa fields could improve pollination and increase seed production. We compared three types of field setups: one without honey bees, one with two colonies, and one with three colonies. We recorded how often insects visited the flowers, how they behaved while feeding, and how many seeds were produced. Honey bees were the most common visitors in the fields where hives were placed, although native wild bees also contributed significantly to pollination. We found that the setup with two honey bee colonies produced more seeds than the setup with three. This suggests that more bees do not always lead to better results. Our results also showed that both honey bees and native wild bees played a role in seed development. This research shows that a balanced use of managed honey bees has the potential to improve seed production in alfalfa.

Abstract: Alfalfa (*Medicago sativa*) is an important fodder crop with 16–25% protein and 20–30% fibre content. It requires pollinators for seed development through a process known as "tripping". Honey bees are preferred for their easy management and mass rearing to enhance pollination. In a field experiment, we tested three treatments: two honey bee colonies (moderately supplemented), three honey bee colonies (highly supplemented), and a control with no honey bee colonies. The colonies European honey bee, *A. mellifera*, were used, starting at the flowering stage. We assessed the abundance, foraging, and tripping behaviour of pollinators, as well as the number of seeds per pod, pod weight, and seed weight. Honey bees were the most abundant in supplemented plots, followed by solitary bees, with increased honey bee visitation in the second year. *Megachile cephalotes*, *M. hera*, and *A. mellifera* were the most efficient pollinators regarding seeds per pod and seed weight. The plot with two honey bee colonies had a higher seed yield than the one with three colonies. Conservation of these pollinators and the introduction of managed *A. mellifera* colonies can lead to sustainable alfalfa seed production. Our findings suggest that honey bee supplementation, along with conservation efforts for solitary bees, can enhance pollination and seed yield in alfalfa.

Keywords: bees; fodder; foraging behaviour; tripping behaviour

1. Introduction

With the increasing human population, the demand for dairy and meat products is increasing; therefore, the livestock industry is expanding. The healthy development of livestock relies on the consistent availability of high-quality fodder. Pakistan produces approximately 55.47 million tons of fodder annually [1]. The main fodder crops in Pakistan include sorghum (*Sorghum bicolor* L.) (Poales: Poaceae), pearl millet (*Pennisetum glaucum* L.) (Poales: Poaceae), maize (*Zea mays* L.) (Poales: Poaceae), cowpeas (*Vigna unguiculata* L.) (Fabales: Fabaceae), guar (*Cyamopsis tetragonoloba* L.) (Fabales: Fabaceae), berseem clover (*Trifolium alexandrinum* L.) (Fabales: Fabaceae), alfalfa (*Medicago sativa* L.) (Fabales: Fabaceae), oats (*Avena sativa* L.) (Poales: Poaceae), and ryegrass (*Lolium perenne* L.) (Poales: Poaceae).

M. sativa is an important fodder crop, containing 20–30% fibre and 16–25% protein [2]. Due to its high nutritional value and yield, alfalfa is known as "the queen of fodders" [3]. It is a perennial forage legume, a deep-rooted crop with multiple cuttings and a fast-growing nature, characterised by purple flowers [4,5]. The inflorescence of alfalfa is trifoliolate, 1.0–2.5 cm long, and 0.3–1.6 cm wide. Papilionaceous flowers remain closed during anthesis and only open after being "tripped" by a pollinator, making alfalfa highly dependent on entomophilous pollination. Furthermore, alfalfa has a reproductive system that is primarily self-incompatible and self-sterile [6]. The fruit of alfalfa is a polyspermous pod. Small, sickle-shaped, or corkscrew-coiled alfalfa pods typically have one or more coils, with a maximum of five. Alfalfa seeds are tiny, reniform, angular, or oval-shaped, with colours ranging from yellow to light brown and greyish brown [7].

In Pakistan, alfalfa is cultivated as a mixed crop for fodder and seed production. After multiple cuttings, the crop is left to mature for seed production [8]. Pakistan has a low output of alfalfa fodder and seeds. The total cultivated area under fodder crops is projected to exceed two million hectares, with an annual fodder production of 55.4 million tons. However, to meet the feeding requirement, this amount of fodder is insufficient [9]. The low production levels can be attributed to the unavailability of high-quality seeds and the farmers' limited awareness of advanced crop production technologies [10].

Effective pollination in alfalfa requires a tripping mechanism, where pollinators must physically force and open flower parts to release pollen grains, enabling successful seed production [11]. There are different pollinators of the alfalfa crop, among which bees are considered the primary pollinators [12–15]. Managed honey bee *Apis mellifera* L. (Hymenoptera: Apidae) ranks as the most frequently visiting single species in many crops, including alfalfa [16]. *A. mellifera* is an important pollinator of alfalfa crops mainly due to its ability to forage for more extended periods up to 11 h per day. This prolonged foraging activity improves its effectiveness in pollinating alfalfa, which makes it particularly valuable for optimising seed yield [17].

Supplementary bee pollination has been used to enhance the productivity and quality of various crops. For example, in melons, supplemental pollination by managed solitary bees and honey bees enhanced the fruit set, fruit size and yield parameters of the crop [18]. Moreover, another study on oil tree peonies has shown that supplementary honey bee pollination can significantly enhance fruit growth rate and yield [19]. In the same study, managed honey bees in controlled pollination settings have increased seed production and oil content [19]. Often, a single pollinator species is insufficient for efficient pollination, therefore, the presence of other pollinators is necessary [20–22].

To increase alfalfa crop yield, farmers use bee colonies and place them near crops for the best pollination [23]. For optimal alfalfa seed production, 4–6 colonies have been recommended per hectare [24]. Although honey bees are highly effective in increasing pollination rates and seed yields due to their ability to be mass-reared and easily managed, they are less efficient at pollinating alfalfa flowers compared to some solitary bees [25,26]. However, the management of solitary bees requires year-round maintenance of floral resources as well as nesting sites [27]. For example, *Megachile* spp. are highly efficient in tripping alfalfa flowers, yet they present unique management challenges due to their nesting needs and lower tolerance to environmental disturbances [27]. Moreover, due to climate change, the number of native solitary bees is expected to decrease [28,29]. In these scenarios, honey bee supplementation could serve as a quick option to improve pollination services in alfalfa crops.

The increased presence of *A. mellifera* can lead to enhanced competition for floral alfalfa resources, potentially affecting foraging behaviour and the population dynamics of local pollinators [30]. Despite this, when combined with conservation efforts for solitary bees, honey bee supplementation can create a balanced approach that enhances overall pollination efficiency and seed production in alfalfa fields [30].

Little is known about the effect of *A. mellifera* supplementation on foraging behaviour and the abundance of local pollinators in alfalfa fields. To our knowledge, only one study has compared the single-visit efficiency of native solitary bees in alfalfa crops in Pakistan [31]. No previous study has reported the role of *A. mellifera* supplementation in increasing the seed yield of alfalfa crops. Therefore, the current study was designed to investigate the optimal number of honey bee colonies for alfalfa seed production, considering the presence of local pollinators. We hypothesise that supplemental honey bee pollination significantly enhances alfalfa seed yield in response to pollination deficits. Based on this hypothesis, we predicted that alfalfa plots with supplemental honey bee pollination would yield higher seed yields than those without supplementation.

2. Materials and Methods

2.1. Study Site

The research was conducted at the experimental farm of Muhammad Nawaz Shareef University of Agriculture, Multan, Punjab, Pakistan (30.1575° N, 71.5249° E) over two seasons, including the years 2021–2022 and 2022–2023. The experimental crop was Alfalfa (*Medicago sativa* L.). Alfalfa cultivar '*Sargodha 2000*' was sown at a rate of 8 kg per acre over a three-acre area on 8 November 2021 (first year) and 10 November 2022 (second year). The climate in Multan is subtropical, characterised by dry and hot summers, mild winters, and an average annual rainfall of 175 mm [32]. The minimum and maximum temperatures in this area in summer are 24 °C and 42 °C, respectively, while in winter, the range is between 4.5 °C and 22 °C [33,34]. The experimental plots and alfalfa sowing were managed homogeneously to exclude any variance that might impair the experiment's analytical comparisons. Standard pest management practices (use of neem-based biopesticides) and agronomic practices (thinning and two cuttings) were implemented for the alfalfa crop, and manual weed removal was also performed.

2.2. Experimental Design and Layout

Three plots (each measuring 4046.86 m^2) were established, which were at least 1 km apart from each other, and named as a highly supplemented block, a moderately supplemented block, and a non-supplemented plot. We used the distance measurement tool available in Google Maps to ensure that all selected experimental fields were at least 1 km apart. Three honey bee colonies were introduced into a highly supplemented block, while two colonies were placed in a moderately supplemented block (Figure 1). No colonies were introduced into the non-supplemented block, which was the control in both years. All honey bee colonies were healthy, with no signs of disease or pest infestation observed during or after the experiment. Throughout the experiment, all colonies were maintained under standard apicultural practices. They were provided with sugar syrup (1:1 ratio of sugar and water) as a supplemental food source. We also conducted routine inspections to monitor colony health, queen presence, and overall brood development. We did observe any absconding or swarming behavior during or after the experiment. Standard agronomic practices were followed for growing the alfalfa crop [35]. *A. mellifera* colonies, each housed in hives containing 10 frames, were used for this experiment. Each hive measured 20 cm (length) × 30 cm (height) × 16 cm (width), and the hive cover measured 20 cm × 16 cm × 4 cm. Each colony had an estimated strength of ~25,000–30,000 worker bees at the start of the experiment. The colonies were purchased from a progressive beekeeper in Layyah, Punjab, Pakistan, and were transported at night on a mobile vehicle due to the low activity of the bees. The bee colonies were then placed on an iron stand under the shade of a tree in the experimental farm of Muhammad Nawaz Shareef University of Agriculture, Multan, Punjab, Pakistan. For moderate honey bee supplementation, two hives were placed diagonally opposite each other, facing southeast. For high honey bee supplementation, three hives were arranged in a triangular formation, with two colonies aligned on one side and the third colony placed in the middle on the opposite side, facing southeast.

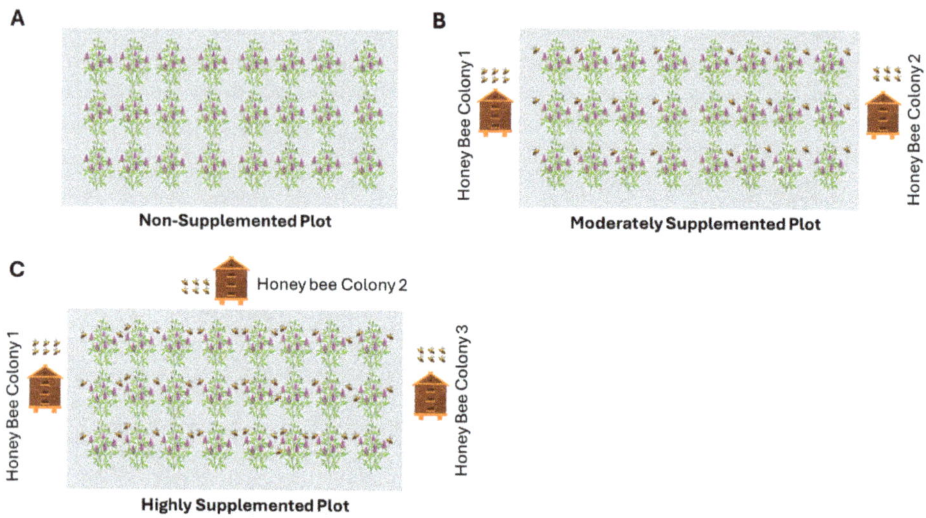

Figure 1. Experimental layout showing three pollination treatments in alfalfa fields: (**A**) Non-supplemented plot with no honey bee colonies, (**B**) Moderately supplemented plot with two *Apis mellifera colonies* placed at the field edges, and (**C**) Highly supplemented plot with a three *A. mellifera* colonies surrounding the field. The illustration represents the relative abundance of honey bees and colony placement across the treatments to assess the impact of honey bee supplementation on pollinator activity and seed set.

2.3. Abundance of Honey Bees and Other Insect Pollinators

The abundance of honey bees and other insect pollinators was measured using a one-square-meter iron frame placed in ten randomly selected quadrats per plot during the flowering period (April–May). The one-square-meter iron frame was randomly tossed to determine the observation area by dividing the experimental plot into units, assigning serial numbers to the units and using a random number generator to select places for the iron frame. After the frame landed, a one-minute pause was observed to ensure that pollinator activity had returned to normal. After the data collection from that area, the frame was tossed again while standing at the previous data point, and this process was repeated. The frame was placed directly on the ground to prevent any movement that could hinder the observation process. Insect pollinator counts were conducted twice weekly at four-hour intervals (06:00, 10:00, 14:00, and 18:00) throughout the flowering season (April–May) in both study years. Abundance was assessed by observing each quadrat for five minutes and recording the number of insect visitors [36].

2.4. Foraging Behavior

The duration of stay for honey bees and other insect pollinators was measured by visual observation of how long each pollinator stayed on a single raceme during a 60-s duration. The duration was recorded by using a digital stopwatch (XL-011, Binloo, China). The observer stood at least 0.3 m away from the alfalfa plant. During the observation period, care was taken to ensure that the observer remained still and avoided any unnecessary movement to prevent disturbance to pollinator activity. The visitation rate was calculated by counting the number of racemes visited by honey bees and other insect pollinators within 60 s. Visual observations were conducted twice weekly throughout the flowering season (April–May) at four different times of day: 06:00, 10:00, 14:00, and 18:00 h, to capture variation in pollinator activity [25,37].

2.5. Tripping Behaviour of Honey Bees and Other Pollinators

Tripping behaviour of insect pollinators was assessed through counting: (1) the number of flowers tripped per raceme (exposure of anthers upon pollinator interaction), and (2) the number of already tripped flowers visited per raceme (already exposed anthers due to a previous pollinator interaction) [38,39]. A tripped alfalfa flower was identified by the visible stigmas and anthers. In contrast, untripped flowers remained closed, with the reproductive structures not yet exposed. Data were collected twice a week throughout the entire flowering season of alfalfa.

2.6. Single-Visit Seed Set Efficiency and Reproductive Success of Alfalfa

To measure the single-visit seed set efficiency, the observer placed himself at least 0.3 m away from an unopened flower (un-tripped) and waited for a pollinator to trip it. After being tripped by a specific insect, the flower was caged using a butter paper bag (3 cm × 3 cm) and tagged with the name of the pollinator; the remaining flowers on that raceme were removed. Ten pollinators of the same species were each allowed a single visit to different flowers. After the seed ripened, the cages were removed, and the pods were harvested at maturity in June. After harvesting, the pod weight was measured using digital weighing balance. After weighing, pod was crushed manually to harvest the seeds. Following hand crushing, the seed weight was measured using digital weighing balance. Following the weight measurements, the number of seeds was counted in each floret. Single visits of seven pollinator species were assessed in the first year of the trial, as the pods visited by the two pollinator species, *Amegilla* sp. and *E. aeneus*, were damaged due to rainfall, and no seed was recovered. Moreover, no data regarding single-visit efficiency were collected during the second year.

2.7. Seed Yield

To assess the effect of honey bee supplementation on seed yield, seed weight per square meter was measured across three treatments: highly supplemented, moderately supplemented, and non-supplemented plots. We selected ten quadrats of 1 m^2 with similar plant densities and marked them across the entire acre. These same plants were then used to assess reproductive success and record seed weight (in grams). Seed weight was then compared across treatments to evaluate the impact of honey bee supplementation on overall seed yield.

2.8. Statistical Analysis

To estimate abundance, the number of different insect pollinators visiting the flowers was analysed in 2022 and 2023. For stay time, the average time spent by each pollinator in seconds was compared between the two years. For the visitation rate, the mean number of flowers visited per minute was analysed for 2022 and 2023. For tripping efficiency, the average number of flowers tripped versus already-tripped flowers was recorded for each bee species and analysed separately for the two years. The mean seed weight per m^2 across three honey bee supplementation treatments was also analysed separately each year. The single-visit efficiency of individual bee species was measured as the mean pod weight, mean seed weight, and mean seed number per visit of each bee species, and these metrics were analysed annually. We used the Shapiro–Wilk test to check whether the data followed a normal distribution and applied a log transformation to achieve normality. Analysis of variance (ANOVA) was performed, and means were compared using the Tukey test (Statistix 8.1).

3. Results

3.1. Abundance of Insect Pollinators

There was a significant difference in the abundance of pollinator species in all the non-supplemented block ($F = 12.2$, df = 5, $p < 0.001$), moderately supplemented block ($F = 16.7$, df = 5, $p < 0.001$), and highly supplemented block ($F = 16.4$, df = 5, $p < 0.001$) during the first year (Figure 1). Similar differences were found in the second year in non-supplemented block ($F = 20.3$, df = 5, $p < 0.001$), moderately supplemented block ($F = 20.6$, df = 5, $p < 0.001$), and highly supplemented block ($F = 46.1$, df = 5, $p < 0.001$) (Figure 2). In the first year, the abundance of other honey bees (*A. dorsata* and *A. florea*) was significantly higher in the highly supplemented block where three honey bee colonies were kept for supplementary pollination and in the non-supplemented block compared to the abundance of other pollinator species (Figure 1A,C). Syrphid flies (*E. aeneus* and *E. arvorum*) were also significantly higher than solitary bees (*M. cephalotes*, *M. hera*, and *Eucera* sp.), butterflies and wasps in the moderately supplemented block where two honey bee colonies were installed for supplementary honey bee pollination. The abundance of European honey bees (*A. mellifera*) was significantly greater in the moderately supplemented block than in other supplementary treatments (Figure 2).

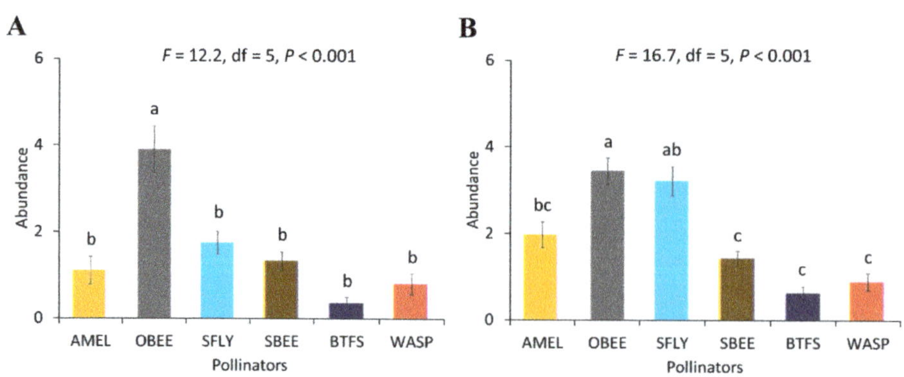

Figure 2. *Cont.*

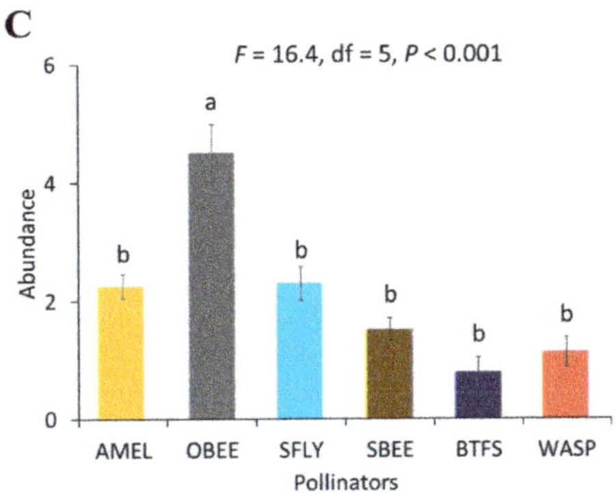

Figure 2. Abundance of different pollinator species visiting lucerne flowers in 2022: (**A**) non-supplemented block, (**B**) moderately supplemented block, and (**C**) highly supplemented block. Standard error bars with the same letters are not significantly different (Tukey–Kramer test, α = 0.05). Pollinator group abbreviations: AMEL—*Apis mellifera*, OBEE—other honey bees; SFLY—syrphid flies; SBEE—solitary bees; BTFS—butterflies; WASP—wasps.

In the second year, the abundance of pollinator species was lowest where no honey bee hives were kept for supplementary pollination. The highest abundance of European honey bees (*A. mellifera*) was recorded in the moderately supplemented block compared to other blocks with varying levels of supplementation. The abundance of syrphid flies, solitary bees, butterflies, and wasps was not significantly different from one another in all of the supplementary pollinated plots in the second year (Figure 3).

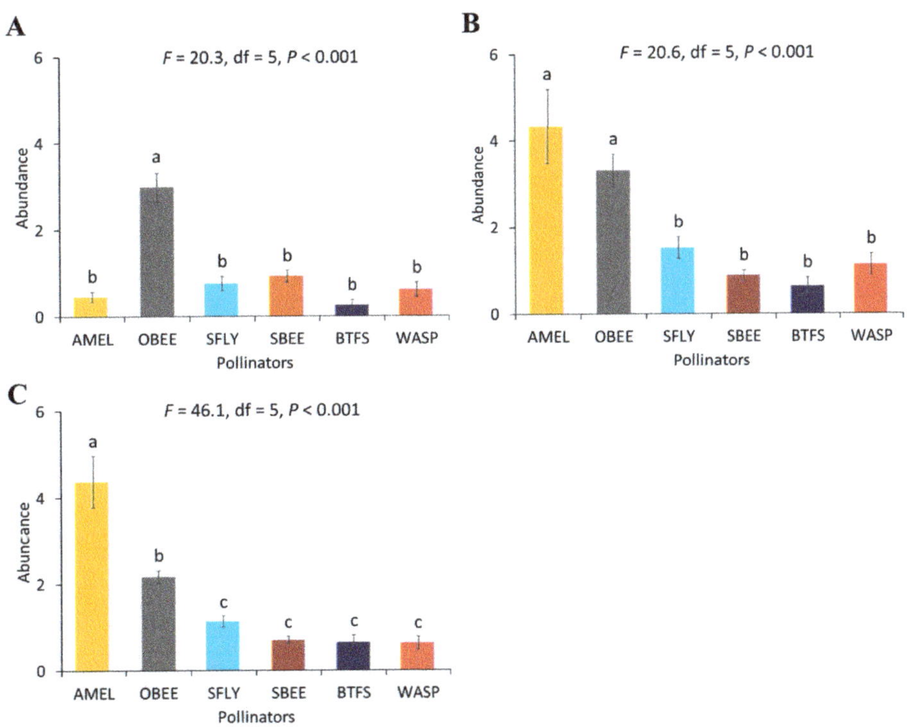

Figure 3. Abundance of different pollinator species visiting lucerne flowers in 2023: (**A**) non-supplemented block, (**B**) moderately supplemented block, and (**C**) highly supplemented block. Standard

error bars with the same letters are not significantly different (Tukey–Kramer test, α = 0.05). Pollinator group abbreviations: AMEL—*Apis mellifera*, OBEE—other honey bees; SFLY—syrphid flies; SBEE—solitary bees; BTFS—butterflies; WASP—wasps.

3.2. Stay Time

The time spent per flower by individual pollinator species differed significantly between the years 2022 ($F = 12.5$, df = 8, $p < 0.001$) and 2023 ($F = 9.9$, df = 8, $p < 0.001$). The stay time of *E. aeneus* was significantly higher than that of all other pollinator species in both years, while *Amegilla* sp. were significantly less frequent visitors in both years. There was no significant difference in the stay time of *A. mellifera*, *A. dorsata*, *M. cephalotes*, *M. hera* and *Eucera* sp. in both years. *A. florea* and *E. arvorum* were significantly lower than all other insect pollinators except *Amegilla* sp. (Figure 4).

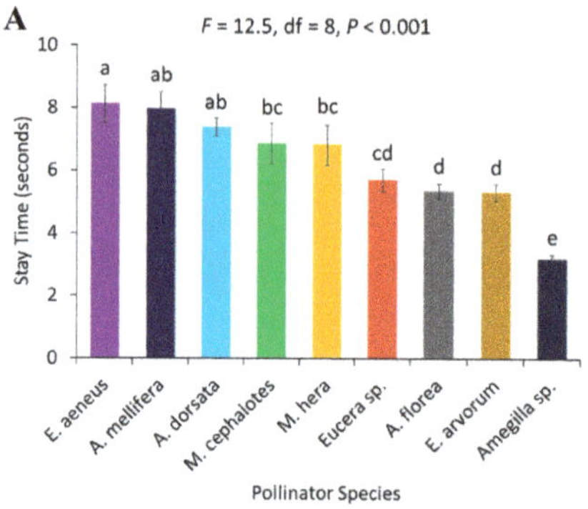

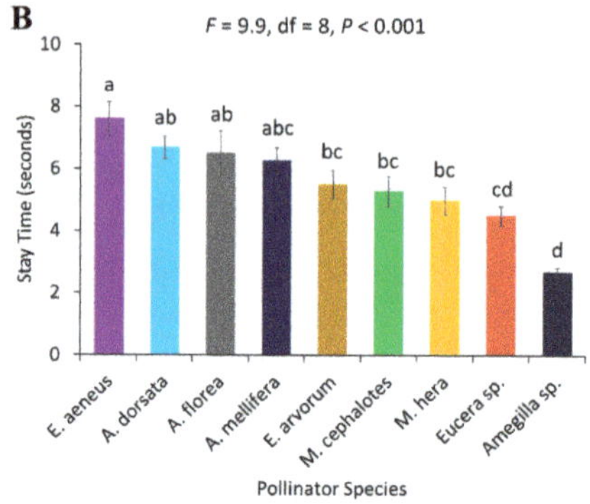

Figure 4. Stay time (time spent per flower) of different pollinator species in two years, (**A**) 2022 and (**B**) 2023. Standard error bars with the same letters are not significantly different (Tukey–Kramer test, α = 0.05).

3.3. Visitation Rate

The number of visited flowers per minute by each pollinator species was significantly different in 2022 ($F = 5.6$, df = 8, $p < 0.001$) and 2023 ($F = 8.7$, df = 8, $p < 0.001$). The visitation rates of *Amegilla* sp., *A. mellifera*, and *M. cephalotes* were significantly higher than all other recorded pollinator species in both years. *Eucera* sp., *A. florea* and *E. aeneus* were not significantly different from each other and had a lower visitation rate as compared to other pollinator species. *E. arvorum* had the significantly lowest visitation rate of all pollinator species in both years (Figure 5).

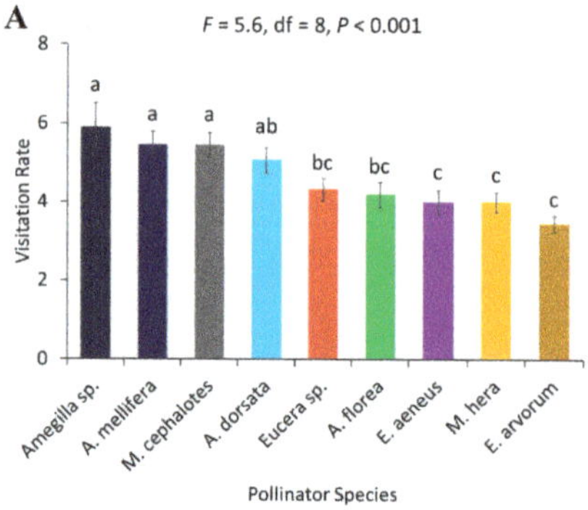

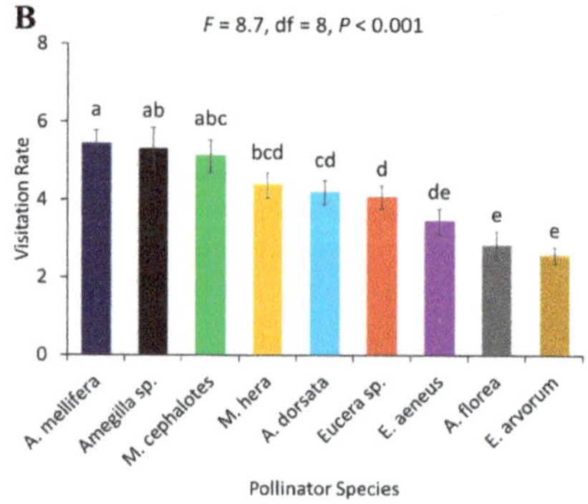

Figure 5. Visitation rate (number of flowers visited per minute) of different pollinator species in two years, (**A**) 2022 and (**B**) 2023. Standard error bars with the same letters are not significantly different (Tukey–Kramer test, α = 0.05).

3.4. Tripping

The tripping behaviour of insect pollinators differed significantly between the two years: 2022 ($F = 11.19$, df = 8, $p < 0.001$) and 2023 ($F = 11.83$, df = 8, $p = 0.004$). *Megachile* sp. was the most effective pollinator in terms of tripping efficiency, with tripping by *M. cephalotes* and *M. hera* being significantly greater than that of all other pollinator species. The number of tripped flowers due to the visits of *A. mellifera* was higher than that of other solitary and honey bees, followed by *Megachille* sp. bees. The number of already visited tripped flowers was higher in *A. florea*, followed by solitary bees (*M. hera*, *M. cephalotes*, *Eucera* sp.) in both years of data recordings (Figure 6).

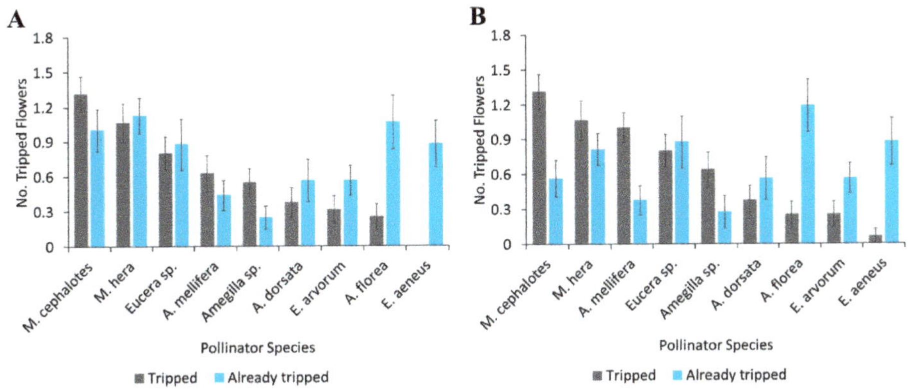

Figure 6. Tripping trends in alfalfa flowers (tripped vs. already-tripped flowers) by different pollinator species in (**A**) 2022 and (**B**) 2023. Error bars indicate standard error.

3.5. Seed Weight

Seed weight per square meter differed significantly across various supplementary treatments in 2022 (F = 5.2, df = 2, $p = 0.031$) and 2023 (F = 11.8, df = 2, $p = 0.002$). Seed weight was highest in the moderately supplemented block, where two honey bee colonies were placed for supplementary pollination. There was no significant difference in seed weight per square meter in the highly supplemented and non-supplemented blocks (Figure 7).

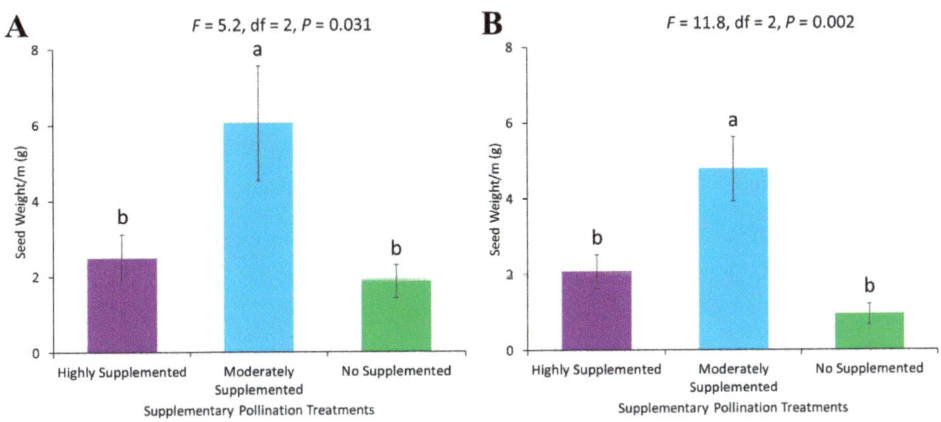

Figure 7. Seed weight (grams) per square meter in different supplementary pollination treatments in (**A**) 2022 and (**B**) 2023. Standard error bars with the same letters are not significantly different (Tukey–Kramer test, $\alpha = 0.05$).

3.6. Single-Visit Seed Set Efficiency

There was a significant difference among the pollinator species in terms of pod weight ($F = 2.2$, df = 7, $p = 0.015$), seed weight ($F = 2.8$, df = 6, $p = 0.017$), and number of seeds produced ($F = 4.3$, df = 6, $p < 0.001$) of the tested pollinator species in a single visit. Single-visit efficacy showed that *M. hera* and *M. cephalotes* were the most effective insect pollinators, followed by *A. mellifera*. Syrphid flies (*E. arvorum*) were the least effective in terms of pod weight, seed weight and seed number (Figure 8). Moreover, the Pearson's correlation analysis showed that flower tripping by pollinators was positively and significantly associated with all measured seed yield parameters (Table 1). Statistically significant correlation was observed between flower tripping and pod weight ($r = 0.2575$, $p = 0.0433$). Similarly, flower tripping showed a stronger correlation with the number of seeds per pod ($r = 0.3605$, $p = 0.0040$) and seed weight ($r = 0.3347$, $p = 0.0078$). Among the seed parameters themselves, significant positive correlations were found. Pod weight was significantly correlated with both the number of seeds ($r = 0.4812$, $p < 0.001$) and seed weight ($r = 0.5286$, $p < 0.001$). The strongest correlation observed was between the number of seeds and seed weight ($r = 0.8907$, $p < 0.001$) (Table 1).

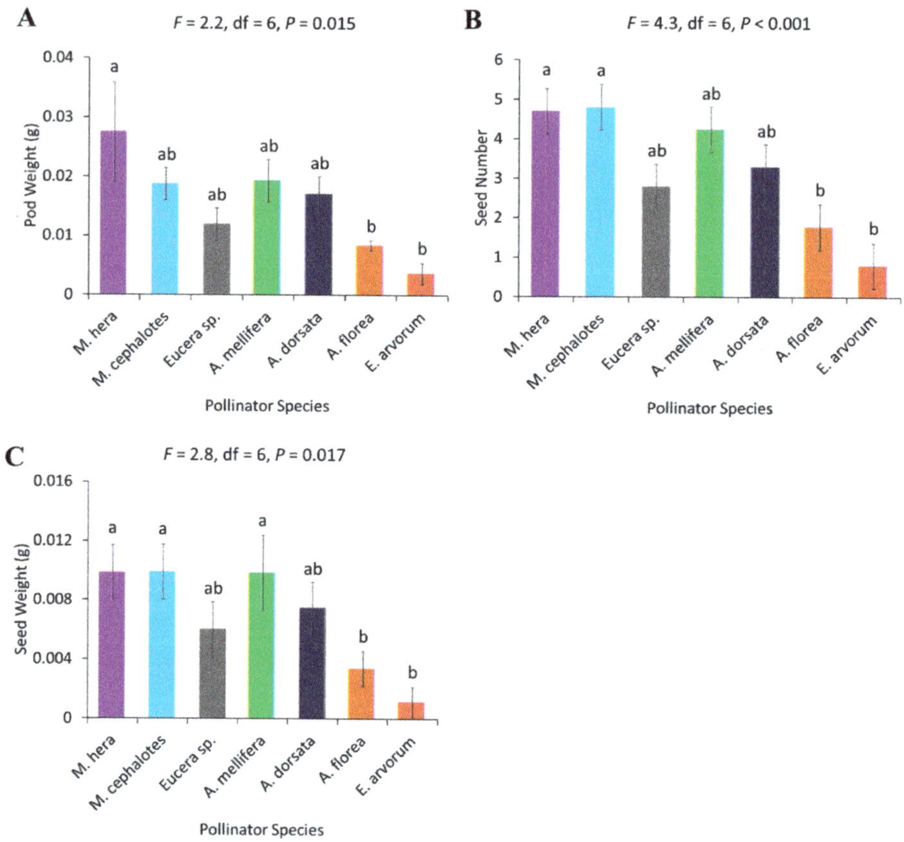

Figure 8. Yield parameters after a single visit by different pollinator species in a lucerne crop: (**A**) pod weight in grams, (**B**) seed weight in grams, and (**C**) seed number per pod. Standard error bars with the same letters are not significantly different (Tukey–Kramer test, $\alpha = 0.05$).

Table 1. Correlation matrix showing Pearson's correlation coefficients (r) and corresponding significance levels *p*-values for showing the relationships between flower tripping by pollinators and seed yield parameters (pod weight, number of seeds, and seed weight).

Variables	Flower Tripping	Pod Weight (g)	No. Seeds	Seed Weight (g)
Flower tripping				
Pod weight (g)	0.2575 ($p = 0.0433$)			
No. Seeds	0.3605 ($p = 0.0040$)	0.4812 ($p < 0.001$)		
Seed weight (g)	0.3347 ($p = 0.0078$)	0.5286 ($p < 0.001$)	0.8907 ($p < 0.001$)	

4. Discussion

The present research showed that *Megachile cephalotes* and *M. hera* were the most prevalent visitors of alfalfa flowers, followed by syrphid flies, solitary bees, honey bees, and wasps, while lepidopterans were sporadically observed in non-supplemented fields. Some previous studies have reported variable results regarding the effects of supplementary honey bee pollination on the yield of different crops. In this study, we observed that the bee population affects pollination efficacy on the crop, and ultimately, yield differences were observed in different supplemented fields. Honey bee colonies were placed around the alfalfa field to test the varying pollination efficacy by *A. mellifera* according to the supplemented treatments, while in one plot, no honey bee hive was installed. It was found that the abundance was higher in honey bee-supplemented plots as compared to non-supplemented plots. Within the supplemented plots, the abundance of *A. mellifera* was higher with the moderately supplemented treatment compared to the higher supplementary treatment. Zhang et al. [19] reported that bees at lower abundances pollinated more efficiently than those at higher abundances. While it is generally assumed that a higher number of supplemented honey bees will enhance crop pollination and production, some studies have shown that this is not always the case [17,40]. Sufficient bee pollination could ensure stable yield and maintenance of genetic variability of crop species [41]. It is also reported that the highest yield was obtained when the optimum number of *A. mellifera* (6–8 bees per 1000 flowers) was present for pollination of kiwifruit (*Actinidia chinensis*) [42]. They also revealed that when there were 60 flowers in a square meter, pollination services by *A. mellifera* were better with 22 honey bees present per square meter, compared to 44 honey bees per square meter [42].

In this study, we found that the highly supplemented alfalfa field yielded less than the moderately supplemented field. This suggests that foraging resources become limited when a high number of supplemented honey bees are introduced to enhance pollination. Similar results have been reported in a study where increased pollen deposition on the stigma of a flower, due to frequent visits by *A. mellifera,* did not affect the drupelet set of raspberries. However, it is suggested that a higher visitation rate of bees can have a detrimental effect on fruit production, ultimately reducing crop yield [43]. The honey bee *A. mellifera* demonstrates a low tripping rate of flowers, but this is compensated by a high frequency of visits, resulting in greater pollination effectiveness. However, a saturation point may be reached where pollination effectiveness no longer improves despite an increase in visit frequency [44]. Contrary results have also been obtained from previous studies, which reported no effects on yield from supplementary treatments, regardless of whether pollinators were present or absent in the soybean crop [45,46]. One reason for the low contribution of supplementary pollination may be the insufficient number of honey bees visiting the crop. Another reason could be the low yield potential of the cultivars used for the experiment. However, installing honey bee colonies in watermelon fields can have a negative impact on native insect pollinators, which are essential for sustainable crop practices [18]. A high abundance of honey bees could reduce native pollinator populations. However, if there is already a low abundance of solitary bees causing a pollination deficit, supplementary pollination using *A. mellifera* is the only feasible option available to increase seed production in alfalfa [44].

A single visit by pollinator species is an essential criterion for assessing the efficacy of native insect pollinators [37]. In our study, *Megachile* sp. gave the highest number of seeds, and seed and pod weights in a single visit, followed by *A. mellifera*, the supplemented bee. A sufficient population of effective pollinators is necessary to pollinate crops properly [26,47]. Previous studies have reported the effectiveness of a single visit of *A. mellifera* [18]. Solitary bees are difficult to manage under agricultural intensification practices due to their complex seasonal behaviour and low availability of suitable nesting sites. However, low abundance of the most efficient pollinators can lead to a pollination deficit [48]. The reduction of pollination in crops can lead to decreased yield production [49,50]. To fulfil this pollination deficit in the crop, supplementary honey bee pollination is necessary to ensure that an adequate number of insect pollinators are available in the field [51].

In this study, bees were identified as the most abundant pollinators in the field, followed by syrphid flies, butterflies, and wasps. In the first year of the study, the abundance of *M. cephalotes* and *M. hera* was higher than that of *A. mellifera*, likely due to the presence of other crops, such as sponge gourd, cucumber, and other cucurbits, which attracted bees from our study fields. In the second year, we observed a significantly higher abundance of *A. mellifera*. This may be due to climatic factors, such as higher rainfall in the second year, which reduced flight activity and led to a low availability of other bees attracted to flowers. The presence of pollinators in sufficient density is essential to achieve adequate pollination, and this can be managed by supplementing with *A. mellifera* [44]. However, honeybees physically disturb and exclude smaller pollinator species from flowers [52]. In several earlier investigations, solitary bees were seen in more significant numbers at alfalfa flowers [39]. It has been noted that other crops, such as canola [37], pumpkin [53], sponge gourd [54], and radish, have larger abundances of *A. mellifera* and solitary bees [14].

In alfalfa, effective pollination is essential for seed production due to its unique floral morphology, which requires a mechanical action known as "tripping" to expose reproductive structures. Our results show a correlation between flower tripping and seed parameters, which proves the dependence of alfalfa on efficient pollinators [26]. However, the ability of a pollinator to trip alfalfa flowers is influenced by several morphological, behavioural, and ecological traits. Specialized bees like *Megachile* species are more efficient as compared to the honey bees, which are less efficient due to sudden strike of anthers on their head during the alfalfa flower tripping process [14]. Taking into consideration the tripping efficiency of different pollinator species, we found that *M. cephalotes* and *M. hera* showed the best performance, while *A. mellifera* demonstrated moderate tripping efficiency. Moreover, *A. florea* showed a nectar-feeding habit without tripping a flower. *A. florea* bees inserted their proboscis to feed on the nectar from the base of a flower, and the flower remained mostly untripped [55]. The low tripping rate of alfalfa by *A. florea* bees was due to this type of behaviour. These findings highlight the importance of selecting efficient pollinator species and optimizing their densities to maximize pollination success and seed yield in alfalfa cultivation.

5. Conclusions

Pollinator management, particularly through the supplementation of honey bee colonies, played a key role in increasing pollinator abundance and consequently improved seed weight and overall reproductive success of alfalfa plants. The lack of significant differences in seed weight between the highly supplemented and non-supplemented blocks draws attention to the nuances of pollination dynamics. This indicates that optimal conditions for pollination might extend beyond the number of pollinators present and could involve the interactions and behaviour of individual species—a critical consideration for future agricultural practices. The visitation rate analyses corroborated these findings, revealing that species like *Amegilla* sp. and *M. cephalotes* were significantly more active, thereby potentially influencing pollination outcomes positively. Furthermore, the tripping behaviour exhibited by different pollinator species provided a fascinating layer of complexity; *M. cephalotes* and *M. hera* demonstrated superior tripping efficacy, indicating their crucial role in ensuring adequate flower fertilisation. The integration of both managed and wild pollinators into agricultural systems could underpin sustainable practices that bolster biodiversity while optimising crop yields.

Author Contributions: Conceptualization, K.E. and M.A.; methodology, K.E.; software, F.Z.A.K.; validation, K.E., M.A. and R.M.; formal analysis, F.Z.A.K.; investigation, K.E.; resources, M.A.; data curation, F.Z.A.K.; writing—original draft preparation, K.E.; writing—review and editing, M.A., F.Z.A.K. and R.M.; visualization, F.Z.A.K.; supervision, R.M.; project administration, M.A.; funding acquisition, R.M. All authors have read and agreed to the published version of the manuscript.

Funding: This research was funded by PARB Project No. 20–276, Government of Punjab, Pakistan, available to M.A., and by the Lithuanian state grant through the State Scientific Research Institute Nature Research Centre, program 2 Climate and Eco-systems, Vilnius, Lithuania, available to R.M.

Institutional Review Board Statement: Not applicable.

Informed Consent Statement: Not applicable.

Data Availability Statement: Data will be available on request.

Acknowledgments: The authors are thankful to the staff at the University farm for crop management.

Conflicts of Interest: The authors declare no conflicts of interest.

References

1. Ayub Agricultural Research Institute (AARI). *AARI Fodder Crop Varieties*; Ayub Agricultural Research Institute: Faisalabad, Pakistan, 2025.
2. Horvat, D.; Viljevac Vuletić, M.; Andrić, L.; Baličević, R.; Kovačević Babić, M.; Tucak, M. Characterization of Forage Quality, Phenolic Profiles, and Antioxidant Activity in Alfalfa (*Medicago sativa* L.). *Plants* **2022**, *11*, 2735. [CrossRef] [PubMed]
3. Ben-Laouane, R.; Ait-El-Mokhtar, M.; Anli, M.; Boutasknit, A.; Oufdou, K.; Wahbi, S.; Meddich, A. Microbial Biotechnologies for Salt Tolerance in Alfalfa: Agro-Nutritional Comparison Between Local and Imported Varieties. *Nitrogen* **2025**, *6*, 27. [CrossRef]
4. Bell, L.W. Relative Growth Rate, Resource Allocation and Root Morphology in the Perennial Legumes, *Medicago sativa*, *Dorycnium rectum* and *D. hirsutum* Grown under Controlled Conditions. *Plant Soil.* **2005**, *270*, 199–211. [CrossRef]
5. Russelle, M.P. Alfalfa: After an 8,000-Year Journey, the "Queen of Forages" Stands Poised to Enjoy Renewed Popularity. *Am. Sci.* **2001**, *89*, 252–261. [CrossRef]
6. Zhang, Y.; Wang, L. Advances in Basic Biology of Alfalfa (*Medicago sativa* L.): A Comprehensive Overview. *Hortic. Res.* **2025**, *12*, uhaf081. [CrossRef]
7. Zhang, S.; Hu, J.; Zhang, Y.; Xie, X.J.; Knapp, A. Seed Priming with Brassinolide Improves Lucerne (*Medicago sativa* L.) Seed Germination and Seedling Growth in Relation to Physiological Changes under Salinity Stress. *Aust. J. Agric. Res.* **2007**, *58*, 811. [CrossRef]
8. Haq, I.U.; Ijaz, S. *Sustainable Winter Fodder: Production, Challenges, and Prospects*; CRC Press: Boca Raton, FL, USA, 2022; ISBN 978-0-367-51798-4.
9. Latif, A.; Sun, Y.; Noman, A. Herbaceous Alfalfa Plant as a Multipurpose Crop and Predominant Forage Specie in Pakistan. *Front. Sustain. Food Syst.* **2023**, *7*, 1126151. [CrossRef]
10. Ahmad, J.; Iqbal, A.; Mahmood, A.; Iqbal, M.A.; Khan, H.Z.; Abbas, R.N.; Akbar, N.; Maqsood, M. Effect of Cutting Management, Seeding Rates and Sowing Method on Seed Yield of Alfalfa (*Medicago sativa* L.). *Pak. J. Bot.* **2020**, *52*, 1449–1454. [CrossRef]

11. Larkin, R.A.; Graumann, H.O. Anatomical Structure of the Alfalfa Flower and an Explanation of the Tripping Mechanism. *Bot. Gaz.* **1954**, *116*, 40–52. [CrossRef]
12. Linsley, G.E. Insect Pollinators of Alfalfa in California. *J. Econ. Entomol.* **1946**, *39*, 18–29. [CrossRef]
13. Bohart, G.E. Pollination of Alfalfa and Red Clover. *Annu. Rev. Entomol.* **1957**, *2*, 355–380. [CrossRef]
14. Batra, S.W.T. Comparative Efficiency of Alfalfa Pollination by Nomia melanderi, Megachile rotundata, Anthidium florentinum and Pithitis smaragdula (Hymenoptera: Apoidea). *J. Kans. Entomol. Soc.* **1976**, *49*, 18–22.
15. Calderone, N.W. Insect Pollinated Crops, Insect Pollinators and US Agriculture: Trend Analysis of Aggregate Data for the Period 1992–2009. *PLoS ONE* **2012**, *7*, e37235. [CrossRef]
16. Garibaldi, L.A.; Steffan-Dewenter, I.; Winfree, R.; Aizen, M.A.; Bommarco, R.; Cunningham, S.A.; Kremen, C.; Carvalheiro, L.G.; Harder, L.D.; Afik, O.; et al. Wild Pollinators Enhance Fruit Set of Crops Regardless of Honey Bee Abundance. *Science* **2013**, *339*, 1608–1611. [CrossRef]
17. Delaplane, K.S.; Mayer, D.F. *Crop Pollination by Bees*; CABI Books; CABI: Wallingford, CT, USA, 2000; ISBN 978-0-85199-783-4.
18. Layek, U.; Kundu, A.; Bisui, S.; Karmakar, P. Impact of Managed Stingless Bee and Western Honey Bee Colonies on Native Pollinators and Yield of Watermelon: A Comparative Study. *Ann. Agric. Sci.* **2021**, *66*, 38–45. [CrossRef]
19. Zhang, K.; Li, Y.; Sun, K.; Bao, J.; He, C.; Hou, X. Supplementary Honey Bee (*Apis mellifera* L.) Pollination Enhances Fruit Growth Rate and Fruit Yield in Paeonia Ostii (Family: Paeoniaceae). *PLoS ONE* **2022**, *17*, e0272921. [CrossRef]
20. Garibaldi, L.A.; Carvalheiro, L.G.; Vaissière, B.E.; Gemmill-Herren, B.; Hipólito, J.; Freitas, B.M.; Ngo, H.T.; Azzu, N.; Sáez, A.; Åström, J.; et al. Mutually Beneficial Pollinator Diversity and Crop Yield Outcomes in Small and Large Farms. *Science* **2016**, *351*, 388–391. [CrossRef]
21. Alomar, D.; González-Estévez, M.A.; Traveset, A.; Lázaro, A. The Intertwined Effects of Natural Vegetation, Local Flower Community, and Pollinator Diversity on the Production of Almond Trees. *Agric. Ecosyst. Environ.* **2018**, *264*, 34–43. [CrossRef]
22. Woodcock, B.A.; Garratt, M.P.D.; Powney, G.D.; Shaw, R.F.; Osborne, J.L.; Soroka, J.; Lindström, S.A.M.; Stanley, D.; Ouvrard, P.; Edwards, M.E.; et al. Meta-Analysis Reveals That Pollinator Functional Diversity and Abundance Enhance Crop Pollination and Yield. *Nat. Commun.* **2019**, *10*, 1481. [CrossRef]
23. Hagler, J.R.; Mueller, S.; Teuber, L.R.; Machtley, S.A.; Van Deynze, A. Foraging Range of Honey Bees, *Apis mellifera*, in Alfalfa Seed Production Fields. *J. Insect Sci.* **2011**, *11*, 1–12. [CrossRef]
24. Gordon, D.M.; Barthell, J.F.; Page, R.E.; Kim Fondrk, M.; Thorp, R.W. Colony Performance of Selected Honey Bee (Hymenoptera: Apidae) Strains Used for Alfalfa Pollination. *J. Econ. Entomol.* **1995**, *88*, 51–57. [CrossRef]
25. Rader, R.; Howlett, B.G.; Cunningham, S.A.; Westcott, D.A.; Newstrom-Lloyd, L.E.; Walker, M.K.; Teulon, D.A.J.; Edwards, W. Alternative Pollinator Taxa Are Equally Efficient but Not as Effective as the Honeybee in a Mass Flowering Crop. *J. Appl. Ecol.* **2009**, *46*, 1080–1087. [CrossRef]
26. Haedo, J.P.; Martínez, L.C.; Graffigna, S.; Marrero, H.J.; Torretta, J.P. Managed and Wild Bees Contribute to Alfalfa (*Medicago sativa*) Pollination. *Agric. Ecosyst. Environ.* **2022**, *324*, 107711. [CrossRef]
27. Nagano, Y.; Wabiko, N.; Yokoi, T. Female Solitary Bees Flexibly Change Foraging Behaviour According to Their Floral Resource Requirements and Foraging Experiences. *Sci. Nat.* **2023**, *110*, 37. [CrossRef]
28. Herrera, C.M.; Núñez, A.; Valverde, J.; Alonso, C. Body Mass Decline in a MEditerranean Community of Solitary Bees Supports the Size Shrinking Effect of Climatic Warming. *Ecology* **2023**, *104*, e4128. [CrossRef]
29. Melone, G.G.; Stuligross, C.; Williams, N.M. Heatwaves Increase Larval Mortality and Delay Development of a Solitary Bee. *Ecol. Entomol.* **2024**, *49*, 433–444. [CrossRef]
30. Rollin, O.; Garibaldi, L.A. Impacts of Honeybee Density on Crop Yield: A Meta-Analysis. *J. Appl. Ecol.* **2019**, *56*, 1152–1163. [CrossRef]
31. Rauf, A.; Saeed, S.; Ali, M.; Nadeem Tahir, M.H. Comparative Efficiency of Native Insect Pollinators in Reproductive Performance of *Medicago sativa* L. in Pakistan. *Insects* **2021**, *12*, 1029. [CrossRef]
32. Ahmed, N.; Umer, A.; Ali, M.A.; Iqbal, J.; Mubashir, M.; Grewal, A.G.; Butt, B.; Rasheed, M.K.; Chaudhry, U.K. Micronutrients Status of Mango (*Mangifera indica*) Orchards in Multan Region, Punjab, Pakistan, and Relationship with Soil Properties. *Open Agric.* **2020**, *5*, 271–279. [CrossRef]
33. Abbas, F. Analysis of a Historical (1981–2010) Temperature Record of the Punjab Province of Pakistan. *Earth Interact.* **2013**, *17*, 1–23. [CrossRef]
34. Abbas, F.; Ahmad, A.; Safeeq, M.; Ali, S.; Saleem, F.; Hammad, H.M.; Farhad, W. Changes in Precipitation Extremes over Arid to Semiarid and Subhumid Punjab, Pakistan. *Theor. Appl. Clim.* **2014**, *116*, 671–680. [CrossRef]
35. Putnam, D.H. Factors Influencing Yield and Quality in Alfalfa. In *The Alfalfa Genome*; Yu, L.-X., Kole, C., Eds.; Compendium of Plant Genomes; Springer International Publishing: Cham, Switzerland, 2021; pp. 13–27. ISBN 978-3-030-74465-6.
36. Mazeed, A.R.; Marey, R.A. Pollinators Activity on Onion Flowers and Its Effect on Seeds Yield at Sohag Governorate, Egypt. *Egypt. J. Agric. Res.* **2018**, *96*, 465–475. [CrossRef]

37. Ali, M.; Saeed, S.; Sajjad, A.; Whittington, A. In Search of the Best Pollinators for Canola (*Brassica napus* L.) Production in Pakistan. *Appl. Entomol. Zool.* **2011**, *46*, 353–361. [CrossRef]
38. Cane, J.H. Pollinating Bees (Hymenoptera: Apiformes) of U.S. Alfalfa Compared for Rates of Pod and Seed Set. *J. Econ. Entomol.* **2002**, *95*, 22–27. [CrossRef]
39. Brunet, J.; Stewart, C.M. Impact of Bee Species and Plant Density on Alfalfa Pollination and Potential for Gene Flow. *Psyche A J. Entomol.* **2010**, *2010*, 1–7. [CrossRef]
40. Morris, W.F.; Vázquez, D.P.; Chacoff, N.P. Benefit and Cost Curves for Typical Pollination Mutualisms. *Ecology* **2010**, *91*, 1276–1285. [CrossRef]
41. Stein, K.; Coulibaly, D.; Stenchly, K.; Goetze, D.; Porembski, S.; Lindner, A.; Konaté, S.; Linsenmair, E.K. Bee Pollination Increases Yield Quantity and Quality of Cash Crops in Burkina Faso, West Africa. *Sci. Rep.* **2017**, *7*, 17691. [CrossRef] [PubMed]
42. Peace, A.; Pattemore, D.; Broussard, M.; Fonseka, D.; Tomer, N.; Bosque-Pérez, N.A.; Crowder, D.; Shaw, A.K.; Jesson, L.; Howlett, B.G.; et al. Orchard Layout and Plant Traits Influence Fruit Yield More Strongly than Pollinator Behaviour and Density in a Dioecious Crop. *PLoS ONE* **2020**, *15*, e0231120. [CrossRef]
43. Sáez, A.; Morales, C.L.; Ramos, L.Y.; Aizen, M.A. Extremely Frequent Bee Visits Increase Pollen Deposition but Reduce Drupelet Set in Raspberry. *J. Appl. Ecol.* **2014**, *51*, 1603–1612. [CrossRef]
44. Haedo, J.P.; Graffigna, S.; Martínez, L.C.; Pérez-Méndez, N.; Torretta, J.P.; Marrero, H.J. Effectiveness Landscape of Crop Pollinator Assemblages: Implications to Pollination Service Management. *Agric. Ecosyst. Environ.* **2023**, *348*, 108417. [CrossRef]
45. Blettler, D.C.; Fagúndez, G.A.; Caviglia, O.P. Contribution of Honeybees to Soybean Yield. *Apidologie* **2018**, *49*, 101–111. [CrossRef]
46. Gazzoni, D.L.; Barateiro, J.V.G.R.P.; Da Rosa Santos, P. Supplemental Bee Pollination Effect on the Productivity of Soybean Grown in a Low Yield Potential Condition. *J. Apic. Res.* **2024**, *63*, 788–800. [CrossRef]
47. Duque-Trujillo, D.; Hincapié, C.A.; Osorio, M.; Zartha-Sossa, J.W. Strategies for the Attraction and Conservation of Natural Pollinators in Agroecosystems: A Systematic Review. *Int. J. Environ. Sci. Technol.* **2023**, *20*, 4499–4512. [CrossRef]
48. Garratt, M.P.D.; De Groot, G.A.; Albrecht, M.; Bosch, J.; Breeze, T.D.; Fountain, M.T.; Klein, A.M.; McKerchar, M.; Park, M.; Paxton, R.J.; et al. Opportunities to Reduce Pollination Deficits and Address Production Shortfalls in an Important Insect-pollinated Crop. *Ecol. Appl.* **2021**, *31*, e02445. [CrossRef] [PubMed]
49. Bukhari, S.F.; Ali, M.; Khan, F.Z.A.; Awan, T.H. From Petals to Seeds: Understanding the Role of Hymenopteran and Dipteran Pollinators in the Reproductive Success of Onion (*Allium cepa* L.). *Plant Bull.* **2024**, *3*, 19–25. [CrossRef]
50. Haider, S.; Khan, F.Z.A.; Gul, H.T.; Ali, M.; Iqbal, S. Assessing the Role of Conservation Strips in Enhancing Beneficial Fauna in the Wheat-Cotton Agricultural System in Punjab, Pakistan. *Pak. J. Zool.* **2024**, *56*, 1–9. [CrossRef]
51. Mashilingi, S.K.; Zhang, H.; Garibaldi, L.A.; An, J. Honeybees Are Far Too Insufficient to Supply Optimum Pollination Services in Agricultural Systems Worldwide. *Agric. Ecosyst. Environ.* **2022**, *335*, 108003. [CrossRef]
52. Gross, C.L.; Mackay, D. Honeybees Reduce Fitness in the Pioneer Shrub *Melastoma affine* (Melastomataceae). *Biol. Conserv.* **1998**, *86*, 169–178. [CrossRef]
53. Ali, M.; Saeed, S.; Sajjad, A.; Bashir, M.A. Exploring the Best Native Pollinators for Pumpkin (*Cucurbita pepo*) Production in Punjab, Pakistan. *Pak. J. Zool.* **2014**, *46*, 531–539.
54. Abdullah, S.; Ali, M.; Khan, F.Z.A.; Sajjad, A.; Qayyum, M.A.; Ahmad, N. Solitary Bees Are More Efficient Pollinators of Sponge Gourd than Giant Honeybees and Syrphid Flies. *Sociobiology* **2024**, *71*, e10279. [CrossRef]
55. Reinhardt, J.F. Some Responses of Honey Bees to Alfalfa Flowers. *Am. Nat.* **1952**, *86*, 257–275. [CrossRef]

biology

MDPI

Article

Seasonal Variation in Flower Traits, Visitor Traits, and Reproductive Success of *Solanum sisymbriifolium* Lamarck (Solanaceae) in the Rarh Region of West Bengal, India

Ujjwal Layek [1], Pappu Majhi [2], Alokesh Das [1], Prakash Karmakar [3] and Arijit Kundu [3,*]

1 Department of Botany, Rampurhat College, Rampurhat 731224, West Bengal, India; layekujjwal@yahoo.co.in (U.L.); apualokesh@gmail.com (A.D.)
2 Department of Agricultural Engineering, Visva-Bharati, Sriniketan 731236, West Bengal, India; majhipappu97@gmail.com
3 Department of Botany & Forestry, Vidyasagar University, Midnapore 721102, West Bengal, India; prakashbot1973@gmail.com
* Correspondence: ksubaikundu@gmail.com

Simple Summary: This study examined the seasonal variations in flower traits, plant–pollinator interactions, and reproductive success of *Solanum sisymbriifolium*. The flowering intensity, flower display size, and pollen and ovule production peaked during spring, summer, and the monsoon, while flower longevity and stigmatic receptivity were the longest in winter. The plant species showed self-compatibility and strong dependency on pollinators. Visitor traits also varied seasonally, with the peak during the spring–monsoon period. Vital pollinators were solitary bees and stingless bees. The fruit and seed sets were higher during the spring–monsoon period and lower in winter. This study will help in the management of the weed species and the conservation of associated bee populations.

Abstract: The wild tomato (*Solanum sisymbriifolium*) is a globally distributed shrubby weed with both negative and positive impacts, including its invasive properties and the potential for pharmaceutical and traditional medicinal uses. Despite its ecological significance, the plant's reproductive biology and pollination ecology remain poorly understood. This study aimed to investigate the floral biology, pollination ecology, and plant reproduction of the weed species. Some flower traits, such as flowering intensity, flower display size, and pollen and ovule production, peaked during spring, summer, and the monsoon, while flower longevity and stigmatic receptivity were the longest in winter. The plant species was self-compatible (ISI = 0.02), heavily depended on pollinators (IDP = 0.72), and experienced minimal pollination limitation (D = 0.10) under open-pollination conditions. Flower visitors' traits (e.g., abundance, diversity, and richness) were higher in the spring, summer, and the monsoon, and these were lower in winter. The vital pollination service was provided by *Amegilla zonata*, *Ceratina binghami*, *Lasioglossum cavernifrons*, *Nomia* (*Curvinomia*) *strigata*, *Tetragonula pagdeni*, *Xylocopa aestuans*, *Xylocopa amethystina*, *Xylocopa fenestrata*, and *Xylocopa latipes*. Reproductive success, as indicated by fruit and seed set, varied seasonally, being higher during the spring–monsoon period and lower in winter. These findings support effective management of this weed species and help conserve the associated bee populations.

Keywords: buzz pollination; flower display size; flowering intensity; flowering phenology; pollination; pollination service index

1. Introduction

Flowering is a key stage in the life cycle of angiosperms, marking the shift from vegetative growth to reproductive maturity and playing a crucial role in the plant's ability to produce seeds. This transition is foundational not only to individual survival and fecundity but also to broader population dynamics, species interactions, and ecosystem functioning [1,2]. Flowering influences ecological networks by determining the timing of resources, which affects pollinators, herbivores, pathogens, seed dispersers, and microbial interactions [3,4]. The characteristics of flowering—its onset, frequency, duration, and intensity—emerge from an intricate balance between internal regulators such as genetic programs and hormonal signals [5,6], and external stimuli including ambient temperature, day length, moisture availability, and soil fertility [7–9]. Across taxa and landscapes, these flowering patterns exhibit remarkable diversity, mirroring both evolutionary legacies and local adaptive responses. For example, Gentry [10] categorised the flowering patterns of Bignoniaceae into five types: steady state, modified steady state, cornucopia, big bang, and multiple bang. Frankie et al. [11] grouped flowering patterns into two categories: seasonal and extended, while Bawa [12] divided them into massive and extended types. These flowering patterns can influence the extent of outcrossing, availability of mates, frequency of near-neighbour matings, and overall reproductive output [13,14].

Floral characteristics lie at the heart of angiosperm reproductive ecology, shaping both the timing and success of pollination and fertilisation. Key attributes—including the flower longevity, moment of anthesis, the quantity of pollen and ovules produced, pollen viability, and the period of stigma receptivity—together govern a plant's reproductive potential. These traits vary widely within and between species, influenced by environmental constraints and the evolution of mating systems [15,16]. For example, precise scheduling of flower opening can align pollen availability with peak pollinator foraging, while extended flower longevity may buffer against unpredictable pollinator visits. Likewise, the allocation of resources to pollen versus ovule production reflects strategic trade-offs in male and female function, and the overlap of pollen viability with stigma receptivity ensures optimal fertilisation.

A wide variety of animals visit the flowers of different plant species to collect resources like nectar and pollen, including insects such as bees, flies, butterflies, and beetles, as well as vertebrates like birds and bats. The species composition of the flower visitors' spectrum is plant species-specific and also varies spatially and temporally [17,18]. The interaction between flowering plants and their animal visitors represents a cornerstone of terrestrial ecosystems, underpinning the reproductive success of most angiosperms. Not all floral visitors are effective pollinators; distinguishing between casual visitors and those that actively transfer pollen is key to understanding the ecological dynamics of plant reproduction. Accurately identifying which species are effective pollinators and understanding how their behaviours affect plant fitness are crucial for conserving biodiversity, sustaining ecosystem services, and ensuring the stability of plant–pollinator networks. The role of a pollinator is evaluated by measuring several parameters, including visitation pattern (legitimate or illegitimate), visitation rate, single-visit pollination efficiency [19], and pollinator importance [20]. As pollinator populations decline globally due to anthropogenic pressures such as habitat loss, pesticide use, and climate change [21,22], understanding the efficiency and specificity of these interactions becomes increasingly urgent.

Plant reproductive fitness, typically measured by fruit and seed set, serves as a vital indicator of a species' capacity to sustain viable populations over time. In flowering plants, successful reproduction hinges on an intricate interplay of biotic elements, such as the presence and behaviour of pollinators, and abiotic factors, including temperature, rainfall, and photoperiod [9,23,24]. Seasonal fluctuations in these variables can significantly alter the timing, frequency, and efficiency of pollination [25,26], resulting in notable differences in both the quantity and quality of reproductive output. For instance, peak pollinator activity during optimal weather conditions may boost fruit set, while adverse conditions in off-peak seasons can suppress seed production or skew resource allocation.

Litchi tomato (*Solanum sisymbriifolium* Lam.) originates from South America and is a shrubby weed currently distributed worldwide, including hotter parts of India [27]. The plant exhibits various pharmaceutical activities, including antimicrobial and antioxidant properties [28,29]. Different plant parts have been widely used to treat numerous diseases, including diarrhoea, hypertension, and respiratory and urinary tract infections [30,31]. The plant is also used as a good trap crop against potato cyst nematodes, root-knot nematodes, and root-lesion nematodes [32]. Limited information exists on the floral and pollination biology of *Solanum sisymbriifolium*, as noted by Saha and Dutta [33] and Hill and Hulley [34]. Additionally, the seasonal variation in flower biology and reproductive ecology is not well understood. Plant–pollinator interactions are also not characterised for the plant species. Understanding the reproductive biology of a plant species is essential for effective population management. Therefore, a comprehensive study of the biology of the flower, plant–pollinator interactions, and reproduction of *Solanum sisymbriifolium* is crucial.

This study intended to expand knowledge on the reproductive biology of *Solanum sisymbriifolium* and assess the seasonal variation in flower traits, visitor traits, and plant reproductive success. Specifically, we investigated the following: (i) flower biology, (ii) floral visitors and pollinators, and (iii) the seasonal variation in flower traits, visitor traits, and plant reproductive success in the Rarh region of West Bengal, India. We hypothesised that flower traits change with the seasons, leading to fluctuations in plant–pollinator interactions and affecting the reproductive fitness of the species.

2. Materials and Methods

2.1. Study Area

We conducted surveys across various sites in the Rarh region (Bankura, Birbhum, and Paschim Medinipur districts) of West Bengal, India. The study areas characterise medium-density, mixed vegetation, with human habitats. The study area experiences seasonal variations—(i) summer (April to June, with daytime temperatures ranging between 35 and 42 °C), (ii) monsoon (July to August, day temperatures ranging 28 to 35 °C, an annual rainfall of 1479.9 mm), (iii) autumn (September to mid-October, day temperatures ranging between 30 and 33 °C), (iv) late autumn (mid-October to November, daytime temperature ranging from 29 to 32 °C), (v) winters (December to January, with daytime temperatures between 7 and 15 °C), and (vi) spring (February to March, daytime temperature ranging from 20 to 30 °C).

2.2. Plant Species

The study was conducted on *Solanum sisymbriifolium* Lam. (Solanaceae) from 2020 to 2023. The plant is known as vila-vila, sticky nightshade, or litchi tomato. It is originally native to Central and South America but is now found worldwide. The small, shrubby plant grows on roadside, uncultivated lands, and wastelands. The plant has sticky stems and branches with orange-yellow spines.

2.3. Floral Biology

The data about flower phenological events (e.g., flowering frequency, timing, and patterns) were recorded following the methods outlined by Gentry [10] and Hopkins [35]. We estimated flowering intensity (FI) and flower display size (FDS) as follows:

$$\text{Flowering intensity} = Fn \quad (1)$$

$$\text{Flower display size} = Ft \quad (2)$$

Here, Fn is the number of freshly open flowers (i.e., 1st day flowers) per day per reproductive unit, such as an inflorescence, flowering twig, or individual plant (in this case, an individual plant was considered). On the other hand, Ft is the number of total open flowers at a time per reproductive unit. For this purpose, individuals of old, medium, and young age classes (all of which had reached the reproductive phase) were chosen (N = 60 × 6 = 360 observations; 6 seasons, data equally derived from each age class). Flower display size was considered during the peak visitation period (i.e., 6:00–8:00 h) of flower visitors.

Flower opening and closing times were documented through field observations on flowers (N = 40 × 6 = 240 observations, 6 seasons) at a two-hour interval throughout the day, and the longevity of selected flowers was recorded.

Flower morphological traits were examined using various microscopes [including (i) a simple microscope, (ii) a compound light microscope (Primo Star, Zeiss), (iii) a stereo microscope (Stemi 508, Zeiss), and (iv) a field emission scanning electron microscope (FE-SEM, GeminiSEM 450, Gemini 2 column)]. The methods used for the SEM study are detailed in Table S1. For pollen count per flower, we collected all anthers of a mature flower bud (before dehiscence of anthers; N = 10 × 6 = 60 buds, 6 seasons) in a vial. We added 5 mL of sucrose solution to the anthers and crushed the anthers. Then, we added 1 mL of *Lycopodium* spore solution (which standardised about 75,600 spores/mL). Through microscopic observation, we counted *Lycopodium* spores and pollen grains from flowers. Then we estimated the number of flower pollens as '[(number of counted flower pollen ÷ number of counted *Lycopodium* spore) × number of *Lycopodium* spores added to solution]' (as described by Layek et al. [36]). We studied pollen morphology (processed pollens: acetosyled according to Erdtman [37], and also unprocessed pollens) using a light microscope and a scanning electron microscope (detailed in Table S1). To count ovules per flower, we took an ovary (N = 40 × 6 = 240 ovaries, 6 seasons) on a glass slide, ruptured the ovary wall, and slightly pressed it to spread out the ovules. Then, we counted the ovules using a simple dissecting microscope.

Pollen viability was assessed using the 2,3,5-triphenyl tetrazolium chloride (TTC) staining method [38]. To prepare a 1% solution, 0.1 g of TTC and 6 g of sucrose were dissolved in 10 mL of distilled water. One drop of this solution was placed on a glass slide, and then pollen grains were added to the TTC solution. The pollen solution was veiled with a cover slip and kept in the dark for two hours. After this incubation period, a minimum of 100 pollen grains per sample were observed under a compound microscope. Grains exhibiting a reddish stain were classified as viable, whereas those displaying no colour change (or other colours like black or yellow) were considered non-viable. We tested pollen viability for all seasons (N = 10 × 6 = 60 flowers for pollen viability test, 6 seasons). Additionally, we assessed viability considering different day times with a 12 h interval starting from 0 h (i.e., during flower opening time) to 48 h after opening (N = 10 × 5 = 50 flowers, 5 time slots).

Pollen germinability was evaluated using an in vitro germination assay, following the protocol of Parfitt and Ganeshan [39], which utilised a germination medium consisting of

0.5% agar, 5 ppm boric acid, and 10% sucrose. A small quantity of the prepared medium was poured into sterile Petri dishes, and bulk pollen grains were then sprinkled onto the surface. The Petri dishes were incubated in the dark at 25 °C for 24 h. Following incubation, the medium containing pollen was gently transferred onto microscope slides for examination under a compound microscope. For each sample, at least 100 pollen grains were counted, and germination was considered successful when the length of the pollen tube was greater than the diameter of the corresponding pollen grain. Like the pollen viability test, the pollen germination test was conducted for each season (N = 10 × 6 = 60 flowers for the pollen germinability test, 6 seasons).

Stigma receptivity was assessed using the hydrogen peroxide test [40]. We collected flowers at different stages of development, spanning their longevity (e.g., 1st day morning, late afternoon; 2nd day morning, late afternoon; 3rd day morning, late afternoon, etc.), and tested stigma receptivity ($n = 10$ stigmas for each interval for each of the six seasons). For the last day of flower longevity, we tested receptivity at two-hour intervals, starting from morning until senescence. Stigmas, along with a small portion of the style, were carefully excised from the flowers using razor blades and placed on a glass slide. A drop of hydrogen peroxide (H_2O_2) solution was then applied, and after 10–15 s, stigma receptivity was examined under dissecting and stereo microscopes. The formation of numerous bubbles indicates receptivity to stigma, and a comparatively greater number of bubbles signifies a higher level of receptivity.

2.4. Mating System

To evaluate the mating system, we performed five pollination treatments during the peak flowering season in summer: (i) open pollination, (ii) pollinator exclusion to test for spontaneous autogamy, (iii) self-pollination (manually), (iv) cross-pollination (manually), and (v) supplementary pollination. In the late afternoon, matured flower buds were randomly selected ($n = 100$ flowers per treatment), marked, and enclosed in nylon mesh bags (except for those in the open and supplementary pollination treatments) to prevent external pollination until the floral parts were senesced. For manual self-pollination, we unbagged the flowers between 8:00 and 10:00 a.m., manually transferred pollen from the same flower to its stigma, and then immediately re-bagged the flowers. Flowers selected for cross-pollination treatment were emasculated prior to anther dehiscence. During the receptive period (8:00–10:00 h), the flowers were unbagged, pollens were added (collected from different individuals) on the stigma, and again re-bagged. In supplementary pollination, pollen (collected from the same or different individuals) was manually applied to the stigmas in addition to the natural pollination by native pollinators. Fruit development was recorded 5–7 days post-pollination, and seed set was assessed approximately 15 days thereafter for each treatment.

We calculated the index of self-incompatibility (ISI) using the method described by Raduski et al. [41], as follows:

$$\mathrm{ISI} = 1 - \frac{\mathrm{Rsp}}{\mathrm{Rcp}} \quad (3)$$

Rsp and Rcp refer to reproductive success in the self-pollination and cross-pollination treatments, respectively. We considered it as the seed set per flower. Based on the ISI value, a plant species can be classified into one of the three categories: (1) self-compatible (ISI ≤ 0.2), (2) partially self-incompatible ($0.2 <$ ISI < 0.8), and (3) self-incompatible (ISI ≥ 0.8).

To determine the plant species' dependency on pollinators, we calculated the index of dependency on pollinators (IDP) following the method described by Layek et al. [36], as outlined below:

$$\mathrm{IDP} = 1 - \frac{\mathrm{Re}}{\mathrm{Rs}} \quad (4)$$

Re and Rs represent the number of seeds produced per flower in the pollinator exclusion and supplementary pollination treatments, respectively. The IDP value ranges from 0 to 1, with higher values indicating a stronger dependence of the plant on pollinators.

To assess whether the plant species experience pollination limitations under open-field conditions, we calculated the pollination deficit coefficient (D) following the method described by Layek et al. [25], as detailed below:

$$D = 1 - \frac{\text{Ro}}{\text{Rs}} \tag{5}$$

Ro represents the number of seeds produced per flower under open pollination. The value of D ranges from 0 to 1, with a higher value ($D \geq 0.1$) indicating a significant pollination deficit in the plant species.

2.5. Floral Visitors

We conducted daytime field surveys to observe flower visitors, dividing the observation period into seven two-hour time slots between 4:00 and 18:00 h. Each survey, based on individual plant sampling, lasted 5 min per plant. One observation was conducted during each time slot per sampling day, totalling seven observations per day. Over the study period, a total of 1260 observations ($N = 30 \times 7 \times 6 = 1260$; 30 observations per time slot per season) were conducted. Visitors were either identified directly in the field or captured for identification at a later time.

Visitor abundance was estimated as the number of individuals observed per plant within a 5 min period. Next, we calculated the relative abundance (RA) for each flower-visiting species following Layek et al. [42]. Visitor richness was assessed using Margalef's index (D) [43], and the diversity of flower visitors was measured using the Shannon-Weaver diversity index (H') [44]. The methodologies are given in Table S2.

We recorded insect visitors collecting floral resources (e.g., nectar, pollen grains, and floral tissues). The flower visitation rate (VR), defined as the number of flowers visited per minute, was measured ($n \geq 20$ observations for an insect species). For visitors with low visitation rates (such as beetles, flies, and stingless bees), the number of flowers visited over a 10 min period was recorded and then standardised to a per-minute rate. Additionally, flower handling time—the duration an insect spent on a single flower during a visit—was recorded ($n \geq 20$ observations for an insect species).

2.6. Pollinating Strategies of Visitors

We recorded the visits made by different insect species on the flowers ($N \geq 300$ visits). Then, we estimated flower visit proportion (FV) for each flower-visiting species as follows:

$$\text{FV} = \frac{\text{V}i}{\text{V}t} \tag{6}$$

Vi is the number of visits recorded for the i species, and Vt is the total number of visits considering all species.

We recorded the visitors' visitation patterns (i.e., legitimate and illegitimate types of visits). We examined the pollen adherence sites on visitors using a stereo microscope and a scanning electron microscope. For this purpose, visitors were captured using a medium-sized plastic container, immobilised by freezing, and then dried in a hot air oven (Digilab). Then, the samples were studied under a stereo microscope and a scanning electron microscope (the methodology is provided in Table S1).

For legitimate floral visitors, we documented various pollination modes (e.g., nototribic, sternotribic, notosternotribic, appendage-mediated, or a special type such as buzz pollination or pulsatory pollination [9]). To identify the vital pollinators of the plant species, we used a composite metric—the pollination service index (PS*i*)—by combining the standardised values (ranging from 0 to 1) of several key parameters.

$$PSi = FV \times FSi \times AR \times SR \quad (7)$$

FV is the flower visit proportion for visitor species *i*. FS*i* is the flower sex type selection index. For hermaphrodite flowers, each visitor is assigned an FS*i* value of 1. If a visitor's FS*i* is greater than 1, all values are rescaled on a 0–1 scale, with the highest value normalised to 1 and the other values adjusted proportionally. AR and SR are the anther contact rate and stigma contact rates, respectively (calculated by dividing the number of visits involving contact by the total number of visits). When flower visit proportion (FV) data are unavailable, an alternative parameter can be used as a substitute [e.g., RA $\times$ VR/Σ(RA $\times$ VR); here RA is relative abundance and VR is flower visitation rate]. The pollination service index (PS*i*) ranges from 0 to 1, with higher values indicating a greater pollination service provided by the pollinator species.

2.7. Seasonal Reproductive Success

Fruit and seed sets were observed under open-field (i.e., unmanipulated) conditions throughout all seasons. In the morning, newly opened flowers (characterised by greenish-yellow anthers) were selected and marked by applying an ink spot to the pedicel and adding tags ($n = 10 \times 10 = 100$ flowers per season, with ten sampling days per season). Subsequently, fruit and seed sets were recorded, following the same procedure as in the previously described pollination treatment experiment.

2.8. Statistical Analysis

We examined the data within each group, assessing key assumptions for parametric tests, including normality (using the Shapiro–Wilk test and Q-Q plots), homoscedasticity (using scatter plots and the Breusch-Pagan test), and the homogeneity of variance (using Levene's test). The non-parametric Kruskal–Wallis H test was used for data that did not follow a normal distribution (e.g., seasonal flower traits: flowering intensity, flower longevity, ovule number, pollen production, pollen viability and germinability, stigma receptivity; seasonal visitor traits: abundance, richness, diversity; reproductive success: fruit and seed sets). When the *p*-value was significant ($p \leq 0.05$), Dunn's test was used as the post hoc analysis. Statistical analyses were conducted using IBM SPSS Statistics version 26.0 and R software 2022 (R Core Team, Vienna, Austria).

3. Results

3.1. Floral Biology

Solanum sisymbriifolium flowered consistently each year, displaying a steady-state flowering pattern. All individuals that had reached the reproductive stage bloomed synchronously. Flowering intensity showed significant seasonal variation (Kruskal–Wallis H test: $\chi^2 = 39.12$, df = 5, $p < 0.001$), with higher intensity observed in spring, summer, and the monsoon, and lower intensity during winter (Table 1). The flower display size also varied according to the seasons (Kruskal–Wallis H test: $\chi^2 = 18.40$, df = 5, $p < 0.01$), as did flowering intensity. The time of flower opening varied seasonally. During warmer seasons, such as summer and the monsoon, flowers opened earlier (between 5:00 and 6:00 a.m.) compared to the colder winter, when they opened later (between 7:00 and 8:00 a.m.). Flower longevity also varied across different seasons (Kruskal–Wallis H test: $\chi^2 = 176.42$, df = 5, $p < 0.001$). Comparatively, a higher flower longevity was observed during cold winter (55.60 $\pm$ 1.37 h) and lower in hot summer (32.60 $\pm$ 1.58 h). On the first day, in the late afternoon, the flowers became flaccid and closed up until the next morning. On the second day morning, closed flowers were reopened, and then closed again in the late afternoon (Figure 1). Anther colour also slightly changed from greenish yellow (in freshly opened flowers) to deep yellow in aged flowers.

Table 1. Flower traits of *Solanum sisymbriifolium* across different seasons in West Bengal, India.

Flower Traits	Summer	Monsoon	Autumn	Late Autumn	Winter	Spring	Statistics
Flowering intensity	$3.85^{a} \pm 4.61$	$4.12^{a} \pm 5.12$	$3.03^{ab} \pm 2.97$	$2.13^{bc} \pm 1.98$	$1.33^{c} \pm 1.46$	$3.47^{a} \pm 2.49$	$\chi^2 = 39.12$, ***
Flower display size	$7.75^{a} \pm 9.21$	$8.18^{a} \pm 9.73$	$6.47^{ab} \pm 6.14$	$5.68^{bc} \pm 5.17$	$4.13^{c} \pm 4.58$	$7.17^{a} \pm 4.81$	$\chi^2 = 18.40$, ***
Flower longevity (hours)	$32.60^{d} \pm 1.58$	$33.10^{d} \pm 1.81$	$33.90^{d} \pm 1.92$	$48.05^{b} \pm 8.98$	$55.60^{a} \pm 1.37$	$38.10^{c} \pm 6.55$	$\chi^2 = 176.42$, ***
Pollen (no. of grains flower^{-1})	$515{,}306.93^{a} \pm 49{,}153.55$	$520{,}167.56^{a} \pm 21{,}340.97$	$428{,}793.87^{b} \pm 60{,}018.43$	$350{,}910.06^{c} \pm 32{,}606.08$	$319{,}852.16^{c} \pm 11{,}734.98$	$439{,}075.23^{b} \pm 62{,}140.65$	$\chi^2 = 44.37$, ***
Ovule (no. of ovules flower^{-1})	$80.25^{a} \pm 15.01$	$83.20^{a} \pm 13.97$	$73.38^{b} \pm 13.40$	$68.32^{bc} \pm 11.57$	$64.95^{c} \pm 11.40$	$74.38^{b} \pm 12.52$	$\chi^2 = 50.35$, ***
Pollen viability	79.98 ± 7.23	80.73 ± 7.02	81.05 ± 7.11	81.77 ± 7.38	82.16 ± 7.31	81.54 ± 7.28	$\chi^2 = 1.25$, $p = 0.94$
Pollen germinability	74.35 ± 6.42	74.76 ± 6.25	75.12 ± 6.32	75.86 ± 6.57	76.82 ± 6.54	75.27 ± 6.51	$\chi^2 = 1.23$, $p = 0.82$
Stigma receptivity (duration in hours)	$51.80^{c} \pm 1.82$	$52.40^{c} \pm 1.90$	$52.90^{c} \pm 1.89$	$57.50^{b} \pm 5.76$	$76.90^{a} \pm 2.29$	$56.30^{b} \pm 4.51$	$\chi^2 = 83.74$, ***

Values are presented as mean ± standard deviation. Different superscript letters within a row (i.e., across seasons) indicate statistically significant differences (Dunn's post hoc test at the 0.05 level). Statistics: df = 5 for each parameter, ***—$p < 0.001$.

Figure 1. Flowers of different ages. **(A)** 1st day morning (freshly opened flower), **(B)** 1st day late afternoon, **(C)** 2nd day morning, **(D)** 2nd day late afternoon, **(E)** 3rd day morning, **(F)** 3rd day late afternoon, **(G)** 4th day morning, and **(H)** 5th day morning.

Flowers were borne on raceme inflorescence (Figure 2). Inflorescences were 82.70 ± 34.78 mm (53.20–134.80 mm) in length. There were two types of flowers: (i) normal fertile bisexual flowers, and (ii) female-sterile flowers with short styles (Figure 3). Most flowers were normal, bisexual flowers, and female-sterile flowers were very rare (approximately 3% of the total flowers). Flowers were actinomorphic, bisexual, hypogynous, and white. The calyx had five sepals, which were gamosepalous (upper parts were free and fused at the base), persistent, and had a dorsal surface with spines of variable length and glandular hairs (Figure 4). The corolla had five petals, gamopetalous, with an acute apex, and was white. The outer surface of the petals bore branched, star-shaped hairs. Five stamens, epipetalous (attached at the base of petals), filaments very short (about 2.5 mm), anthers elongated (about 11 mm), basifixed, yellow-coloured, and dehisced by apical pores. The palynological analysis revealed that the pollen grains were monads with a spheroidal shape, measuring approximately 25.96 μm in diameter. Their ambs were sub-triangular, and they had trizonocolporate apertures. The exine of a pollen grain was about 2.5 μm thick and exhibited micro-verrucate ornamentation (Figure 5). Gynoecium had two carpels, syncarpous; stigma capitate, greenish, and apart from the anthers; style long (about 18 mm), whitish; ovary dome-shaped, and whitish. Many glandular hairs were on the style base and the upper part of the ovary. Ovules were many; placentation was the axile type. Fruit was berry type, globose, bright red when mature, and covered by an enlarged calyx. Each fruit contained many reniform seeds (Figure 6).

Figure 2. Flowering twig of *Solanum sisymbriifolium*.

Figure 3. Two types of flowers of *Solanum sisymbriifolium*. (**A**) Fertile bisexual flower and (**B**) female-sterile flower. Scale bars = 20 mm.

Figure 4. (**A**,**B**) Calyx; (**C**) outer surface of sepal; (**D**) inner surface of sepal; (**E**,**F**) a portion of sepal showing glandular hairs; (**G**) corolla; (**H**) outer surface of petal showing branched, star-shaped hairs; (**I**) androecium; (**J**) epipetalous stamens; (**K**) basifixed anther; (**L**) openings at the top of anther lobes; (**M**) gynoecium; (**N**) upper part of ovary showing hairs; (**O**) style surface showing glandular hairs; (**P**) t. s. of ovary showing axile placentation.

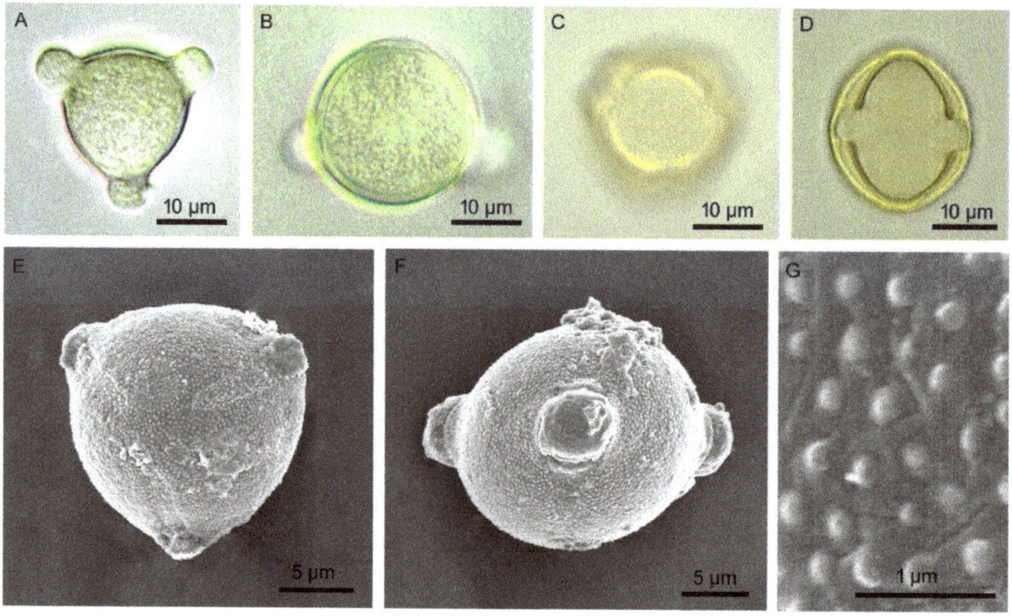

Figure 5. Microphotographs of pollen grains of *Solanum sisymbriifolium*. (**A–D**) Light microscopy (**A**,**B**: unprocessed pollens; **C**,**D**: processed pollens). (**E–G**) Scanning electron microscopy (**E**: polar view, **F**: equatorial view, **G**: exine ornamentation).

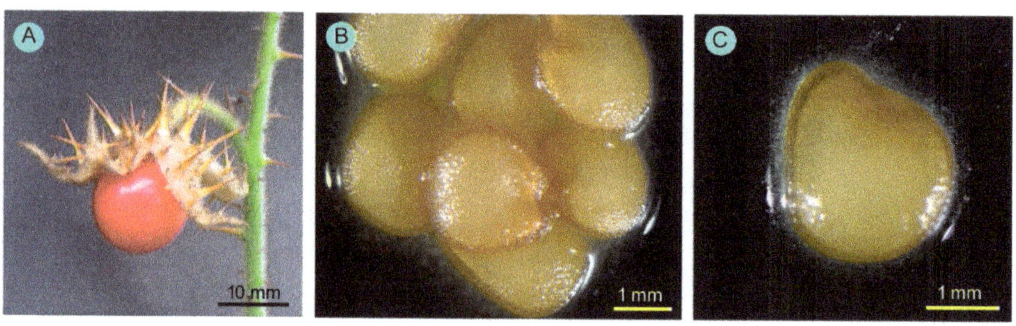

Figure 6. (**A**) Fruit and (**B**,**C**) seeds of *Solanum sisymbriifolium*.

Each flower produced an average of 429,017.64 ± 86,608.11 pollen grains (mean ± SD, $n = 60$), while the number of ovules per flower averaged 74.08 ± 14.37 (mean ± SD, $n = 240$). The number of pollen grains and ovules per flower varied significantly across seasons (pollen grains: $\chi^2 = 44.37$, df = 5, $p < 0.001$; ovules: $\chi^2 = 50.35$, df = 5, $p < 0.001$). Pollen and ovule production were higher in the summer and monsoon seasons and lower during the cold winter (Table 1). The ovule to pollen ratio was 1:5791.27.

Anther dehiscence began 2 to 3 h after the flower had fully opened. Anther dehiscence was through apical pores (formation of a small opening at the top of the anther lobe). At the time of flower opening, pollen showed a viability of 82.16 ± 7.31% and a germinability of 76.82 ± 6.54%. Pollen viability and germinability showed no seasonal variation. However, pollen viability and germinability decreased with time elapsed (Table S3), being highest during opening time and very low by the 3rd day morning (i.e., at the end of flower longevity). Stigma remained receptive for the entire lifespan of the flower, from the moment it opened until it withered. Therefore, the flowers showed the protogynous type of dichogamy. During the time of flower opening (1st day flowers in the morning), stigma receptivity was lower, gradually reached its peak on the 2nd day flowers, and then decreased gradually. The duration of stigma receptivity varied significantly across seasons ($\chi^2 = 83.74$, df = 5, $p < 0.001$), being longest in winter (76.90 ± 0.23 h) and shortest in summer (51.80 ± 1.82 h). Figure 7 illustrates viable pollen and receptive stigmas.

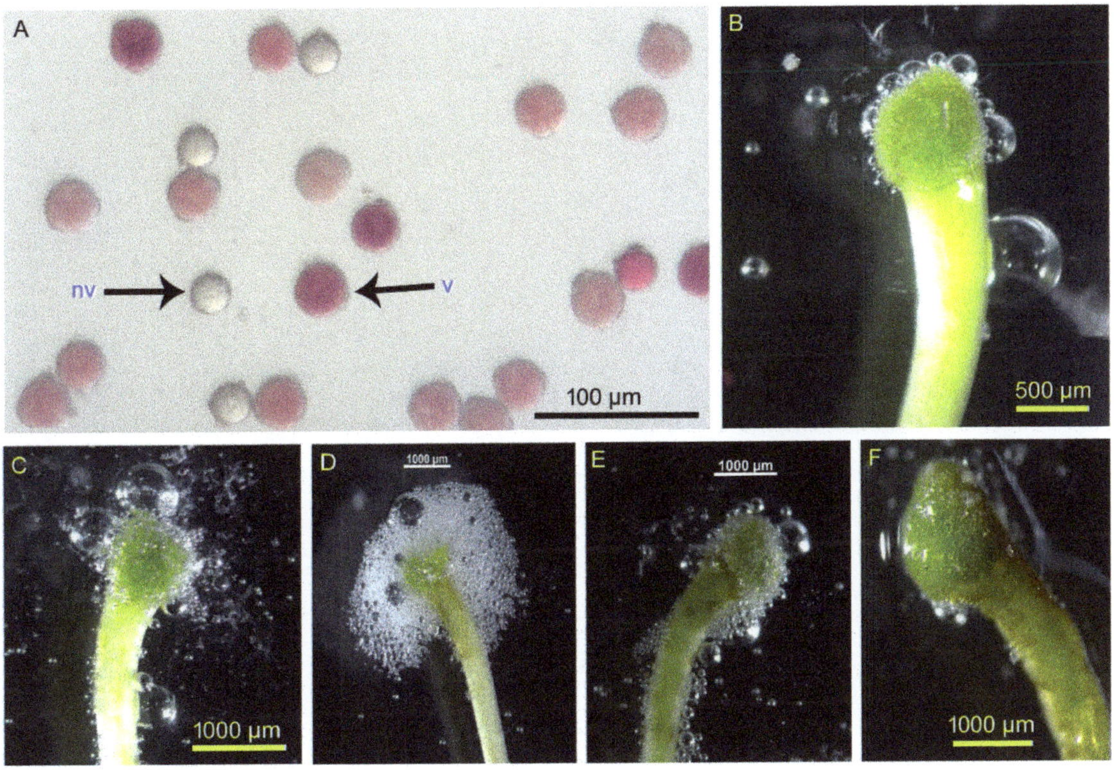

Figure 7. (**A**) Pollen viability test using the TTC staining method (v—viable, nv—non-viable), (**B–F**) stigma receptivity test using H_2O_2 (different ages of flowers: (**B**)—bud condition, (**C**)—1st day, (**D**)—2nd day, (**E**)—3rd day, (**F**)—4th day).

3.2. Mating System

All five pollination treatments led to both fruit and seed formation (Table S4). Fruit and seed sets differed from pollinator exclusion treatments with other treatments (fruit set: $\chi^2 = 27.01$, df = 4, $p < 0.001$; seed set: $\chi^2 = 145.54$, df = 4, $p < 0.001$), with the lowest in pollinator exclusion treatment (fruit set: 45 ± 10.80%; seed set: 20.06 ± 23.87 seeds per flower). Fruit and seed sets did not differ among open pollination, manual selfing, manual crossing, and supplementary pollination treatments. The index of self-incompatibility (ISI) was calculated as 0.02, indicating that the plant species is fully self-compatible. The species exhibited a strong reliance on pollinators, with an index of dependency on pollinators (IDP) of 0.72. The coefficient of pollination deficit was very low (D = 0.10), suggesting that the plant species experienced minimal to no pollination limitation under open conditions.

3.3. Floral Visitors

A total of 21 insect species were recorded as floral visitors of *Solanum sisymbriifolium* in West Bengal, India (Table 2, Figure 8). The majority belonged to the order Hymenoptera, comprising 15 species. Within this group, most were from the family Apidae (12 species), followed by Halictidae (3 species).

Table 2. Flower visitors of *Solanum sisymbriifolium* in West Bengal, India.

Flower Visitor	RA	VR	HT	Resource	Pollination Strategies							
					FV	Visitation Pattern	Buzzing Activity	Pollinating Mode	Pollen Adhering Parts	AR	SR	PS*i*
Coleoptera												
Epuraea luteola	16.21	0.11 ± 0.03	-	P	0.011	IL	No	-	H, VA, VT, L	-	-	-
Hycleus phalarantha	0.19	0.10 ± 0.0	-	FT	0.001	IL	No	-	-	-	-	-
Diptera												
Eristalinus megacephalus	0.51	0.34 ± 0.13	-	P	0.003	IL	No	-	H, L	-	-	-
Helophilus peregrinus	0.33	0.32 ± 0.14	-	P	0.002	IL	No	-	H, L	-	-	-
Lucilia sericata	1.17	0.28 ± 0.10	-	P	0.002	IL	No	-	H, L	-	-	-
Paragus serratus	1.03	1.40 ± 0.57	-	P	0.003	L	No	-	H, L, VA, VT	1	0.53	0.002
Hymenoptera												
Amegilla zonata	6.45	5.13 ± 1.20	0.74 ± 0.23	P	0.096	L	Yes	S, A	VA, VT, L	1	0.92	0.088
Apis cerana	1.31	3.43 ± 1.20	15.54 ± 22.17	P	0.014	L	No	S, A	VA, VT, L	1	0.44	0.006
Apis dorsata	1.45	3.73 ± 1.38	14.38 ± 20.26	P	0.016	L	No	S, A	VA, VT, L	1	0.45	0.007
Apis florea	0.56	3.27 ± 1.07	16.63 ± 21.75	P	0.005	L	No	S, A	VA, VT, L	1	0.42	0.002
Brounsapis mixta	3.69	2.43 ± 1.13	12.04 ± 13.38	P	0.029	L	No	S, A	VA, VT, L	1	0.40	0.012
Ceratina binghami	7.90	2.87 ± 1.21	20.52 ± 29.81	P	0.068	L	No	S, A	VA, VT, L	1	0.42	0.029
Ceratina hieroglyphica	2.43	3.51 ± 1.49	15.85 ± 24.64	P	0.026	L	No	S, A	VA, VT, L	1	0.41	0.011
Halictus (Seladonia) lucidipennis	3.36	3.33 ± 1.47	15.73 ± 23.80	P	0.024	L	No	S, A	VA, VT, L	1	0.41	0.010
Lasioglossum cavernifrons	14.63	3.17 ± 1.39	17.68 ± 28.44	P	0.143	L	Yes	S, A	VA, VT, L	1	0.48	0.069
Nomia (Curvinomia) strigata	13.60	5.97 ± 1.71	8.18 ± 10.23	P	0.224	L	Yes	S, A	VA, VT, L	1	0.52	0.116
Tetragonula pagdeni	12.34	0.74 ± 0.26	100.78 ± 82.15	P	0.075	L	No	S, A	VA, VT, L	1	0.37	0.028
Xylocopa aestuans	4.49	6.53 ± 1.53	1.38 ± 0.50	P	0.088	L	Yes	S, A	VA, VT, L	1	0.94	0.083
Xylocopa amethystina	2.10	7.27 ± 1.64	0.78 ± 0.24	P	0.045	L	Yes	S, A	VA, VT, L	1	0.91	0.041
Xylocopa fenestrata	4.30	6.97 ± 1.38	1.29 ± 0.48	P	0.089	L	Yes	S, A	VA, VT, L	1	0.96	0.085
Xylocopa latipes	1.96	6.10 ± 1.63	1.42 ± 0.53	P	0.036	L	Yes	S, A	VA, VT, L	1	0.92	0.033

RA—relative abundance (%); VR—flower visitation rate; HT—flower handling time (s); Resource: FT—floral tissue, P—pollen; FV—flower visit proportion; Visitation pattern: IL—illegitimate, L—legitimate; Pollinating mode: S—sternotribic, A—appendage-mediated; Pollen adhering body parts: H—head, L—legs, VA—ventral side of abdomen, VT—ventral side of thorax; AR—anther contacting rate; SR—stigma contacting rate; PS*i*—pollination service index.

The abundance, richness, and diversity of floral visitors fluctuated across seasons (Table 3). The greatest abundance, richness, and diversity of visitors were observed during summer (abundance: 2.40 ± 2.05 visitors/plant/5 min; richness, D = 0.52 ± 0.55; Shannon–Weaver diversity index, $H' = 0.38 \pm 0.38$), while the lowest values occurred in winter (abundance: 0.88 ± 1.13 visitors/plant/5 min; richness, D = 0.23 ± 0.50; diversity index, $H' = 0.13 \pm 0.28$). Visitor traits such as abundance, richness, and diversity showed significant variation across different times of the day (Table S5). The abundance, richness and diversity of flower visitors remained higher during 6.00–10.00 h for most seasons (except in winter, traits were higher at 10.00–12.00 h) and lower at 16.00–18.00 h.

Table 3. Abundance, richness, and diversity of floral visitors on *Solanum sisymbriifolium* in West Bengal.

Season	Abundance	Richness	Diversity
Summer	$2.40^{a} \pm 2.05$	$0.52^{a} \pm 0.55$	$0.38^{a} \pm 0.38$
Monsoon	$2.07^{ab} \pm 2.05$	$0.48^{ab} \pm 0.55$	$0.34^{ab} \pm 0.39$
Autumn	$1.60^{cd} \pm 1.71$	$0.39^{bc} \pm 0.56$	$0.26^{cd} \pm 0.36$
Late autumn	$1.33^{d} \pm 1.41$	$0.30^{cd} \pm 0.50$	$0.19^{de} \pm 0.31$
Winter	$0.88^{e} \pm 1.13$	$0.23^{e} \pm 0.50$	$0.13^{e} \pm 0.28$
Spring	$1.91^{bc} \pm 1.83$	$0.43^{ab} \pm 0.54$	$0.30^{bc} \pm 0.38$
Throughout year	1.70 ± 1.80	0.39 ± 0.54	0.27 ± 0.36
Statistical analysis	$\chi^2 = 87.45$, df = 5, $p < 0.001$	$\chi^2 = 51.72$, df = 5, $p < 0.001$	$\chi^2 = 62.99$, df = 5, $p < 0.001$

Abundance: number of visitors/individual/5 min; richness: Margalef's index D; diversity: index of Shannon-Weaver, H'; Data are presented as mean ± standard deviation. Different superscript letters within the same column denote statistically significant differences (Dunn's post hoc test, $p < 0.05$).

Figure 8. Flower visitors of *Solanum sisymbriifolium* in West Bengal. (**A**) *Amegilla zonata*, (**B**) *Brounsapis mixta*, (**C**) *Ceratina binghami*, (**D**) *Epuraea luteola*, (**E**) *Halictus* (*Seladonia*) *lucidipennis*, (**F**) *Hycleus phalarantha*, (**G**) *Lasioglossum cavernifrons*, (**H**) *Lucilia sericata*, (**I**) *Nomia* (*Curvinomia*) *strigata*, (**J**) *Paragus serratus*, (**K**) *Tetragonula pagdeni*, and (**L**) *Xylocopa fenestrata*. Scale bars = 10 mm.

The flower-visiting species with greater abundance were *Ceratina binghami* (relative abundance = 7.90%), *Epuraea luteola* (relative abundance = 16.21%), *Lasioglossum cavernifrons* (relative abundance = 14.63%), *Nomia (Curvinomia) strigata* (relative abundance = 13.60%), and *Tetragonula pagdeni* (relative abundance = 12.34%) (Table 2).

All the flower visitors collected pollen grains. Beetles also feed on floral tissues, especially the anther wall. The visitation rates for beetles and flies were very low. Among the hymenopteran members, the visitation rates were higher for *Amegilla zonata*, *Nomia (Curvinomia) strigata* and *Xylocopa* spp., while the visitation rate remained very low for stingless bees. Flower handling time was higher for beetles, flies, and stingless bees. The handling time for *Amegilla zonata* and *Xylocopa* spp. was very low.

The weed flowers received a higher amount of visits by *Nomia (Curvinomia) strigata* (FV = 0.224), *Lasioglossum cavernifrons* (FV = 0.143), *Amegilla zonata* (FV = 0.096), *Xylocopa fenestrata* (FV = 0.089), *Xylocopa aestuans* (FV = 0.088), *Tetragonula pagdeni* (FV = 0.075), and *Ceratina binghami* (FV = 0.068) (Table 2). The hymenopteran members (honeybees, solitary bees, and stingless bees) legitimately visited *Solanum sisymbriifolium* flowers (Table 2). Beetles and some flies visited flowers in an illegitimate manner. However, beetles and flies' bodies adhered to *Solanum sisymbriifolium* pollens (Figure 9). Pollen was primarily attached to the legs—particularly the scopae in solitary bees and the corbiculae in honeybees and stingless bees—as well as to the ventral surfaces of the thorax and abdomen. Some solitary bees (e.g., *Amegilla zonata*, *Nomia (Curvinomia) strigata* and *Xylocopa* spp.) showed buzz pollination. They primarily contacted the stigma with the ventral side of their abdomen. Sometimes, they touched the stigmatic surface via the lateral sides of the thorax. Honeybees and stingless bees did not exhibit buzz pollination; instead, they employed the sternotribic pollination mode via the ventral surface of their abdomen and thorax. Bees also conducted appendage-mediated pollination. The anther touching rates were higher for all bee species. Stigma touching rates were higher for blue banded bees (*Amegilla zonata*) and carpenter bees (*Xylocopa* spp.). Based on the pollination service index (PS*i*), vital pollinators were *Nomia (Curvinomia) strigata* (PS*i* = 0.116), *Amegilla zonata* (PS*i* = 0.088), *Xylocopa fenestrata* (PS*i* = 0.085), *Xylocopa aestuans* (PS*i* = 0.083), *Lasioglossum cavernifrons* (PS*i* = 0.069), *Xylocopa latipes* (PS*i* = 0.033), *Ceratina binghami* (PS*i* = 0.029), and *Tetragonula pagdeni* (PS*i* = 0.028).

3.4. Seasonal Plant Reproduction

Reproductive success, measured by fruit and seed set, varied significantly across seasons (fruit set: $\chi^2 = 11.19$, df = 5, $p < 0.05$; seed set: $\chi^2 = 46.22$, df = 5, $p < 0.001$). The fruit set remained higher during spring (82 $\pm$ 9.19%), summer (86 $\pm$ 9.66%) and monsoon (85 $\pm$ 8.50%) (Figure 10). Seed set was the highest during summer (63.86 $\pm$ 34.20 seeds/flower) and the lowest in winter (38.68 $\pm$ 28.59 seeds/flower). During summer, the plant produced a higher number of seeds (63.86 $\pm$ 34.20 seeds per flower), while seed set was lower in winter (38.68 $\pm$ 28.59 seeds per flower).

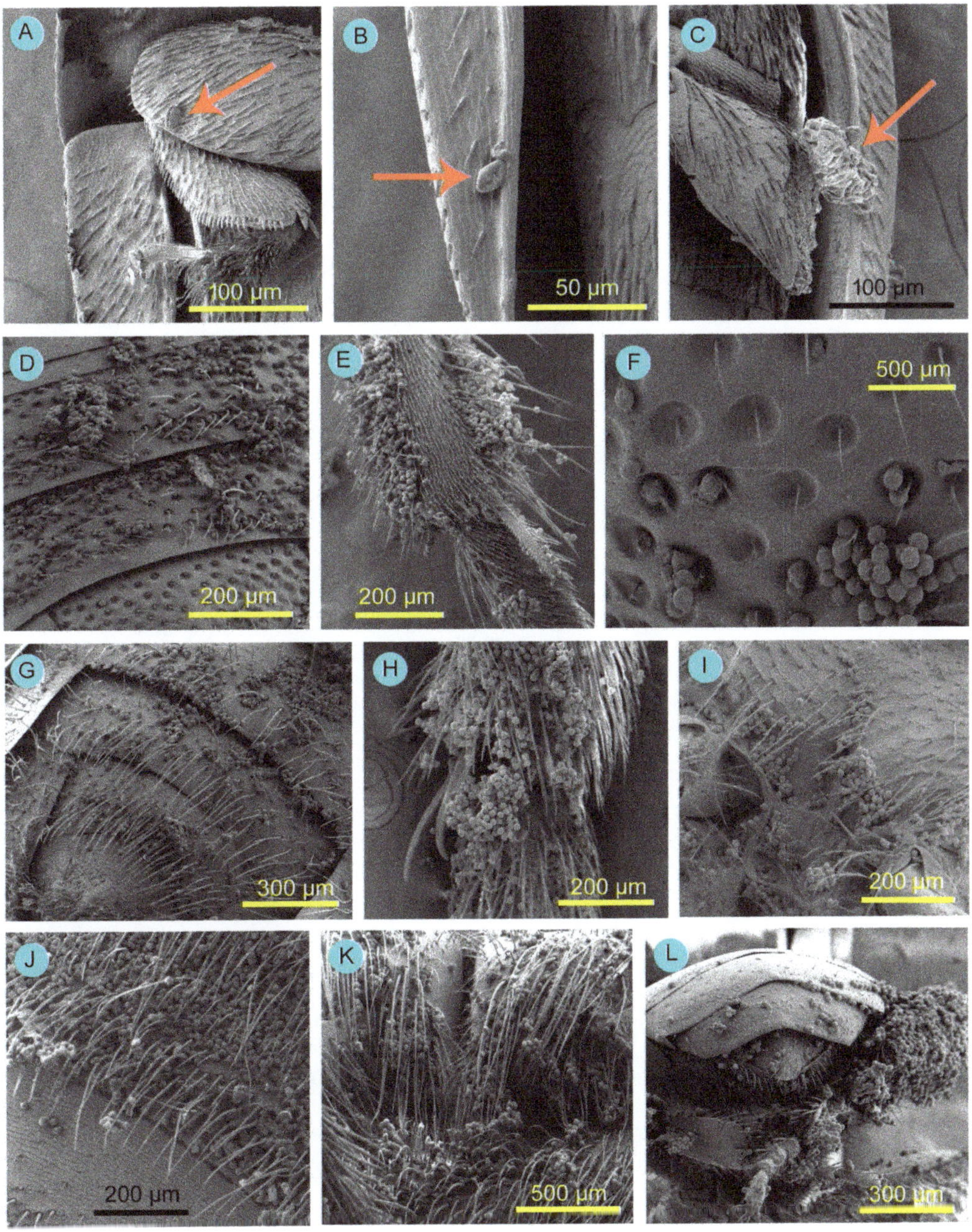

Figure 9. Showing attachment of pollen grains on insect body parts. (**A–C**) Illegitimate visitor, beetle *Epuraea luteola*. (**A**) Pollen attached to the segment of the leg, and (**B**,**C**) the margin of the wing. Arrows are showing attached pollen grains. (**D–L**) Legitimate visitors. (**D–F**) *Ceratina binghami* (ventral abdomen, leg, and ventral thorax, respectively), (**G–I**) *Lasioglossum cavernifrons* (ventral abdomen, leg, and ventral thorax, respectively), (**J**,**K**) *Nomia* (*Curvinomia*) *strigata* (ventral abdomen and ventral thorax), (**L**) *Tetragonula pagdeni* (pollen grains on legs and ventral abdomen and thorax).

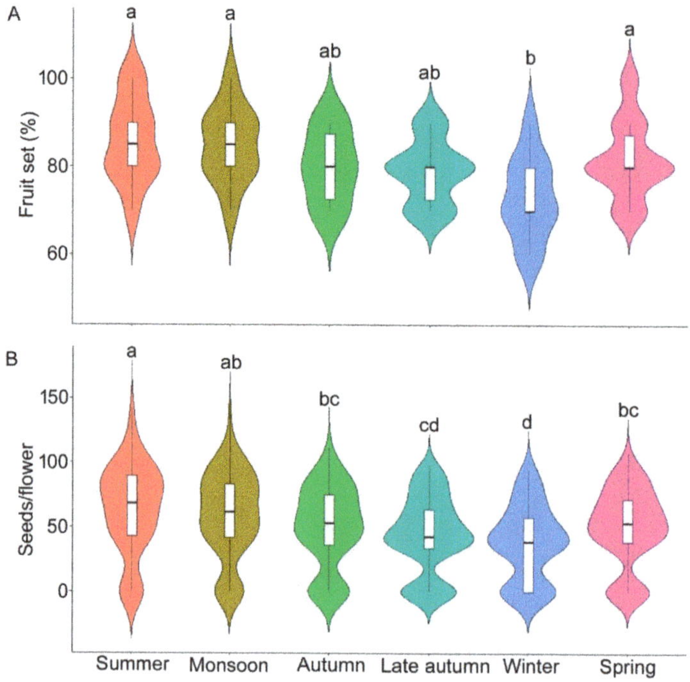

Figure 10. Season-wise fruit set (**A**) and seed set (**B**) of *Solanum sisymbriifolium* in West Bengal, India. Different superscript letters within a row (i.e., season-wise) indicate significant differences (Kruskal–Wallis H test and Dunn's post hoc test at 0.05% level).

4. Discussion

The classification of flowering patterns in angiosperms has been conducted in various ways by several researchers. For example, Gentry [10] mentioned five types of flowering patterns (cornucopia, big bang, multiple bang, steady state, and modified steady state), Bawa [12] divided into two groups (i.e., extended and massive), and Frankie et al. [11] classified flowering patterns into two groups (i.e., extended and seasonal). The plant species under study displayed steady-state flowering year-round in the Rarh region of West Bengal. A similar flowering pattern has been observed in many tropical plant species (e.g., *Turnera ulmifolia* [9]). Steady-state flowering enhances cross-pollination and may ensure reproductive success under adverse weather. Synchronous blooming among individuals allows widespread gene exchange within the population, boosting genetic diversity [45,46]. The majority of flower traits, including flowering intensity, flower longevity, ovule number, pollen production, and duration of stigma receptivity, varied seasonally in Rarh Bengal. This may be due to variations in seasonal atmospheric factors. The responsiveness of floral traits to environmental factors has been well documented in many plant species (e.g., *Hordeum vulgare* [47], *Theobroma cacao* [48], and *Turnera ulmifolia* [9]). Certain floral traits—such as flowering intensity, flower display size, and ovule and pollen production—were enhanced during the summer and monsoon season and declined in winter. In contrast, some others (e.g., flower longevity and duration of stigma receptivity) showed a reverse trend. In winter, an extended period of stigma receptivity may help offset the low flowering frequency, thereby improving reproductive success [49]. This trait likely aids biotic pollination under harsh conditions. The duration of female receptivity is also influenced by the occurrence of pollination [50,51], and long-lasting stigmas increase the chances of cross-pollination. The pollen viability and germinability did not vary seasonally but significantly decreased with the ageing of flowers. During senescence (especially 3rd day old flowers), viability and germinability were retained very little.

In terms of anther and stigma maturation, the plant species exhibited protogyny, a condition less common in angiosperms compared to protandry. It has also been reported for some other Solanaceae members, such as *Anthocercis gracilis* [52], *Jaltomata sinuosa* [53], and *Mandragora caulescens* [54]. This form of dichogamy is highly effective in promoting outcrossing in the species. Based on its mating system, fruit and seed development in bagged flowers suggested that the plant species is capable of spontaneous self-pollination. This strategy provides reproductive assurance in challenging environments where pollinator activity is limited. Additionally, similar seed set levels between self-pollination and cross-pollination treatments suggested that the species was self-compatible. Among different species of *Solanum*, researchers found both phenomena: self-compatibility [55,56] and self-incompatibility [52,57,58]. The reproductive assurance is considered a major selective force driving the evolution of self-fertilisation [59,60]. Self-fertilisation enables individuals to reproduce even when the success of cross-fertilisation is constrained by limited access to compatible mates or pollinators. Additionally, selfing may enhance the colonisation ability of a plant species [61]. In contrast, cross-fertilisation is thought to be superior to selfing, considering it reduces inbreeding, which is often associated with the production of offspring of lower genetic quality [62]. Seed set in the pollinator exclusion treatment was relatively lower than in the supplementary pollination treatments, indicating that the plant species has a moderate reliance on biotic pollinators. The pollinator dependence of fruit and seed sets has also been reported for various species of *Solanum* (e.g., *Solanum aethiopicum* [63], *Solanum anguivi* [63], and *Solanum melongena* [64]). In open field conditions, the plant species had a negligible or very low pollination deficit (coefficient of pollination deficit, D = 0.10). Pollination limitation in a plant species is influenced by floral traits, pollinator availability, and weather conditions during the flowering period [9,36,65].

Only a limited amount of work has been performed on the flower visitors of the weed species. Saha and Dutta [33] worked from Tripura, India, and Hill and Hulley [34] reported flower visitors for the weed from South Africa. We provided detailed information about flower visitors and their interactions on *Solanum sisymbriifolium*, for the first time from the Rarh region of West Bengal, India. The weed flowers were visited by beetles, flies, honeybees, solitary bees, and stingless bees, with beetles, solitary bees, and stingless bees being the most dominant. The most solitary bees (e.g., *Amegilla zonata*, *Nomia* (*Curvinomia*) *strigata*, *Xylocopa* spp.) showed buzzing activity on the weed species. The buzzing activity of visitors is well established on the Solanaceae members [66,67]. The effective pollinators (based on pollination service index, PS*i*) of the weed species were solitary bees (e.g., *Amegilla zonata*, *Lasioglossum cavernifrons*, *Nomia* (*Curvinomia*) *strigata*, *Xylocopa* spp.) and stingless bees (*Tetragonula pagdeni*). Previous studies (e.g., Saha and Dutta [33]; Hill and Hulley [34]) have also shown the importance of blue-banded bees (*Amegilla* spp.) and carpenter bees (*Xylocopa* spp.) as pollinators of the weed. However, the composition of insect species varied across studies. The composition of flower visitors depends not only on the plant species but also varies across time and space [18]. Regarding temporal variation, visitor traits, including abundance, richness, and diversity, were higher in summer and lower in winter. The greater flowering intensity and flower display size during summer positively influenced visitor traits. The influence of floral traits on pollinator characteristics has been demonstrated in several plant species by numerous researchers (e.g., McCall and Primack [68]; Ebeling et al. [69]; Layek et al. [9]). The pollinator traits are also influenced by the surrounding vegetation [36,70], as it deters the availability of food and habitat for native pollinators.

The fruit and seed set of the weed species showed seasonal variation, with higher levels observed in spring, summer, and the monsoon season, and lower levels during winter. Reproductive success (i.e., fruit and seed set) of a plant species depends on several factors, including nutrient availability [71], the physiological condition of the individuals [72], and pollination services [36,73]. The pollination service is also linked to flowering intensity, pollinator traits and weather conditions [9,74]. During the spring to monsoon season, the higher flowering intensity and floral display size of the weed species remained more attractive to pollinators, resulting in higher pollinator abundance, increased pollination services, and ultimately, higher reproductive success. Additionally, flower traits, especially a higher number of ovule production during these seasons, were also associated with higher seed set. A positive correlation between ovule production and seed set has been well documented by several researchers (e.g., Strelin and Aizen [75]; Layek et al. [9]).

5. Conclusions

The wild tomato (*Solanum sisymbriifolium*) bloomed throughout the year with a steady-state pattern. Flower traits varied seasonally, with some (e.g., flowering intensity, flower display size, ovule and pollen production) peaking during spring, summer, and the monsoon, while values remained lowest during winter. Some other flower traits, such as flower longevity and the duration of stigma receptivity, exhibited a reverse seasonal variation. The plant species exhibited full self-compatibility (ISI = 0.02), showed a high reliance on pollinators (IDP = 0.72), and faced only slight pollination limitation under open conditions (pollination deficit coefficient, D = 0.10). Considering flower visitors, beetles, flies, honeybees, solitary bees, and stingless bees visited *Solanum sisymbriifolium* flowers. The most abundant visitors were *Epuraea luteola*, *Lasioglossum cavernifrons*, *Nomia* (*Curvinomia*) *strigata*, and *Tetragonula pagdeni*. Based on the pollination service index (PS*i*), effective pollinators were *Amegilla zonata*, *Ceratina binghami*, *Lasioglossum cavernifrons*, *Nomia* (*Curvinomia*) *strigata*, *Tetragonula pagdeni*, *Xylocopa aestuans*, *Xylocopa amethystina*, *Xylocopa fenestrata*, and *Xylocopa latipes*. The reproductive success of the plant, as measured by fruit and seed set, varied seasonally. The fruit and seed set remained higher during hot seasons (i.e., spring–monsoon) and comparatively lower in cold winter. The present work provided information about flower biology, visitors and plant–pollinator interactions on the weed species, and these data will help in the management of the weed and also in the conservation of the associated wild bee fauna.

Supplementary Materials: The following supporting information can be downloaded at: https://www.mdpi.com/article/10.3390/biology14070865/s1. Table S1. The methodologies for the microscopic study of flower parts and pollen grains; Table S2. Methodologies for determining relative abundance, richness and diversity of flower visitors; Table S3. Flower-age-wise pollen viability and germinability of *Solanum sisymbriifolium* in West Bengal, India; Table S4. Fruit and seed sets in different pollination treatments on *Solanum sisymbriifolium* in West Bengal, India; Table S5. Daytime-wise, visitor traits (abundance, richness, and diversity) of *Solanum sisymbriifolium* in West Bengal, India.

Author Contributions: Conceptualization: U.L. and P.K.; methodology: U.L., P.M., A.D. and A.K.; formal analysis: U.L. and A.K.; investigation: U.L., P.M. and A.D.; data curation: U.L. and A.K.; writing—original draft: U.L.; writing—review and editing: P.M., A.D., A.K. and P.K.; visualisation: U.L.; supervision: P.K. All authors have read and agreed to the published version of the manuscript.

Funding: There is no funding for this research.

Institutional Review Board Statement: Not applicable.

Informed Consent Statement: Not applicable.

Data Availability Statement: Data were available within the article and Supplementary Files.

Acknowledgments: The authors express their gratitude to the authorities of various institutions, including Rampurhat College, Vidyasagar University, and Visva-Bharati (West Bengal, India), for providing laboratory facilities. The authors thank the authorities of several institutions (e.g., Rampurhat College, Vidyasagar University, and Visva-Bharati, West Bengal, India) for providing laboratory facilities, including SEM studies. They express their gratitude to the entomologists at ZSI, Kolkata, West Bengal, for their assistance in identifying the flower visitors. They are also thankful to Shashidhar Virakamath (Emeritus Scientist, University of Agricultural Sciences, Bengaluru, India) for helping in identifying the stingless bee species.

Conflicts of Interest: The authors declare that they have no conflicts of interest to disclose.

References

1. Fornoff, F.; Klein, A.M.; Hartig, F.; Benadi, G.; Venjakob, C.; Schaefer, H.M.; Ebeling, A. Functional flower traits and their diversity drive pollinator visitation. *Oikos* **2017**, *126*, 1020–1030. [CrossRef]
2. Gillespie, M.A.; Baude, M.; Biesmeijer, J.; Boatman, N.; Budge, G.E.; Crowe, A.; Davies, N.; Evans, R.; Memmott, J.; Morton, R.D.; et al. Flowering plant communities mediate the effects of habitat composition and configuration on wild pollinator communities. *Funct. Ecol.* **2024**, *38*, 2576–2594. [CrossRef]
3. Rering, C.C.; Vannette, R.L.; Schaeffer, R.N.; Beck, J.J. Microbial co-occurrence in floral nectar affects metabolites and attractiveness to a generalist pollinator. *J. Chem. Ecol.* **2020**, *46*, 659–667. [CrossRef] [PubMed]
4. Sponsler, D.; Iverson, A.; Steffan-Dewenter, I. Pollinator competition and the structure of floral resources. *Ecography* **2023**, *2023*, e06651. [CrossRef]
5. Zeevaart, J.A. Physiology of Flowering: Flowering is hormonally controlled, but the nature of the hormone remains to be elucidated. *Science* **1962**, *137*, 723–731. [CrossRef] [PubMed]
6. Chailakhyan, M.K.; Timiriazev, K.A. Genetic and hormonal regulation of growth, flowering, and sex expression in plants. *Am. J. Bot.* **1979**, *66*, 717–736. [CrossRef]
7. Shillo, R.; Halevy, A.H. The effect of various environmental factors on flowering of gladiolus. III. Temperature and moisture. *Sci. Hortic.* **1976**, *4*, 147–155. [CrossRef]
8. Lee, Z.; Kim, S.; Choi, S.J.; Joung, E.; Kwon, M.; Park, H.J.; Shim, J.S. Regulation of flowering time by environmental factors in plants. *Plants* **2023**, *12*, 3680. [CrossRef] [PubMed]
9. Layek, U.; Das, N.; Samanta, A.; Karmakar, P. Impact of seasonal atmospheric factors and photoperiod on floral biology, plant–pollinator interactions, and plant reproduction on *Turnera ulmifolia* L. (Passifloraceae). *Biology* **2025**, *14*, 100. [CrossRef] [PubMed]
10. Gentry, A.H. Flowering phenology and diversity in tropical Bignoniaceae. *Biotropica* **1974**, *6*, 64–68. [CrossRef]
11. Frankie, G.W.; Baker, H.G.; Opler, P.A. Comparative phenological studies of trees in tropical wet and dry forests in the lowlands of Costa Rica. *J. Ecol.* **1974**, *62*, 881–919. [CrossRef]
12. Bawa, K.S. Patterns of flowering in tropical plants. In *Handbook of Experimental Pollination Biology*; Jones, C.E., Little, R.J., Eds.; Van Nostrand Reinhold: New York, NY, USA, 1983; pp. 394–410.
13. Koptur, S. Outcrossing and pollinator limitation of fruit set: Breeding systems of neotropical *Inga* trees (Fabaceae: Mimosoideae). *Evolution* **1984**, *38*, 1130–1143. [CrossRef] [PubMed]
14. Bawa, K.S.; Kang, H.; Grayum, M.H. Relationships among time, frequency, and duration of flowering in tropical rain forest trees. *Am. J. Bot.* **2003**, *90*, 877–887. [CrossRef] [PubMed]
15. Delph, L.F.; Ashman, T.L. Trait selection in flowering plants: How does sexual selection contribute? *Integr. Comp. Biol.* **2006**, *46*, 465–472. [CrossRef] [PubMed]
16. Goodwillie, C.; Sargent, R.D.; Eckert, C.G.; Elle, E.; Geber, M.A.; Johnston, M.O.; Kalisz, S.; Moeller, D.A.; Ree, R.H.; Vallejo-Marin, M.; et al. Correlated evolution of mating system and floral display traits in flowering plants and its implications for the distribution of mating system variation. *New Phytol.* **2010**, *185*, 311–321. [CrossRef] [PubMed]
17. Price, M.V.; Waser, N.M.; Irwin, R.E.; Campbell, D.R.; Brody, A.K. Temporal and spatial variation in pollination of a montane herb: A seven-year study. *Ecology* **2005**, *86*, 2106–2116. [CrossRef]
18. Layek, U.; Das, A.D.; Das, U.; Karmakar, P. Spatial and temporal variations in richness, diversity and abundance of floral visitors of curry plants (*Bergera koenigii* L.): Insights on plant-pollinator interactions. *Insects* **2024**, *15*, 83. [CrossRef] [PubMed]
19. Spears Jr, E.E. A direct measure of pollinator effectiveness. *Oecologia* **1983**, *57*, 196–199. [CrossRef] [PubMed]
20. Layek, U.; Kundu, A.; Bisui, S.; Karmakar, P. Impact of managed stingless bee and western honey bee colonies on native pollinators and yield of watermelon: A comparative study. *Ann. Agric. Sci.* **2021**, *66*, 38–45. [CrossRef]
21. Potts, S.G.; Biesmeijer, J.C.; Kremen, C.; Neumann, P.; Schweiger, O.; Kunin, W.E. Global pollinator declines: Trends, impacts and drivers. *Trends Ecol. Evol.* **2010**, *25*, 345–353. [CrossRef] [PubMed]

22. LeBuhn, G.; Luna, J.V. Pollinator decline: What do we know about the drivers of solitary bee declines? *Curr. Opin. Insect Sci.* **2021**, *46*, 106–111. [CrossRef] [PubMed]
23. Bakker, J.C. The effects of air humidity on flowering, fruit set, seed set and fruit growth of glasshouse sweet pepper (*Capsicum annuum* L.). *Sci. Hortic.* **1989**, *40*, 1–8. [CrossRef]
24. Tutin, C.E.; Fernandez, M. Relationships between minimum temperature and fruit production in some tropical forest trees in Gabon. *J. Trop. Ecol.* **1993**, *9*, 241–248. [CrossRef]
25. Abrahamczyk, S.; Kluge, J.; Gareca, Y.; Reichle, S.; Kessler, M. The influence of climatic seasonality on the diversity of different tropical pollinator groups. *PLoS ONE* **2011**, *6*, e27115. [CrossRef] [PubMed]
26. Lawson, D.A.; Rands, S.A. The effects of rainfall on plant–pollinator interactions. *Arthropod-Plant Interact.* **2019**, *13*, 561–569. [CrossRef]
27. Chakravarty, A.K.; Mukhopadhyay, S.; Saha, S.; Pakrashi, S.C. A neolignan and sterols in fruits of *Solanum sisymbrifolium*. *Phytochemistry* **1996**, *41*, 935–939. [CrossRef]
28. Gupta, V.K.; Simlai, A.; Tiwari, M.; Bhattacharya, K.; Roy, A. Phytochemical contents, antimicrobial and antioxidative activities of *Solanum sisymbriifolium*. *J. Appl. Pharm. Sci.* **2014**, *4*, 75–80. [CrossRef]
29. Gebrewbet, G.H.; Hndeya, A.G. Phytochemical screening and antibacterial activity studies on the crude leaf extract of *Solanum sisymbriifolium*: Traditional Ethiopian medicinal plant. *Adv. Gut Microbiome Res.* **2023**, *2023*, 5525606. [CrossRef]
30. Hellión-lbarrola, M.C.; Montalbetti, Y.; Heinichen, O.; Alvarenga, N.; Figueredo, A.; Ferro, E.A. Isolation of hypotensive compounds from *Solanum sisymbriifolium* Lam. *J. Ethnopharmacol.* **2000**, *70*, 301–307. [CrossRef] [PubMed]
31. Ferro, E.A.; Alvarenga, N.L.; Ibarrola, D.A.; Hellión-Ibarrola, M.D.C.; Ravelo, A.G. A new steroidal saponin from *Solanum sisymbriifolium* roots. *Fitoterapia* **2005**, *76*, 577–579. [CrossRef] [PubMed]
32. Perpétuo, L.S.; da Cunha, M.J.; Batista, M.T.; Conceição, I.L. *Solanum linnaeanum* and *Solanum sisymbriifolium* as a sustainable strategy for the management of Meloidogyne chitwoodi. *Sci. Rep.* **2021**, *11*, 3484. [CrossRef] [PubMed]
33. Saha, M.; Dutta, B.K. Reproductive biology of *Solanum sisymbriifolium* Lamk. (Solanaceae) in Tripura, North-east India. *Int. J. Plant Reprod. Biol.* **2017**, *9*, 59–62.
34. Hill, M.P.; Hulley, P. Aspects of the phenology and ecology of the South American weed, *Solanum sisymbriifolium*, in the Eastern Cape province of South Africa. *Afr. Plant Prot.* **2000**, *6*, 53–59.
35. Hopkins, H.C. Floral biology and pollination ecology of the neotropical species of *Parkia*. *J. Ecol.* **1984**, *72*, 1–23. [CrossRef]
36. Layek, U.; Bhagira, N.K.; Das, A.; Kundu, A.; Karmakar, P. Dependency of crops on pollinators and pollination deficits: An approach to measure them in considering the influence of various reproductive traits. *Agriculture* **2023**, *13*, 1563. [CrossRef]
37. Erdtman, G. The acetolysis method. A revised description. *Sven. Bot. Tidskr.* **1960**, *54*, 561–564.
38. Norton, D.J. Testing of plum pollen viability with tetrazolium salts. *J. Am. Soc. Hortic. Sci.* **1966**, *89*, 132–134.
39. Parfitt, E.D.; Ganeshan, S. Comparison of procedures for estimating viability of *Prunus* pollen. *HortScience* **1989**, *24*, 354–356. [CrossRef]
40. Dafni, A.; Maués, M.M. A rapid and simple procedure to determine stigma receptivity. *Sex. Plant Reprod.* **1998**, *11*, 177–180. [CrossRef]
41. Raduski, A.R.; Haney, E.B.; Igic, B. The expression of self-incompatibility in angiosperms is bimodal. *Evolution* **2012**, *66*, 1275–1283. [CrossRef] [PubMed]
42. Layek, U.; Das, U.; Karmakar, P. The pollination efficiency of a pollinator depends on its foraging strategy, flowering phenology, and the flower characteristics of a plant species. *J. Asia-Pac. Entomol.* **2022**, *25*, 101882. [CrossRef]
43. Margalef, R. Temporal succession and spatial heterogeneity in phytoplankton. In *Perspectives in Marine Biology*; Buzzati-Traverso, A.A., Ed.; University of California Press: Berke ley, CA, USA, 1958; pp. 323–347.
44. Shannon, C.E.; Weaver, W. *The Mathematical Theory of Communication*; University of Illinois Press: Champaign, IL, USA, 1949; p. 117.
45. Augspurger, C.K. Reproductive synchrony of a tropical shrub (*Hybanthus prunifolius*): Influence on pollinator attraction and movement. *Ecology* **1981**, *62*, 774–788. [CrossRef]
46. Martínez-Sánchez, J.J.; Segura, F.; Aguado, M.; Franco, J.A.; Vicente, M.J. Life history and demographic features of *Astragalus nitidiflorus*, a critically endangered species. *Flora* **2011**, *206*, 423–432. [CrossRef]
47. Karsai, I.; Kőszegi, B.; Kovács, G.; Szűcs, P.; Mészáros, K.; Bedő, Z.; Veisz, O. Effects of temperature and light intensity on flowering of barley (*Hordeum vulgare* L.). *Biol. Futura* **2008**, *59*, 205–215. [CrossRef]
48. Omolaja, S.S.; Aikpokpodion, P.; Oyedeji, S.; Vwioko, D.E. Rainfall and temperature effects on flowering and pollen productions in cocoa. *Afr. Crop Sci. J.* **2009**, *17*, 41–48. [CrossRef]
49. Aronne, G.; Buonanno, M.; De Micco, V. Reproducing under a warming climate: Long winter flowering and extended flower longevity in the only Mediterranean and maritime *Primula*. *Plant Biol.* **2015**, *17*, 535–544. [CrossRef] [PubMed]

50. Torres-Díaz, C.; Gomez-Gonzalez, S.; Stotz, G.C.; Torres-Morales, P.; Paredes, B.; Pérez-Millaqueo, M.; Gianoli, E. Extremely long-lived stigmas allow extended cross-pollination opportunities in a high Andean plant. *PLoS ONE* **2011**, *6*, e19497. [CrossRef] [PubMed]
51. Vesprini, J.; Pacini, E. Temperature-dependent floral longevity in two *Helleborus* species. *Plant Syst. Evol.* **2005**, *252*, 63–70. [CrossRef]
52. Stace, H.M. Protogyny, self-incompatibility and pollination in *Anthocercis gracilis* (Solanaceae). *Aust. J. Bot.* **1995**, *43*, 451–459. [CrossRef]
53. Mione, T.; Plourd, K.C.; Leiva González, S.; Yacher, L.; Philbrick, C.T.; Blume, D. Pollination decreases longevity of the protogynous flowers of *Jaltomata sinuosa* (Solanaceae). *J. Biol. Nat.* **2017**, *7*, 123–129.
54. Wan JinPeng, W.J.; Zhu XingFu, Z.X.; Li QingJun, L.Q. Breeding system of protogynous *Mandragora caulescens* (Solanaceae). *Plant Divers. Resour.* **2011**, *33*, 565–570.
55. Easlon, H.M.; Richards, J.H. Drought response in self-compatible species of tomato (Solanaceae). *Am. J. Bot.* **2009**, *96*, 605–611. [CrossRef] [PubMed]
56. Behling, W.L.; Douches, D.S. The effect of self-compatibility factors on interspecific compatibility in *Solanum* Section Petota. *Plants* **2023**, *12*, 1709. [CrossRef] [PubMed]
57. Vallejo-Marín, M.; O'Brien, H.E. Correlated evolution of self-incompatibility and clonal reproduction in *Solanum* (Solanaceae). *New Phytol.* **2007**, *173*, 415–421. [CrossRef] [PubMed]
58. Miller, J.S.; Kostyun, J.L. Functional gametophytic self-incompatibility in a peripheral population of *Solanum peruvianum* (Solanaceae). *Heredity* **2011**, *107*, 30–39. [CrossRef] [PubMed]
59. Schoen, D.J.; Morgan, M.T.; Bataillon, T. How does self-pollination evolve? Inferences from floral ecology and molecular genetic variation. *Philos. Trans. R. Soc. Lond. B Biol. Sci.* **1996**, *351*, 1281–1290.
60. Pannell, J.; Barrett, S. Baker's Law revisited: Reproductive assurance in a metapopulation. *Evolution* **1998**, *52*, 657–668. [CrossRef] [PubMed]
61. Baker, H. Self-compatibility and establishment after 'long distance' dispersal. *Evolution* **1955**, *9*, 347–349. [PubMed]
62. Charlesworth, D. Evolution of plant breeding systems. *Curr. Biol.* **2006**, *16*, 726–735. [CrossRef] [PubMed]
63. Oyelana, O.A.; Ogunwenmo, K.O. Floral biology and the effects of plant-pollinator interaction on pollination intensity, fruit and seed set in *Solanum*. *Afr. J. Biotechnol.* **2012**, *11*, 14967–14981.
64. Jayasinghe, U.S.R.; Silva, T.S.E.; Karunaratne, W.I.P. Buzzing wild bee visits enhance seed set in eggplant, *Solanum melongena*. *Psyche* **2017**, *2017*, 4624062. [CrossRef]
65. Webber, S.M.; Garratt, M.P.; Lukac, M.; Bailey, A.P.; Huxley, T.; Potts, S.G. Quantifying crop pollinator-dependence and pollination deficits: The effects of experimental scale on yield and quality assessments. *Agric. Ecosyst. Environ.* **2020**, *304*, 107106. [CrossRef]
66. Tayal, M.; Kariyat, R. Examining the role of buzzing time and acoustics on pollen extraction of *Solanum elaeagnifolium*. *Plants* **2021**, *10*, 2592. [CrossRef] [PubMed]
67. Vallejo-Marín, M.; Pereira Nunes, C.E.; Russell, A.L. Anther cones increase pollen release in buzz-pollinated *Solanum* flowers. *Evolution* **2022**, *76*, 931–945. [CrossRef] [PubMed]
68. McCall, C.; Primack, R.B. Influence of flower characteristics, weather, time of day, and season on insect visitation rates in three plant communities. *Am. J. Bot.* **1992**, *79*, 434–442. [CrossRef]
69. Ebeling, A.; Klein, A.M.; Schumacher, J.; Weisser, W.W.; Tscharntke, T. How does plant richness affect pollinator richness and temporal stability of flower visits? *Oikos* **2008**, *117*, 1808–1815. [CrossRef]
70. Vrdoljak, S.M.; Samways, M.J.; Simaika, J.P. Pollinator conservation at the local scale: Flower density, diversity and community structure increase flower visiting insect activity to mixed floral stands. *J. Insect Conserv.* **2016**, *20*, 711–721. [CrossRef]
71. Moulton, C.A.; Gough, L. Effects of soil nutrient availability on the role of sexual reproduction in an Alaskan tundra plant community. *Arct. Antarct. Alp. Res.* **2011**, *43*, 612–620. [CrossRef]
72. Sato, S.; Peet, M.M.; Thomas, J.F. Physiological factors limit fruit set of tomato (*Lycopersicon esculentum* Mill.) under chronic, mild heat stress. *Plant Cell Environ.* **2000**, *23*, 719–726. [CrossRef]
73. Falque, M.; Vincent, A.; Vaissiere, B.E.; Eskes, A.B. Effect of pollination intensity on fruit and seed set in cacao (*Theobroma cacao* L.). *Sex. Plant Reprod.* **1995**, *8*, 354–360. [CrossRef]
74. Albrecht, M.; Schmid, B.; Hautier, Y.; Müller, C.B. Diverse pollinator communities enhance plant reproductive success. *Proc. R. Soc. B Biol. Sci.* **2012**, *279*, 4845–4852. [CrossRef] [PubMed]
75. Strelin, M.M.; Aizen, M.A. The interplay between ovule number, pollination and resources as determinants of seed set in a modular plant. *Peer J.* **2018**, *6*, e5384. [CrossRef] [PubMed]

biology MDPI

Article

Mathematical Modeling of Population Dynamics of Pollinators: A Survey

Fernando Huancas [1,*], Anibal Coronel [2,*], Esperanza Lozada [2] and Jorge Torres [2]

1 Departamento de Matemática, Facultad de Ciencias Naturales, Matemáticas y del Medio Ambiente, Universidad Tecnológica Metropolitana, Ñuñoa, Santiago 7750000, Chile
2 Departamento de Ciencias Básicas, Facultad de Ciencias, Universidad del Bío-Bío, Campus Fernando May, Chillán 3780000, Chile; elozada@ubiobio.cl (E.L.); jotorres@ubiobio.cl (J.T.)
* Correspondence: fhuancas@utem.cl (F.H.); acoronel@ubiobio.cl (A.C.)

Simple Summary: Pollinators are crucial for diverse biological processes, but it is recognized that there has been a decline in their populations in recent years. Hence, the study of the population dynamics of pollinators is a relevant topic for research. In this study, to contribute to the state of the art of mathematical modeling of population dynamics, we searched the relevant literature in two databases. This review explores the different contributions, develops a summary and classification, and states some future work to understand the behavior of pollinator interactions.

Abstract: In this paper, we develop a systematic review of the existing literature on the mathematical modeling of several aspects of pollinators. We selected the MathSciNet and Wos databases and performed a search for the words "pollinator" and "mathematical model". This search yielded a total of 236 records. After a detailed screening process, we retained 107 publications deemed most relevant to the topic of mathematical modeling in pollinator systems. We conducted a bibliometric analysis and categorized the studies based on the mathematical approaches used as the central tool in the mathematical modeling and analysis. The mathematical theories used to obtain the mathematical models were ordinary differential equations, partial differential equations, graph theory, difference equations, delay differential equations, stochastic equations, numerical methods, and other types of theories, like fractional order differential equations. Meanwhile, the topics were positive bounded solutions, equilibrium and stability analysis, bifurcation analysis, optimal control, and numerical analysis. We summarized the research findings and identified some challenges that could inform the direction of future research, highlighting areas that will aid in the development of future research.

Keywords: pollinators; plant–pollinator interaction; pesticides

1. Introduction

In the last few decades, the study of pollinators has attracted the attention of several researchers, as it is known that pollinators play a crucial role as ecosystem regulators in nature [1–5]. It is known that there are several types of pollinators, including birds, bats, butterflies, moths, flies, beetles, wasps, small mammals, and, most importantly, insects like bees. These animals are responsible for the bulk of pollination, which significantly affects our daily lives. Some important facts about pollinators are that three out of four crops depend on pollinators; in the extreme case of total disappearance of pollinators, this would lead to a decrease in world food production; and the causes of pollinator decline include disease, climate change, and pesticides. The problem associated with pollinators is complex and should be analyzed from multiple scientific perspectives, particularly biology, chemistry, and mathematics.

Pollination is a crucial event in the reproductive cycle of flowering plants. In this context, several characteristics associated with the evolutionary process of species help maintain and optimize the functioning of various ecosystems [6]. Two widely studied phenomena are flowers that produce nectar and those that do not. First, we consider flowers that produce nectar to be a food source. Some plants provide nectar to pollinators as a reward for their assistance in pollination. In this context, a notable aspect is the fact that plants conceal their nectar, which prevents pollinators from detecting its presence without first entering the flower. Second, we know that there are plants that do not produce nectar. Nectar production requires a considerable amount of energy; some flowers can employ deceptive strategies by not producing nectar. Despite lacking nectar, these flowers can still be pollinated by pollinators. An example of this type of plant is found in certain orchid species, which are pollinated through a phenomenon known as Batesian mimicry. Nectarless flowers are likely the result of evolutionary optimization. Additionally, other pollination-related phenomena include the fact that flowers can also attract pollinators by producing large floral displays, even if they provide no reward. There are pollinators skilled at extracting nectar without pollinating the flower, known as nectar robbers. There are also indirect pollinators, such as ants, which seek other plant nutrients or prey on insects living on the plant, rather than directly seeking nectar or pollen.

In this paper, we aim to elucidate the existing studies on pollinators from a mathematical perspective. Several phenomena related to pollinators can be analyzed using mathematical modeling, such as the dynamics of pollinator populations, plant–pollinator interactions, the effects of climate change on pollinator decline, the impact of pesticides on pollinator populations, and the spread of infectious diseases among pollinators. A recent review developed by Chen et al. introduced the framework of different mathematical models related to the dynamics of honeybee populations [7]. However, to the extent of our knowledge, there is no comprehensive review of the state of the art in mathematical modeling of pollinators and related topics. Therefore, we conducted a systematic literature review using bibliometric methods and following the methodology detailed in [8] (see also [9]).

We surveyed the MathSciNet and WoS databases and examined the topics of each work. We obtained a set of 107 works, comprising 105 journal articles, 1 PhD thesis, and 1 book chapter related to the mathematical modeling of pollinators. The retained list of articles ranged from 1978 to 2025. We analyzed the papers and established a classification based on the mathematical theory involved in the mathematical modeling formulation. The classification introduced considers four groups: ordinary, differential equation models, partial differential equation models, network-based models, and other methodologies. In the case of other methodologies, we found discrete mathematical-based models, stochastic models, and others. We outlined some key contributions of the papers and compiled a list of topics that highlight potential challenges and perspectives for further research on the topic.

This paper is outlined as follows. In Section 2, we describe the methodology, including the list of selected relevant works that were identified and analyzed, as well as the bibliometric analysis. In Section 3, we report the results of the main findings arising from analysis of the existing literature on the mathematical modeling of pollinators. In Section 4, we discuss some biological issues of the retained list. In Section 5, we collect some aspects which are not included in the previous sections but are relevant for the completeness of the work. Finally, in Section 6, we present the conclusions of the paper and also outline some possible future research directions.

2. Methodological Framework

The methodology supporting the present work combines two approaches to developing a literature review: a systematic review and a bibliometric analysis. To be more precise, we adopted the methodology presented in [8], which consists of the five steps given in [9]: (1) framing questions for a review, (2) identifying relevant work, (3) assessing the quality of studies, (4) summarizing the evidence, and (5) interpreting the findings. The results for steps (1) and (2), step (3), and steps (4) and (5) are presented below in Section 2.1, Section 2.2, and Section 3, respectively. A synthesized visualization is presented in Figure 1.

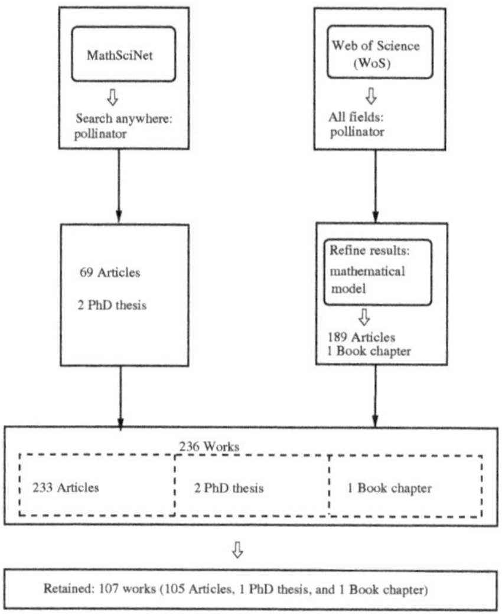

Figure 1. Schematic summary of the process used for identifying the relevant work (see Section 2).

2.1. Framing Questions for a Review and Identifying Relevant Work

We considered the following two questions:

Question 1: What are the studies developed for mathematical modeling of the pollinator population's dynamics?
Question 2: What types of modeling approaches were used in those studies?

Meanwhile, related to the step of identification of the relevant work, we selected two databases, MathSciNet and the Web of Science (WoS), with the following details:

- *MathSciNet* (https://mathscinet.ams.org/mathscinet/, accessed on 14 April 2025): We searched for the word "pollinator" using the option "search term: anywhere" and found that the response reported a total of 71 items: 69 journal articles and 2 PhD theses.
- *WoS* (https://www.webofscience.com/, accessed on 14 April 2025). We used the option "all fields" for the platform's search engine to search for the word "pollinator", obtaining 26,938 items. Then, by using the keyword "mathematical model" in the option "refine results", we found 199 items: 198 journal articles and 1 book chapter.

When combining the two lists, we found that there were 34 duplicated items. Then, we obtained a list of 236 works: 233 journal articles, 2 PhD theses, and 1 book chapter. Here, we note that the search in both databases was not limited to the keyword "pollinator" being specified in the works.

We performed an examination of the 236 works and retained those which were related, namely with mathematical modeling as the topic of the paper, obtaining a list of 107 works [10–116]. In [19], there is a book chapter in conference proceedings, while [31] is a PhD thesis, and the other works are journal articles. We note two additional facts: the present review is registered in OSF (https://doi.org/10.17605/OSF.IO/3DWSR , accessed on 5 September 2025)), and for the inclusion criteria, we considered a work to be about mathematical modeling of pollinators when there was a proposal to research the population dynamics of pollinators.

2.2. Assessing the Quality of the Studies

A graphical distribution of the list, by year and by decade from 1978 to 2025, is shown in Figure 2. Here, we observe that the oldest reference was from 1978, and the most recent one was from 2025. Also, we noticed a clear increase in articles over the decades, even though there were slight decreases in some years. Additionally, we also noted the geographic location declared by each of the authors in the corresponding affiliation for each article; the results are graphically presented in Figure 3. The affiliations of the authors were counted in the 236 works. The details for the retained list are given in Appendix A and more specifically in Table A1. The regions with the highest number of records were the United States of America (USA), the United Kingdom (UK), and China, with 205, 102, and 100 records, corresponding to 27%, 10%, and 10% of the works, respectively. These rankings were followed by regions with less than 4% representation as detailed below:

- Brazil (45) and Australia (35) with 4% each;
- India (29), Japan (29), Canada (28), France (28), Mexico (28), the Czech Republic (26), and Spain (25) with 3% each;
- Brazil (22), Hungary (21), and Serbia (16) with 2% each;
- Italy (14), South Africa (13), Denmark (11), Israel (10), Sweden (10), Taiwan (10), the Netherlands (9), New Zealand (9), Chile (8), Russia (8), Norway (7), Argentina (6), and Belgium (5) with 1% each;
- Bulgaria (4), Finland (4), Poland (4), the Republic of Korea (4), Switzerland (4), the Philippines (3), Slovakia (3), Greece (2), Kenya (2), Estonia (2), Ecuador (1), Indonesia (1), Ireland (1), Pakistan (1), Portugal (1), Saudi Arabia (1), Slovenia (1), and Thailand (1) with 0% each.

Here, the number in parentheses is the number of records for the region. A graphical interpretation is given in Figure 3, with The oldest article being [5].

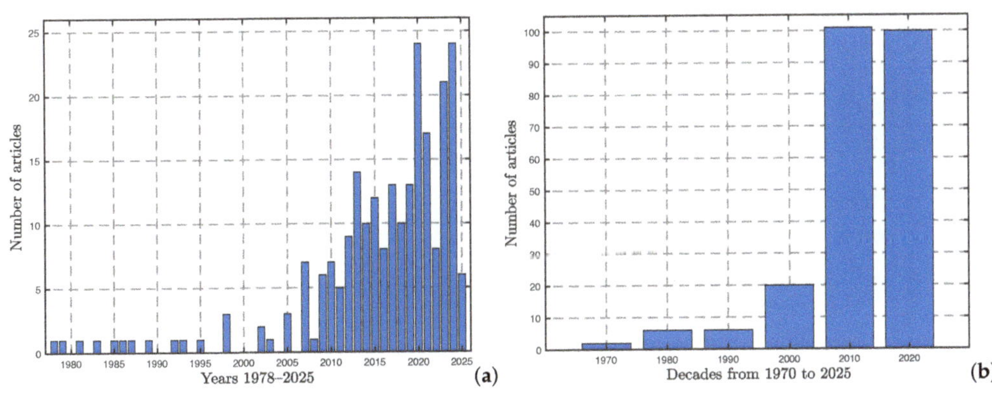

Figure 2. Number of works in MathSciNet and the WoS related to the keywords "pollinators" and "mathematical models". (**a**) Number of articles by year from 1978 to 2025. (**b**) Number of articles by decade.

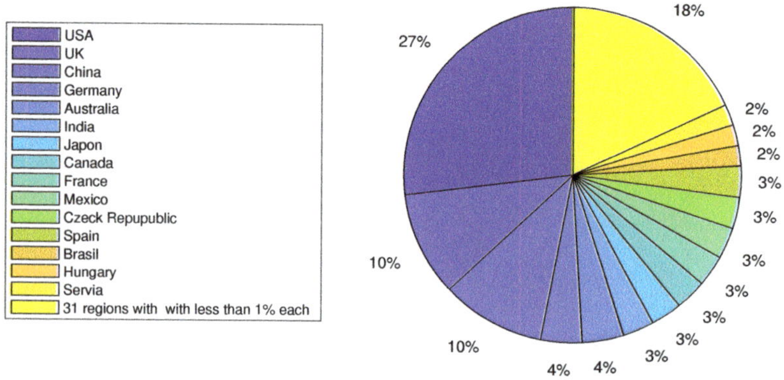

Figure 3. Percentages of the number of authors according to the geographic locations declared by the authors. We rounded off all percentages to their integers.

The indicators for journals and authors in the list are presented below. The retained articles in the list were published in 59 journals. Table 1 shows the nine journals which were in the first four positions according the published articles. We found that there were 13 journals with 2 publications and 37 journals with 1 publication. In Table 2, we show the top 10 journals according to the H index of the SCImago Journal & Country Rank (https://www.scimagojr.com/ accessed on 4 May 2025), and the SJR 2023 indicators, quartiles, and subject areas of those journals were obtained from SJR (https://jcr.clarivate.com/ accessed on 4 May 2025). Refer to Appendix B for more details. Moreover, in Table 3, we present the top four prolific authors. Meanwhile, Tables 4 and 5 provide an extensive and structured synthesis of the retained literature. In Table 4, we detail the main findings of [10–116], and in Table 5, we organized the results by the seven major modeling domains identified (see Section 3.5). Each domain is characterized by its underlying biological assumptions, mathematical formulation, parametrization strategies, validation or calibration procedures, and key ecological insights. The table also highlights the potential policy implications derived from each modeling approach, thereby bridging theoretical contributions with applied relevance. This typological classification facilitates comparative analysis across studies and supports interdisciplinary integration of mathematical ecology, conservation planning, and empirical calibration. Representative references are included to exemplify each category and guide further exploration of methodological trends and thematic priorities.

Table 1. The nine journals in the first four positions, considering the number of articles published.

Rank	Journal	Record Count	% of 106
1°	*Journal of Theoretical Biology*	12	11.32
2°	*Bulletin of Mathematical Biology*	8	7.55
3°	*Ecology Letters*	4	3.7
	International Journal of Biomathematics	4	3.77
	Theoretical Ecology	4	3.77
4°	*Journal of Mathematical Biology*	3	2.83
	Oikos	3	2.83
	Plos One	3	2.83
	Theoretical Population Biology	3	2.83

Table 2. The top 10 journals based on the H index, SJR index, and quartile. The information was obtained from Scimago https://www.scimagojr.com/ (accessed on 20 April 2025). Refer to Table A2 for the complete list of journals.

Rank	Journal	H Index	SJR 2024	Quartile SJR	Subject Area and Category
1°	*Proceedings of the National Academy of Sciences of the United States of America*	896	10.8	Q1	Multidisciplinary Sciences
2°	*Plos One*	467	3.3	Q1	Multidisciplinary Sciences
3°	*Scientific Reports*	347	4.3	Q1	Multidisciplinary Sciences
4°	*Ecology*	343	5.5	Q1	Ecology
5°	*Ecology Letters*	330	9.8	Q1	Ecology
6°	*Plos Pathogens*	260	5.5	Q1	Microbiology, Parasitology
7°	*American Naturalist*	236	3.3	Q2	Ecology
8°	*Evolution*	227	3.0	Q2	Ecology/Evolutionary Biology
9°	*Journal of Ecology*	219	6.1	Q1	Ecology, Plant Sciences
10°	*Journal of Applied Ecology*	216	6.2	Q1	Biodiversity Conservation, Ecology

Table 3. The top four authors with the highest number of articles in the selected list.

Author	Institution	Number of Articles
Yuanshi Wang	Sun Yat-sen University, P. R. China	16
Hong Wu	Sun Yat-sen University, P. R. China	7
Faustino Sánchez-Garduño	Universidad Nacional Autónoma de México, Mexico	6
Fernanda S. Valdovinos	University of California Davis, USA	4

Table 4. Summary of mathematical model topics and phenomena studied in the retained list of papers. Here, RR stands for retained reference.

RR	Model Type	Assumptions	Parameterization	Validation	Key Findings
[10]	Ordinary differential equations of Lotka–Volterra type	Two-species, mutualistic interaction, closed system, constant environment	Interaction coefficients derived from the ecological literature; intrinsic growth rates assumed to be constant	Analytical exploration of equilibrium points and stability, qualitative phase plane analysis	Demonstrated conditions for mutualistic coexistence and thresholds for collapse due to partner dependency
[11]	Ordinary differential Lotka–Volterra equations with ecological feedback	Mutualistic interaction of pollinators and plants, continuous population dynamics, homogeneous environment	Growth rates and interaction, coefficients estimated from empirical observations and the ecological literature	Local stability analysis, numerical simulations exploring population trajectories under varying scenarios	Identification of stable equilibria, limit cycles, and bifurcation points explaining persistence or collapse of mutualistic systems
[12]	Discrete patch-based mutualism model	Species occupy spatially distinct patches; mutualistic benefit depends on local density; extinction and colonization are patch-specific	Parameters included colonization rate, extinction probability, and mutualistic benefit per patch; values treated generically for theoretical exploration	Analytical derivation of equilibrium conditions; stability assessed via local perturbation analysis	Mutualism enhances patch occupancy and persistence; spatial structure stabilizes interactions; coexistence possible without obligate dependence
[13]	Patch-based metapopulation model	Species occupy discrete habitat patches; extinction risk decreases with population size; colonization depends on local density	Extinction and colonization rates modeled as functions of population size; parameters derived from ecological theory	Analytical derivation of equilibrium distributions; stability analysis of patch ensemble	Demonstrated that spatial structure stabilizes mutualistic systems and intra-patch dynamics critically influence metapopulation persistence
[14]	Compartmental disease model with pollination-mediated transmission	Vector-borne fungal disease spreads via pollinators; host recruitment and pollinator visitation drive transmission	Transmission potential and recruitment rates estimated from empirical data; no explicit density dependence	Analytical threshold conditions for disease invasion; bifurcation analysis of host–pathogen dynamics	Identified conditions for disease-induced host extinction; showed that high disease incidence suppresses pollination and drives collapse
[15]	Three-species ordinary differential equations model (herbivore, plant, and pollinator)	Non-obligate mutualism; herbivory reduces pollinator visitation; population dynamics are continuous and deterministic	Functional responses and interaction coefficients derived from the ecological literature; visitation rates modeled explicitly	Stability analysis of equilibria; numerical simulations of population trajectories	Showed that herbivory can indirectly promote pollinator persistence; identified conditions for coexistence and oscillatory dynamics
[16]	Adaptive dynamics with piecewise smooth bifurcation structure	Evolution of mutualistic traits constrained by physiological costs; trait space bounded; bifurcations occur at viability borders	Trait-dependent fitness functions and bifurcation parameters derived from evolutionary stability conditions	Analytical and numerical bifurcation analysis; detection of border collision bifurcations	Revealed abrupt evolutionary transitions in mutualism; identified critical thresholds for trait viability and coexistence
[17]	Cellular automata with pair approximation	Pollination and reproduction modeled as separate processes; spatial clumping affects contact rates; Allee effects are context-dependent	Local and global interaction rules encoded in probabilistic automata; pair approximation used to derive ODE caricatures	Comparison of pair approximation predictions with full simulation outcomes	Demonstrated that spatial structure can eliminate Allee effects; local pollination and colonization enhance reproductive success in small populations
[18]	Individual-based spatial model with evolutionary dynamics	Obligate mutualism between plant and pollinator; parasitoid antagonist and dispersal evolves under selection; homogeneous landscape	Dispersal kernels, visitation rates, and mortality probabilities, parameters derived from ecological data and theoretical distributions	Simulation-based pattern formation; comparison with non-spatial ordinary differential equation model; sensitivity to obligacy constraints	Identified evolutionarily stable dispersal distances; showed that obligacy promotes spatial patterning and coexistence and weak obligacy leads to collapse of spatial structure

Table 4. *Cont.*

RR	Model Type	Assumptions	Parameterization	Validation	Key Findings
[19]	Reaction, diffusion, advection partial differential equation system	Three-species system: plant, pollinator, herbivore; nonlinear interactions; spatial movement and local reactions; continuous time and space	Diffusion and advection coefficients; interaction terms inspired by physical analogies; parameters selected for dynamical richness	Analytical treatment of reduced subsystems; numerical simulations of full spatiotemporal model	Demonstrated emergence of spatial heterogeneity and traveling waves; showed that mutualism and antagonism interact to produce complex spatial patterns
[20]	Three-species ordinary differential equation model with asymmetric mutualism	One plant species with two pollinators differing in efficiency and genetic impact; mutualism and competition coexist; inbreeding affects seed viability	Growth rates, nectar consumption rates, and pollination success calibrated from biological traits; inbreeding modeled as a dynamic feedback	Stability analysis and bifurcation exploration; numerical simulations under varying asymmetry and inbreeding levels	Found that pollinator asymmetry enhances system stability; high inbreeding paradoxically increases population persistence; oscillatory regimes linked to trait divergence
[21]	Stochastic metacommunity network model	Pollination network includes native and alien plants; mutualistic interactions structured by network topology; species abundances evolve over time	Degree distribution, nestedness, and modularity used to define network structure; demographic parameters estimated from field data	Simulations of network dynamics under species removal; comparison of full vs. native-only networks	Removal of alien plants destabilizes network structure and reduces species persistence; network topology strongly influences long-term dynamics and resilience
[22]	Multi-scale ordinary differential equation model with individual-to-population extrapolation	Pollinator foraging occurs at multiple temporal scales; mutualism affects reproduction and survival; population-level dynamics derived from individual-level interactions	Parameters derived from empirical foraging behavior and floral handling times; interaction terms approximated via Beddington–DeAngelis functional response	Analytical derivation of equilibrium conditions; numerical simulations of long-term dynamics	Demonstrated bi-stability and threshold effects in plant-pollinator systems; highlighted the role of individual-level behavior in shaping population-level outcomes
[23]	Conceptual coevolutionary framework with tolerance–virulence trade-offs	Host fitness is reduced by parasite virulence; tolerance evolves independent of resistance; coevolution affects both host and parasite traits	Virulence modeled as a function of parasite density and per-parasite damage; host fitness expressed as a linear function of tolerance and infection burden	Theoretical synthesis of existing models; conceptual validation through comparative analysis of empirical cases	Proposed that tolerance can drive parasite counter-adaptation; emphasized the need for integrated models of host–parasite coevolution, including mutualistic analogs
[24]	Game theoretic model of aggression in multi-partner mutualism	Plants interact simultaneously with ants and pollinators; ant aggression affects pollinator survival; fitness interests are misaligned across partners	Aggression modeled as a continuous trait; payoff matrices constructed for plant, ant, and pollinator strategies; parameters derived from ecological scenarios	Analytical derivation of evolutionary stable strategies; threshold analysis of aggression levels	Identified conditions under which pollinators are excluded due to ant aggression; facultative mutualisms more vulnerable than obligate ones; proposed empirical tests for aggression thresholds
[25]	Network-based dynamical model with topological asymmetry	Mutualistic networks are asymmetric; specialists interact with generalists, habitat destruction removes nodes and links, and extinction cascades depend on network structure	Network topology defined by degree distribution and nestedness; demographic parameters estimated from empirical pollination networks	Simulations of network collapse under progressive habitat loss; entropy metrics used to assess differential susceptibility	Found that disassortative networks buffer specialist species against extinction; asymmetry promotes resilience under habitat fragmentation; validated predictions with real-world network data
[26]	Reaction, diffusion, and advection system	Three-species system (plant, pollinator, herbivore); nonlinear interactions; spatial movement via diffusion and advection; Holling type II functional response	Diffusion and advection coefficients selected for dynamical richness; interaction terms derived from ecological analogies	Analytical proof of existence, positivity, and boundedness; numerical simulations under Dirichlet and Neumann boundary conditions	Showed emergence of spatial patterns and traveling waves; mutualism and antagonism jointly shape spatial heterogeneity and coexistence

Table 4. *Cont.*

RR	Model Type	Assumptions	Parameterization	Validation	Key Findings
[27]	Hybrid dispersal model with intensive and extensive search modes	Foraging population divided into two behavioral states—intensive (local) and extensive (long range)—with movement modeled via diffusion and advection, respectively	Search mode parameters calibrated from empirical movement data; dispersal kernels constructed from behavioral observations	Comparison with traditional diffusion models; simulation-based fit to spatial distribution data	Demonstrated superior predictive power of hybrid model; emphasized behavioral heterogeneity as key to dispersal dynamics and ecological forecasting
[28]	Three-species ordinary differential equation model with unidirectional consumer–resource interaction	Plant provides nectar to both pollinator and nectar robber; pollinator offers mutualistic service, but robber is exploitative; interaction asymmetry drives coexistence	Functional responses tailored to mutualist and exploiter; parameters derived from empirical studies and theoretical ecology	Stability and persistence analysis; numerical simulations of invasion scenarios and equilibrium dynamics	Identified conditions for stable coexistence despite exploitation; showed that pollinator's dual role (beneficial and costly) mediates system resilience
[29]	Eco-evolutionary model of mutualism under climate change	Plant and pollinator phenologies evolve in response to climate shifts; mutualism persistence depends on synchrony and community composition	Evolutionary rates and phenological traits modeled explicitly; alternative partner availability incorporated as dynamic variables	Simulation of evolutionary trajectories under varying climate scenarios; sensitivity analysis of Allee thresholds	Found that mutualism robustness depends on partner diversity and temporal overlap; identified thresholds beyond which climate-induced asynchrony leads to collapse
[30]	Three-species ordinary differential equation model with mutualism and exploitation	Plants interact with pollinators (mutualism) and nectar robbers (parasitism); functional responses differ across interactions and population-level dynamics	Beddington–DeAngelis and Holling type II responses; parameters derived from the ecological literature and theoretical constraints	Analytical conditions for persistence and extinction; bifurcation analysis of coexistence regimes	Identified mechanisms for stable coexistence in presence of robbers; showed that mutualism can persist despite exploitation under specific parameter thresholds
[31]	Cross-diffusion partial differential equation model with empirical calibration	Honey bees and solitary bees forage in almond orchards; movement influenced by environmental favorableness and interspecific interactions	Shigesada–Kawasaki–Teramoto framework; parameters calibrated from field data and canopy structure; spectral Galerkin method used for numerical approximation	Numerical simulations of spatial redistribution; empirical validation via bee visitation data and pollen movement patterns	Demonstrated that cross-diffusion leads to increased inter-tree movement and enhanced cross-pollination; model supports use of diverse pollinator assemblages to improve yield
[32]	Two-species SKT model with habitat choice and productivity feedback	Pollinators choose habitats based on floral density and competition; movement modeled via cross-diffusion; productivity linked to pollen transfer efficiency	SKT model applied to almond trees; favorableness gradients derived from floral distribution; interspecific effects modeled explicitly	Spectral Galerkin simulations; sensitivity analysis of pollinator redistribution and productivity outcomes	Found that interspecific competition drives honey bees into less favorable zones, increasing cross-pollination; spatial heterogeneity enhances productivity in mixed-variety orchards
[33]	Three-species ordinary differential equation model with cheater invasion	Plants classified as nectar secretors or non-secretors; pollinators interact mutualistically with secretors and parasitically with cheaters; fitness depends on nectar availability	Beddington–DeAngelis functional responses for both interactions; efficiency thresholds defined for cheater invasion	Global stability analysis; threshold conditions for persistence and extinction; numerical simulations of invasion dynamics	Showed that nectarless flowers can invade and persist under specific efficiency conditions; identified scenarios where cheaters drive mutualists to extinction, leading to system collapse
[34]	Ordinary differential equation model with mutualism and parasitism	Plants and pollinators form a mutualistic pair; nectar robbers exploit plants without providing pollination; pollinators and robbers share a limiting resource without direct interference	Functional responses include Beddington–DeAngelis and Holling type II; parameters varied to explore invasion and persistence scenarios	Analytical derivation of equilibria and stability conditions; numerical simulations of coexistence, extinction, and invasion dynamics	Mutualism can persist despite robber invasion; coexistence possible under intermediate robber efficiency and favorable initial conditions; pollinators not necessarily driven to extinction by robbers

Table 4. *Cont.*

RR	Model Type	Assumptions	Parameterization	Validation	Key Findings
[35]	Consumer–resource ordinary differential equation model with adaptive foraging	Pollinators adjust foraging efforts based on floral reward availability; network structure influences interaction strength; mutualism is dynamic and plastic	Empirical pollination networks used to calibrate interaction matrices; adaptive foraging implemented via optimization routines	Simulations of network dynamics under species loss; comparison of static vs. adaptive foraging scenarios	Adaptive foraging enhances biodiversity and network robustness and reduces secondary extinctions by promoting niche partitioning among pollinators and plants
[36]	Compartmental model of colony impairment under sublethal stress	Sublethal pesticide exposure impairs individual bees without causing mortality; colony function depends on cumulative impairment; positive density dependence drives collapse	Parameters derived from empirical exposure experiments; impairment modeled as transition between healthy and dysfunctional states	Model fitted to experimental data from bumblebee colonies; comparison with non-impairment models	Sublethal stress leads to colony failure via feedback loops; impairment thresholds generate Allee effects and bistable dynamics; model explains enigmatic patterns of collapse
[37]	Three-species consumer–resource model with indirect interactions	Plant provides nectar to mutualist (pollinator) and parasite (robber); indirect interactions emerge via shared resources; feedback loops modulate coexistence	Functional responses calibrated from empirical nectar consumption rates; interaction strengths derived from ecological theory	Analytical derivation of equilibrium and limit cycles; simulations of indirect interaction strength and persistence	Mutualism stabilizes food web module despite parasitism; (+,–) indirect interactions promote coexistence; density-dependent feedback enhances resilience
[38]	Compartmental population model of honey bee colony dynamics	Colony composed of brood, hive bees, and foragers; food availability regulates development and foraging onset; mortality and recruitment are stage-specific	Parameters derived from empirical data on bee life stages and foraging behavior; food dynamics modeled via differential equations	Simulations of colony trajectories under varying food and mortality rates; equilibrium and threshold analysis	Identified critical thresholds for forager mortality beyond which colonies collapse; food availability buffers colony resilience; model supports predictive management of hive health
[39]	Stage-structured ordinary differential equation model with facultative obligate mutualism	Plant is facultative; pollinator is obligate with life stage structure; external stressors affect larval development and adult survival	Life stage transitions modeled explicitly; demographic rates derived from insect biology and the ecological literature	Analytical stability analysis; bifurcation exploration under varying demographic dominance	Identified hysteresis and collapse thresholds; pollination service vulnerable to shifts in pollinator structure; recovery requires large demographic compensation
[40]	Three-dimensional autonomous ordinary differential equation system with cooperative and competitive interactions	Two plant species compete, one pollinator interacts cooperatively with both, and biodiversity emerges from interaction topology	Growth rates, competition coefficients, and mutualistic terms defined via mean-field approximations; handling time neglected	Analytical and numerical exploration of attractors and limit cycles; comparison with reduced competitive system	Demonstrated that cooperative species enhance biodiversity even when driven to extinction; proposed structural vulnerability of key mutualists
[41]	Hybrid dynamical model with seasonal and intra-seasonal phenology	Plant and pollinator phenologies respond differently to climate change; demographic outcomes depend on synchrony and lifespan	Zonneveld-type non-autonomous ordinary differential equations for within-season dynamics and discrete-time equations for seasonal transitions	Zonneveld-type non-autonomous ordinary differential equations for within-season dynamics and discrete-time equations for seasonal transitions	Found that short-lived species are highly sensitive to mismatching; hybrid models capture demographic consequences of climate-driven phenological shifts
[42]	Reaction, diffusion, and advection system with nonlinear functional responses	Three interacting populations: plant, pollinator, and herbivore; spatial movement via diffusion and advection; non-monotonic response functions	Parameters selected for dynamical richness; functional responses derived from ecological analogies and prior models	Analytical reduction to autonomous ordinary differential equation system; numerical simulations of full spatiotemporal model	Herbivore stabilizes mutualistic dynamics; coexistence enhanced by spatial heterogeneity; limit cycles and attractors depend on interaction strength

Table 4. *Cont.*

RR	Model Type	Assumptions	Parameterization	Validation	Key Findings
[43]	Lotka–Volterra cooperative system with delay	One hyper-connected mutualistic species interacts with multiple peripheral species; delays represent interaction lags; no inter-peripheral interactions	Birth rates and interaction coefficients assumed to be positive; delays modeled via distributed kernels	Lyapunov functionals used to prove global stability; analytical derivation of coexistence conditions	Demonstrated global asymptotic stability of coexistence equilibrium; extended classical Lotka—Volterra framework to nested mutualistic networks
[44]	Three-species ordinary differential equation model with dual mutualism and interference	Plant interacts mutually with both pollinator and ant; ant interferes with pollinator access; functional responses differ across interactions	Beddington–DeAngelis for mutualisms and Holling type II for interference; parameters derived from the ecological literature	Stability analysis and threshold exploration; numerical simulations of coexistence and extinction regimes	Identified threshold in ant aggressiveness; coexistence possible under weak interference; strong interference leads to pollinator extinction and mutualism collapse
[45]	Extended Beddington–DeAngelis model with asymmetric mutualism	Pollinators and ants interact with plants; ants interfere with pollinators but depend on mutualism for survival, with feedback loops included	Functional responses extended to capture interference; parameters derived from empirical studies and theoretical constraints	Global dynamics analyzed via persistence theory; bifurcation analysis of extinction thresholds	Showed that mutualism can persist under moderate interference; strong ant dependence on pollination stabilizes coexistence; extinction cascades occur under high interference
[46]	Three-species dynamical system with extended functional responses	Plant–pollinator and plant—ant interactions are mutualistic; ant interference modeled explicitly; system includes indirect effects	Extended Beddington—DeAngelis responses; interaction strengths calibrated from ecological theory and prior models	Analytical derivation of boundary equilibria; numerical simulations of persistence and extinction dynamics	Defined threshold for ant interference; weak interference promotes mutualism synergy; strong interference destabilizes entire system, including ant population
[47]	Ordinary differential equation model with pollinator learning and expertise differentiation	Pollinators are divided into novice and expert classes; learning improves pollination efficiency; plant growth benefits from expert visitation	Logistic growth for plants; mutualistic benefit modeled via saturating functional response; learning encoded via efficacy parameter σ	Analytical derivation of equilibrium and stability; numerical simulations of coexistence dynamics	Showed stable coexistence of plants, novices, and experts; learning enhances mutualistic benefit and system resilience; expertise evolution supports pollination service
[48]	Three-species ordinary differential equation model with nectar robbing	Pollinators provide mutualistic service; nectar robbers consume floral resources without pollination; robbers compete indirectly with pollinators	Beddington–DeAngelis functional responses for both mutualism and exploitation; parameters derived from ecological theory	Stability and persistence analysis; bifurcation exploration of coexistence and extinction regimes	Identified conditions for coexistence of robbers, pollinators, and plants; robbers can destabilize mutualism or persist without collapsing the system
[49]	Network-based dynamical model with structural variation	Biodiversity depends on network topology (e.g., nestedness, modularity); interaction strength varies across species; life history traits influence persistence	Network structure varied systematically; interaction matrices parameterized across plausible ecological regimes	Simulations across parameter space; biodiversity measured as fraction of surviving species under different topologies	Found multiple regimes linking nestedness to biodiversity; network structure alone can promote or hinder persistence depending on trait configuration
[50]	Reaction and diffusion model with positive steady-state analysis	Plant and pollinator populations diffuse spatially; mutualism modeled via Beddington–DeAngelis response; steady states represent ecological coexistence	Diffusion coefficients and growth rates selected for analytical tractability; functional response includes saturation and interference terms	Leray–Schauder degree theory used to prove existence of positive steady states; stability explored via monotone dynamical systems	Demonstrated existence and uniqueness of positive steady states; coexistence depends on growth–mortality balance and spatial diffusion rates
[51]	Adaptive dynamics model with trait-mediated trade-offs	Plants evolve interaction traits under nutrient enrichment; traits affect both mutualistic and antagonistic interactions; ecological trade-off is convex	Trait values modulate interaction strength; nutrient enrichment modeled as external forcing; parameters derived from ecological theory	Analytical exploration of evolutionary equilibria; simulations of community assembly under enrichment gradient	Evolution modifies community structure and alleviates priority effects; nutrient enrichment promotes plant diversification into attractive and defensive phenotypes

Table 4. *Cont.*

RR	Model Type	Assumptions	Parameterization	Validation	Key Findings
[52]	Eco-evolutionary network model with trait similarity	Invader success depends on trait similarity to resident species; mutualistic interactions are trait-mediated; network structure influences invasibility	Trait distributions and propagule pressure varied systematically; network metrics (nestedness, modularity) used to define recipient community	Simulations of invasion scenarios; robustness and resilience metrics used to assess impact	Trait dissimilarity enhances invasiveness; network stability better predicts invasibility than topology; multiple introductions increase invasion success
[53]	Age-structured partial differential equation model for unidirectional mutualism	Consumer species structured by age; resource species provides benefit without reciprocal cost; interaction includes both positive and negative effects	Age-dependent interaction kernels, Michaelis—Menten saturation for resource uptake; parameters derived from biological life history traits	Hopf bifurcation and stability analysis; numerical simulations of periodic solutions	Identified conditions for oscillatory dynamics; age structure induces complex feedbacks; coexistence depends on balance between benefit and exploitation
[54]	Ordinary differential equation system with attractor geometry in mutualistic networks	Bipartite network of plants and pollinators; mutualistic and competitive interactions; attractor architecture governs long-term dynamics	Growth rates and competition and cooperation coefficients defined via network topology; nestedness encoded in interaction matrices	Morse decomposition and Lyapunov function construction; numerical simulations of attractor transitions	Attractor structure determines biodiversity outcomes; nestedness enhances coexistence; topological robustness linked to dynamical stability
[55]	Two-species ordinary differential equation model with behavioral learning	Plants attract pollinators deceptively without offering rewards; pollinators learn to avoid deceptive flowers over time; learning affects visitation rates	Learning encoded via a dynamic cost function; interaction terms modeled with saturating responses; parameters derived from behavioral ecology	Stability and bifurcation analysis; numerical simulations of coexistence and extinction regimes	Learning reduces pollinator visitation to deceptive plants; coexistence depends on deception cost and learning rate; oscillatory dynamics emerge under intermediate conditions
[56]	Delay differential equation model for bumblebee colonies	Bumblebee population structured by colony stages; time delays represent developmental lags; external pressures affect reproduction and survival	Life history parameters derived from Bombus terrestris data; delays calibrated from empirical colony development timelines	Numerical simulations using spline approximations; sensitivity analysis under resource and pesticide stress	Delay structure captures seasonal dynamics and vulnerability; model predicts colony collapse under combined stressors; useful for evaluating conservation strategies
[57]	Genetic hybridization model with Allee effect mitigation	Small populations suffer from pollen limitation and genetic Allee effects; hybridization with co-flowering species improves pollination quality	Single-locus, two-allele genetic model; pollinators modeled as catalytic agents; parameters derived from kinetic reaction theory	Stability analysis of trivial and hybrid equilibria; phase-plane exploration of invasion dynamics	Neutral hybridization removes or reduces Allee thresholds; hybridization facilitates invasion and persistence; implications for conservation and invasion biology
[58]	Fractional-order diffusion model with Lévy flights	Bee-mediated pollen dispersal follows truncated Lévy flight patterns; long-distance dispersal events drive transgene spread; Brownian motion underestimates risk	Dispersal kernels fitted to empirical pollen movement data; fractional diffusion operator used to interpolate between Brownian and Lévy regimes	Numerical solution of fractional PDEs; comparison with classical diffusion predictions; threshold analysis for isolation distances	Lévy-based models predict significantly larger dispersal ranges; isolation distances must be revised upward; model improves risk assessment for GM pollen escape
[59]	Reaction, diffusion, delay system model with Hopf bifurcation	Plant–pollinator system with spatial diffusion and time delay; unidirectional consumer–resource interaction; periodic patterns emerge from instability	Diffusion coefficients and delay terms derived from ecological reasoning; functional responses include saturation and interference	Hopf bifurcation analysis via center manifold and normal form theory; numerical simulations of spatially homogeneous and inhomogeneous solutions	Identified conditions for temporal and spatial oscillations; delay and diffusion jointly drive pattern formation; bifurcation structure predicts ecological transitions

Table 4. *Cont.*

RR	Model Type	Assumptions	Parameterization	Validation	Key Findings
[60]	Trait-based evolutionary model with directional selection	Mutualism evolves via trait matching; regulation of trait variation includes homeostasis, developmental stability, and partner acceptability; selection is directional	Trait space modeled as continuous; mutation bias and environmental noise incorporated; parameters derived from evolutionary theory	Analytical exploration of joint evolution; stability analysis of trait distributions under different regulatory regimes	Developmental stability promotes mutualism evolution; trait regulation affects partner specificity; coevolutionary feedback shapes mutualistic trait architecture
[61]	Hybrid deterministic and agent-based model for oil palm pollination	Pollination by *Elaeidobius* spp. weevils depends on male inflorescence availability; fruit set linked to pollinator dynamics; spatial heterogeneity matters	Deterministic model uses metapopulation dynamics; agent-based model simulates individual weevil behavior; parameters from plantation data	Comparative simulations of both models; fruit set estimates validated against field observations; sensitivity analysis of inflorescence ratios	Agent-based model captures fine-scale dynamics; deterministic model predicts population thresholds; both approaches inform optimal pollinator management for yield improvement
[62]	Adaptive consumer–resource model with trait-mediated facilitation	Two plants compete for pollinators; pollinators adapt preferences based on plant abundance; facilitation and competition co-occur via trait mediation	Functional responses include adaptive foraging; trait distributions influence interaction strength; parameters derived from optimal foraging theory	Isoleg analysis and ideal free distribution framework; numerical simulations of coexistence and exclusion regimes	Adaptive preferences reduce niche overlap and promote coexistence; trait-mediated facilitation alters community structure; coexistence depends on balance between facilitation and competition
[63]	Ordinary differential equation model with behavioral learning and cost-benefit trade-offs	Pollinators learn to avoid deceptive plants; plant population includes rewarding and non-rewarding individuals; learning affects visitation rates	Learning encoded via dynamic feedback; cost-benefit parameters derived from ecological theory and behavioral studies	Hopf bifurcation analysis; numerical simulations of periodic and damped oscillations	Learning induces sustained or damped oscillations; coexistence depends on deception cost and learning rate; behavioral adaptation stabilizes mutualism
[64]	Three-species dynamical system with competition, parasitism, and mutualism	Plants interact with pollinators (mutualism) and nectar robbers (parasitism); pollinators and robbers compete indirectly for floral resources	Functional responses include Beddington—DeAngelis and Holling type II; parameters derived from the ecological literature and invasion theory	Global stability and persistence analysis; bifurcation diagrams of coexistence regimes	Coexistence possible under intermediate parasitism and competition; mutualism can persist despite robbers; extinction thresholds depend on interaction efficiency
[65]	Dimension-reduced model of mutualistic network collapse	High-dimensional mutualistic networks exhibit tipping points; dimension reduction captures essential dynamics; stochastic perturbations affect resilience	Reduction to 2D system using weighted averaging of empirical networks; parameters derived from 59 real-world datasets	Comparison of reduced model predictions with full network simulations; robustness tested under structural perturbations	Reduced model accurately predicts tipping points; resilience depends on network structure and interaction strength; framework generalizable to other complex systems
[66]	Compartmental population model for pyrophite shrub dynamics	Ulex parviflorus population structured by age and reproductive status; fire regimes influence biomass and regeneration; Mediterranean ecosystem context	Growth, flowering, and seed dispersal modeled via nonlinear differential equations; parameters calibrated from field data in Castellón, Spain	Numerical simulations of post-fire recovery and reproductive cycles; sensitivity analysis of biomass and seed bank dynamics	Pyrophite shrubs exhibit structured recovery under fire disturbance; reproductive success depends on spatial distribution and ecological thresholds
[67]	Three-species nonlinear ordinary differential equation model with limit cycle emergence	Plant–pollinator mutualism coupled with herbivory; herbivores reduce plant biomass and indirectly affect pollinator visitation; functional responses are nonlinear	Type IV functional responses for herbivory and mutualism; parameters selected for dynamical richness and ecological realism	Hopf—Andronov bifurcation theorem applied; Lyapunov coefficient used to confirm stability of limit cycle; numerical simulations support analytical results	Demonstrated existence of a stable limit cycle; herbivory can destabilize mutualism and induce oscillatory dynamics; coexistence depends on interaction strength and saturation effects

Table 4. *Cont.*

RR	Model Type	Assumptions	Parameterization	Validation	Key Findings
[68]	Eco-evolutionary simulation model with genetic algorithm	Obligate pollination mutualism; plant and pollinator traits evolve under fitness trade-offs; genetic algorithm simulates adaptive dynamics	Heuristic Lotka–Volterra-type model; fitness landscapes and trait distributions encoded in algorithm; parameters varied across simulations	Zero-isocline analysis and trait distribution mapping; robustness tested across multiple evolutionary runs	Trade-offs between cost and benefit shape mutualist niches; genetic algorithm reveals multiple stable eco-evolutionary regimes; obligate mutualism can persist under constrained trait evolution
[69]	Network-based model with exploitative competition	Pollinators compete for shared plant resources; exploitative competition affects network topology and species abundance; mutualistic links evolve adaptively	Interaction matrix constructed from rewiring rules; competition strength varied systematically; parameters derived from network theory	Analytical inversion of interaction matrix; simulations of network evolution under competition pressure	Exploitative competition increases plant abundance; pollinator hubs emerge asymmetrically; network rewiring enhances mutualistic benefit while minimizing competition cost
[70]	Two-patch ordinary differential equation model with dispersal and mutualism	Pollinators and plants interact in two spatial patches; dispersal affects persistence and abundance; mutualism is patch-dependent	Resource-service exchange modeled via nonlinear terms; dispersal rates and survival thresholds calibrated from theoretical ecology	Stability analysis of equilibria; numerical simulations of dispersal scenarios and population trajectories	Dispersal enhances pollinator abundance even under low plant density; patch quality influences mutualistic outcomes; small dispersal can outperform non-dispersal in both persistence and productivity
[71]	Three-species ordinary differential equation model with intermediary nectar dynamics	Nectar acts as intermediary resource between plant and pollinator; nectar dynamics influence mutualism persistence; nectar rapidly reaches quasi-steady state in reduced model	Nectar decay, production, and consumption rates derived from the ecological literature; reduced model assumes fast nectar dynamics	Analytical comparison of full and reduced models; bifurcation analysis and numerical simulations	Initial nectar density critically affects pollinator survival; reduced model captures long-term dynamics; nectar-mediated feedback shapes coexistence thresholds
[72]	Conceptual synthesis of mutualistic network theory	Network structure and species traits jointly determine mutualism dynamics; adaptive foraging and trait matching improve predictive capacity	Parameters drawn from empirical datasets and theoretical models; emphasis on trait-based and mechanistic approaches	Comparative review of modeling frameworks; integration of empirical validation strategies	Advocates for biologically grounded models; trait-based and adaptive mechanisms enhance prediction of network responses to perturbations
[73]	Two-plant, one-animal model with adaptive foraging	Animal mediator (mutualist or exploiter) adapts foraging preferences; plant coexistence mediated by behavioral feedback; animal density fixed	Preferences evolve to maximize fitness; generalized isocline framework used; competition strength varied systematically	Differential inclusion and sliding mode analysis; numerical simulations of coexistence regimes	Exploiter generalism promotes coexistence under strong competition; mutualist specialization yields alternative stable states; adaptive behavior reshapes competitive outcomes
[74]	Three-species ordinary differential equation model with food and toxin production	Plant produces both nectar (mutualism) and toxin (defense); pollinator and herbivore interact with plant via distinct pathways; trade-offs govern coexistence	Functional responses include saturation and inhibition; toxin production modeled as dynamic trait; parameters derived from ecological experiment	Hopf bifurcation and persistence analysis; numerical simulations of oscillatory and steady-state regimes	Toxin production modulates herbivore suppression and pollinator survival; coexistence possible via intermediate defense levels; excessive defense leads to collapse
[75]	Three-species ordinary differential equation model with pesticide-induced mortality	Pesticides reduce pollinator survival and indirectly affect plant reproduction; mutualism depends on energetic reward; extinction thresholds exist	Mortality rates and reward thresholds derived from ecological theory; pesticide effects modeled as additive mortality terms	Stability analysis of equilibria; numerical simulations under varying pesticide intensities	High energetic reward can buffer pesticide impact; low reward leads to plant extinction; mutual dependence is sensitive to pesticide pressure

Table 4. *Cont.*

RR	Model Type	Assumptions	Parameterization	Validation	Key Findings
[76]	Ordinary differential equation model with nonlinear mutualism and pesticide feedback	Pollinator mortality increases with pesticide exposure; mutualism modeled via Beddington—DeAngelis response; plant growth depends on pollination service	Pesticide toxicity modeled as a dynamic variable; parameters derived from empirical studies and theoretical ecology	Global stability and persistence analysis; bifurcation diagrams of extinction and coexistence regimes	Mutualism persists under moderate pesticide levels; excessive toxicity leads to collapse; coexistence thresholds depend on pollinator resilience and plant reward rates
[77]	Replicator and ordinary differential equation hybrid model with floral deception	Plants produce nectar-rich or nectarless flowers; pollinators adopt selective or non-selective foraging strategies; evolutionary game dynamics shape population structure	Nectar cost and cheater efficiency encoded in replicator equations; interaction terms derived from behavioral ecology	Bifurcation analysis of periodic and steady-state regimes; numerical simulations of invasion and persistence scenarios	Nectarless flowers can persist via cyclic dynamics; pollinator learning affects strategy evolution; coexistence depends on cost-benefit asymmetry and foraging discrimination
[78]	Three-species ordinary differential equation model with intermediary nectar dynamics	Nectar acts as intermediary resource between plant and pollinator; nectar dynamics influence persistence; plant cannot survive without pollination	Nectar decay, production, and consumption rates derived from ecological experiments; reduced model assumes fast nectar equilibrium	Analytical comparison of full and reduced models; global dynamics and persistence conditions derived rigorously	Initial nectar density determines persistence; low decay rates favor coexistence; intermediary resource mediates survival thresholds and system resilience
[79]	Nonlinear ordinary differential equation model with temperature-dependent mortality	Rising environmental temperature increases pollinator mortality and reduces plant reproduction; mutualism is sensitive to thermal stress	Temperature effects modeled via exponential mortality terms; ecological parameters derived from the climate and pollination literature	Local and global stability analysis; numerical simulations under varying temperature regimes	Elevated temperature reduces pollinator persistence and plant biomass; mutualism collapses under extreme warming; mitigation requires cooling interventions
[80]	Network-based co-adaptation model with dynamic link weights	Mutualistic networks adapt both structurally and dynamically; co-adaptation enhances resilience under perturbations; link weights evolve with species abundance	Link weights updated via feedback rules; empirical networks used for calibration; heterogeneity and connectance preserved	Comparative simulations of static, adaptive, and co-adaptive models; robustness tested under species loss scenarios	Co-adaptation increases resilience without altering connectance; dynamic feedback buffers against coextinction; model generalizable to other complex systems
[81]	Three-species ordinary differential equation model with antagonism between mutualists	Two mutualists share a partner species but interact antagonistically; antagonism may be consumptive or non-consumptive; life stages explicitly modeled	Interaction strengths and specialization levels varied systematically; antagonism encoded as direct negative feedback	Stability and persistence analysis; bifurcation exploration of oscillatory and extinction regimes	Antagonism reduces mutualist persistence; indirect effects dominate at high antagonism rates; specialization modulates system resilience
[82]	Empirical network analysis with epidemiological modeling	Landscape simplification alters plant-pollinator network structure; pathogen prevalence shaped by diet breadth and connectance; dilution effect emerges in complex networks	Eleven empirical networks analyzed; pathogen prevalence measured via molecular assays; network metrics computed from field data	Structural equation modeling and simulation of disease dynamics; robustness tested across landscape gradients	Simplified landscapes increase pathogen prevalence; higher connectance reduces outbreak risk; dominant species' diet breadth mediates community-level infection patterns
[83]	Delay differential equation model with empirical calibration	Fruit yield in dioecious crops depends on orchard layout, flower sex ratio, and pollinator density; pollinator behavior has diminishing returns at high density	Empirical data from kiwifruit orchards in New Zealand; Latin hypercube sampling used for sensitivity analysis	Simulations of fruit set under varying orchard configurations; model validated against field observations	Plant traits and layout more influential than pollinator density; optimal yield achieved with 65–75% female flowers and $\geq$6 bees per 1000 flowers
[84]	Fractional-order differential model with Atangana—Baleanu derivative	Nectar acts as intermediary resource; fractional calculus captures memory effects and non-locality in pollination dynamics	Fractional order $\alpha \in (0,1)$ varied systematically; stability analyzed via Picard—Lindelöf method	Numerical simulations using Adams–Bashforth scheme; stability tested across fractional orders	Fractional models outperform classical ODEs in capturing system memory; persistence depends on nectar dynamics and fractional order

Table 4. *Cont.*

RR	Model Type	Assumptions	Parameterization	Validation	Key Findings
[85]	Mechanistic transmission model with empirical viral assays	Deformed wing virus (DWV) transmits between bee species via shared flowers; transmission is bidirectional and density-dependent	Laboratory experiments with *Apis mellifera* and *Bombus impatiens*; viral load quantified via molecular assays	Mathematical simulations of transmission dynamics; dilution effect tested via floral abundance scenarios	DWV spreads via shared floral resources; increasing floral abundance reduces transmission; managing *Varroa* mites in honeybees mitigates spillover
[86]	Ordinary differential equation model with adaptive foraging by herbivores and pollinators	Two plants compete for shared mutualists and exploiters; animal preferences adapt to plant density; indirect interactions shape coexistence	Trait-mediated interactions modeled via isolegs and ideal free distribution; parameters varied across consumer abundance	Stability analysis of alternative states; simulations of coexistence under adaptive behavior	Adaptive preferences promote coexistence at high consumer abundance; low abundance leads to specialization and exclusion; insect decline alters plant community structure
[87]	Adaptive evolutionary model of nectar provisioning traits	Plants evolve nectar traits (production rate and reservoir volume) under selective pressures from pollinator interactions; traits co-evolve as a suite	Cost-benefit trade-offs modeled explicitly; ecological constraints include pollinator conversion efficiency and plant productivity	Analytical exploration of evolutionary equilibria; simulations of trait dynamics under varying ecological conditions	Higher nectar provisioning evolves under pollinator limitation, compensatory investment across traits stabilizes mutualism; indirect selection shapes trait architecture
[88]	Stochastic–deterministic hybrid model for hoverfly pollination control	Tomato flowers lack nectar; hoverfly density must be supplemented via feeding; pollination success depends on adult density and feeding strategy	Stochastic model estimates required hoverfly density; deterministic optimal control model minimizes feeding cost; parameters derived from greenhouse crop data	Simulation of pollinator dynamics and fruit yield; theoretical validation of control strategy under economic constraints	Supplementary feeding maintains economically viable hoverfly density; optimal control reduces cost; model supports hoverfly-based pollination in nectar-deficient crops
[89]	Epidemiological synthesis with genotype replacement modeling	DWV-A and DWV-B genotypes co-circulate in honeybee populations; DWV-B exhibits higher transmission and virulence; genotype interference affects prevalence	Global dataset (2008–2021) analyzed; mathematical model incorporates genotype competition and host co-infection dynamics	Empirical prevalence data from Germany, Italy, and the UK; model predictions compared to observed genotype shifts	DWV-B is replacing DWV-A globally; genotype interference drives replacement; implications for wild pollinators and beekeeping practices
[90]	Dose–response transmission model with pathogen transport	Pathogen transport via mechanical vectors (e.g., pollinators) alters exposure distribution; transmission depends on host dose–response curve	Two transmission scenarios modeled: amplification and dilution; dose–response functions derived from empirical infection thresholds	Analytical derivation of infection risk under varying transport regimes; simulations of disease spread in pollinator networks	Transport amplifies or dilutes transmission depending on dose–response shape; oversimplified models misestimate risk; framework improves epidemiological predictions
[91]	Consumer–resource ordinary differential equation models with reproductive benefit mechanisms	Plant reproductive benefits arise via pollination or seed dispersal; benefits affect seed set, germination, or recruitment; mutualism may be obligate or facultative	Foraging rate functions and benefit pathways modeled explicitly; parameters varied across ecological scenarios	Stability and bifurcation analysis; simulations of low-density thresholds and Allee effects	Pollination and seed dispersal mutualisms exhibit distinct dynamics; bistability and collapse thresholds depend on benefit mechanism and partner density
[92]	Network-based resilience analysis with dimension reduction	Hybrid ecological networks include mutualism, herbivory, and antagonism; resilience assessed via species contribution and extinction vulnerability	Interaction matrices constructed from hybrid network topology; resilience evaluated via reduced-dimensional metrics	Simulation of species removal and perturbation scenarios; resilience mapped across species categories	Strong contributors to resilience are more extinction-prone; plants are most vulnerable; network structure influences robustness under perturbation
[93]	Non-autonomous nonlinear ordinary differential equation model with seasonal forcing and parasitism	Honey bee population dynamics influenced by seasonal egg-laying and parasitism (e.g., *Varroa* mites); seasonality modulates colony resilience	Seasonality encoded via time-dependent birth rates; parasitism modeled as density-dependent mortality; parameters derived from empirical data	Hopf bifurcation analysis; simulations of collapse and recovery under seasonal and parasitic stress	Seasonality can stabilize or destabilize colonies depending on timing; parasitism induces collapse via bifurcation; synergistic effects shape colony survival

Table 4. *Cont.*

RR	Model Type	Assumptions	Parameterization	Validation	Key Finding
[94]	Delay differential equation model with dual time lags	Plant–pollinator interactions subject to two distinct delays (e.g., response and maturation); delays influence stability and oscillatory behavior	Characteristic equations derived from delay structure; delays treated as bifurcation parameters; ecological rates assumed to be constant	Center manifold and normal form theory applied; numerical simulations of periodic solutions and stability regions	Multiple delays induce Hopf bifurcations and periodic dynamics; delay asymmetry affects system persistence; explicit conditions derived for oscillation onset
[95]	Analytical framework for structural stability in mutualistic–competitive networks	Mutualistic networks include interspecific competition among plants and pollinators; structural stability defined as coexistence feasibility under perturbation	Total of 50 empirical networks used; competition encoded via weighted adjacency matrices; mutualism modeled with saturating responses	Analytical derivation of feasibility domains; numerical simulations of stability under network rewiring	Competitive structure strongly influences coexistence; mutualism alone insufficient for stability; new metric links network architecture to resilience thresholds
[96]	Eco-evolutionary model with adaptive disinvestment	One-sided population decline triggers adaptive reduction in mutualistic investment; feedback loops can accelerate collapse or stabilize interaction	Trade-off between independent growth and mutualistic investment modeled as concave function; adaptation rates varied systematically	Simulations of co-evolutionary trajectories; bifurcation analysis of collapse and recovery regimes	Disinvestment by undisturbed partner precedes collapse; slow adaptation or high initial investment delays extinction; co-evolution essential for recovery
[97]	Socio-mutualistic network model with optimal conservation strategy	Pollinator dynamics coupled with human conservation norms; tipping points emerge from structural and behavioral feedbacks	Network topology varied across nestedness levels; conservation norms applied selectively to pollinator nodes	Dynamical analysis of reduced model; simulations across empirical and synthetic networks	Optimal conservation strategy prevents collapse with minimal intervention; intermediate nestedness most responsive; social norms amplify resilience
[98]	Generalized stressor model for hive and forager bees	Stressors affect bees via transmissibility, lethality, impairment, and timing; colony collapse emerges from labor destabilization and precocious foraging	Stressor attributes generalized across multiple dimensions; model integrates prior CCD frameworks; parameters derived from empirical studies	Validated against emergent colony behaviors; simulations of collapse thresholds under stressor variation	Sublethal stressors sufficient to trigger collapse; timing and impairment level critical; model supports holistic stressor management in conservation
[99]	Ordinary differential equation based community model with antagonism–mutualism continuum	Herbivores reduce pollinator visitation via indirect limitation; mutualistic and antagonistic interactions co-occur; stability assessed across interaction gradients	Interaction strengths varied systematically; network architecture encoded via adjacency matrices; parameters derived from ecological literature	Temporal and compositional stability analyzed; simulations of species persistence and network robustness	Pollinator limitation enhances both temporal and compositional stability; herbivory indirectly promotes persistence; network architecture–stability relationship shifts under limitation
[100]	Behaviorally modified preda-tor–prey–mutualism model	Predation on pollinators alters foraging behavior; behavioral avoidance stabilizes mutualism; direct predation alone insufficient for stability	Pollinator behavior modeled via adaptive response functions; predation rates and avoidance thresholds varied across scenarios	Stability analysis of modified mutualism; simulations of predator-induced behavioral shift	Behavioral modification stabilizes mutualism under predation; avoidance behavior critical for persistence; predator pressure reshapes interaction dynamics
[101]	Network-based thermal response model with tipping point detection	Climate warming alters species-level physiological traits; mutualistic networks exhibit tipping points under thermal stress; generalists play stabilizing roles	Total of 139 empirical networks analyzed; thermal sensitivity encoded via trait-dependent growth and mortality rates	Reduced two-dimensional model used for bifurcation analysis; simulations across temperature gradients	Rising temperature induces rapid transitions in low-strength networks; generalists delay collapse; network structure and thermal traits jointly determine resilience

Table 4. *Cont.*

RR	Model Type	Assumptions	Parameterization	Validation	Key Finding
[102]	Partial differential equation system with chemotaxis-enhanced reaction rates	Chemotaxis enhances biological reaction efficiency; organisms move toward chemical gradients; surface chemotaxis considered for biological realism	Chemotactic flux modeled via Keller–Segel-type terms; reaction rates compared with pure diffusion; parameters derived from biological signaling contexts	Analytical estimates of convergence rates; sharp inequalities derived for Fokker–Planck operators	Chemotaxis significantly enhances reaction success; convergence to equilibrium accelerated; framework applicable to immune signaling, reproduction, and pollination
[103]	Trait-based evolutionary model with trophic structure	Plants and pollinators co-evolve under mutualistic and competitive pressures; trait polymorphisms emerge via Darwinian dynamics; niche breadth influences diversification	Trait matching and competition encoded via continuous trait space; ecological rates derived from trophic interactions	Analytical derivation of evolutionary branching conditions; simulations of polymorphism emergence	Broad plant niches promote phenotypic diversification; mutualistic generalism triggers cascading trait divergence; co-diversification driven by bottom-up control
[104]	Delay differential equation model with brood mortality	Brood deaths induced by insecticide exposure affect colony dynamics; time delay represents developmental lag; collapse thresholds explored	Mortality and delay parameters derived from empirical bee life cycle data; insecticide effects modeled as external forcing	Stability and sensitivity analysis of equilibria; numerical simulations with real data	Brood mortality significantly reduces colony resilience; no Hopf bifurcation within biologically plausible delay range; model supports policy design for insecticide regulation
[105]	Stochastic branching process model with behavioral resistance	Grooming and hygienic behaviors reduce DWV outbreak probability in *Varroa*-infested colonies; transmission occurs via vector and contact routes	Transmission probabilities and behavioral efficacy derived from empirical studies; branching process theory applied to outbreak dynamics	Analytical computation of outbreak probabilities; simulations across behavioral scenarios	Hygienic behavior most effective at reducing DWV outbreaks; grooming reduces vector transmission to near zero; behavioral selection enhances colony survival
[106]	Nonlinear dynamical system with bifurcation analysis in parameter space	Honeybee—mite interactions exhibit multistability, chaos, and bifurcation structures; parameter variation reveals complex attractor landscapes	Interaction rates and mortality parameters varied systematically; bifurcation parameters mapped across biologically relevant ranges	Numerical exploration of phase space; identification of Arnold tongues, jellyfish structures, and saddle regions	System exhibits rich dynamical regimes including chaos and multistability; parameter tuning critical for colony persistence; visual structures aid in ecological interpretation
[107]	Three-species nonlinear ordinary differential equation model with limit cycle dynamics	Pollinators and plants form a mutualistic pair; herbivores consume plants and indirectly reduce pollinator visitation; functional responses are of type IV	Saciety and saturation effects encoded via nonlinear terms; ecological rates selected for dynamical richness	Hopf–Andronov bifurcation theorem applied; Lyapunov coefficient used to confirm stability of limit cycle; numerical simulations support analytical results	Herbivory destabilizes mutualism and induces oscillatory dynamics; coexistence depends on interaction strength and saturation thresholds
[108]	Consumer–resource network model with adaptive foraging	Floral resource availability constrains pollinator specialization; network structure emerges from adaptive foraging and resource depletion	Lotka–Volterra framework with dynamic floral resource pool; parameters varied across resource and animal density gradients	Generalized linear models used to explain nestedness, modularity, and specialization; simulations across 3812 networks	Low floral resource availability reduces specialization; resource dynamics distinct from plant density; adaptive foraging mediates network structure
[109]	Adaptive dynamics model of floral handedness evolution	Mirror-image flowers evolve from mixed to fixed handedness; stylar orientation affects pollination efficiency and mating patterns; inbreeding depression penalizes geitonogamy	Trait evolution modeled via adaptive dynamics; genetic architecture assumed plausible; ecological parameters include pollinator efficiency and plant density	Stability analysis of dimorphic vs. monomorphic states; population genetics simulations validate transitions	Dimorphic enantiostyly evolves under moderate inbreeding and high pollination efficiency; ecological shifts may reverse dimorphism; model supports convergence in floral asymmetry

Table 4. *Cont.*

RR	Model Type	Assumptions	Parameterization	Validation	Key Findings
[110]	Population dynamics model of lethal deceptive pollination	Plants trap and kill pollinators without reward; coexistence depends on vegetative reproduction and sex transition rates; disturbances affect demographic stability	Empirical data from *Arisaema* species used to calibrate model; disturbance scenarios include deer abundance, habitat loss, and plant theft	Stability analysis and extinction thresholds explored; simulations under multiple disturbance regimes	Lethal deception maintained under high vegetative reproduction and slow sex transition; co-extinction risk elevated under strong attraction and anthropogenic disturbance
[111]	Impulsive reaction–diffusion model on a periodically evolving domain	Plant–pollinator system subject to periodic habitat changes and impulsive disturbances; domain evolution affects dispersal and persistence	Ecological reproduction index defined; impulsive effects modeled via discontinuous terms; domain evolution rate treated as bifurcation parameter	Upper–lower solution method applied; numerical simulations of extinction and persistence scenarios	High domain evolution rate promotes pollinator survival; impulsive effects can override spatial benefits; coexistence depends on timing and magnitude of impulses
[112]	Stochastic partial differential equation model with degenerate diffusion	Plant–pollinator dynamics influenced by environmental noise; degenerate diffusion captures spatial heterogeneity and dispersal limitations	Diffusion coefficients vary with population density; stochastic terms modeled via Brownian motion; permanence conditions derived analytically	Stochastic comparison principle and Lyapunov function used; extinction and persistence regions identified	Degenerate diffusion enhances extinction risk under low density; stochasticity can stabilize or destabilize dynamics; permanence requires bounded noise intensity
[113]	Mechanistic ordinary differential equation model with separate pollen and nectar dynamics	Pollinators consume nectar and transfer pollen independently; interspecific pollen transfer reduces pollination efficiency; mutualism may shift to antagonism	Pollen and nectar modeled as distinct resources; pollinator efficiency and interference encoded via saturating functions	Analytical derivation of pollination efficiency; simulations of mutualism-antagonism transitions	Separation of pollen and nectar dynamics reveals hidden antagonism; interspecific pollen transfer reduces plant fitness; ecological context determines net interaction outcome
[115]	Nonlinear dynamical system with predation on pollinators	Predators consume pollinators and destabilize mutualism; plant—pollinator interaction saturates at high density; predator–pollinator dynamics coupled	Saturated mutualism modeled via Holling-type functions; predation rate and initial density varied across simulations	Stability and bifurcation analysis; numerical exploration of transcritical and Hopf bifurcations	High predator density leads to pollinator extinction; mutualistic strength buffers predation impact; coexistence possible under low predation and strong mutualism
[114]	Tripartite ordinary differential equation model with facultative and obligate mutualism and parasitism	Plant–pollinator–parasite system; pollinators may be facultative or obligate; parasites affect pollinator fitness and system stability	Facultative and obligate behavior encoded via growth and mortality terms; bifurcation parameters include parasite conversion and death rates	Bifurcation analysis of subsystems and full model; numerical simulations of multistability and oscillatory regimes	Facultative pollinators enhance resilience via multistability; obligate systems prone to collapse; novel bifurcation patterns (e.g., wavebow) characterize amplitude transitions
[116]	Delayed reaction, diffusion model with memory-based diffusion	Pollinator movement influenced by spatial memory; delay affects diffusion and pattern formation; periodicity emerges from delay–diffusion interplay	Memory delay incorporated via modified Fick's law; diffusion coefficients and delay terms varied systematically	Hopf bifurcation and Lyapunov–Schmidt reduction applied; simulations of spatial and temporal periodic solution	Memory-based diffusion induces spatial heterogeneity and oscillations; delay triggers stability switch; model captures biologically realistic movement biases

Table 5. Summary of retained studies according to biological topics and other characteristics.

Modeling Domain	Assumptions	Parametrization	Validation	Findings	Policy	Works
Biological consistency and population viability	Non-negative, bounded population variables.	Logistic or saturating growth; constrained initial conditions.	Analytical consistency.	Avoids spurious extinction or explosion.	Supports viability thresholds.	[15,19,20]
Long-term dynamics and species persistence	Equilibrium-based persistence or extinction.	Reproduction, mortality, and interaction sensitivity.	Stability via Lyapunov function and linearization.	Identifies resilience thresholds.	Informs sustainability planning.	[10–12]
Regime shifts and critical transitions	Threshold-driven qualitative change.	Bifurcation parameters.	Bifurcation diagrams.	Reveals tipping points.	Enables adaptive management.	[16,30,53]
Mutualism and network structure	Mutualistic coexistence and spatial structure.	Patch connectivity and trait evolution.	Network simulations.	Biodiversity maintenance.	Corridor design and zoning.	[13,16,18]
Seasonal and oscillatory behavior	Seasonal forcing and delay effects.	Time-dependent coefficients.	Periodicity analysis.	Captures seasonal fluctuations.	Supports crop planning.	[30,47,65]
Simulation and empirical calibration	Empirical realism.	Optimization and sensitivity analysis.	Simulation observed data.	Enhances credibility.	Enables data-driven decisions.	[19,25]
Intervention and optimization	Ecological outcomes modifiable via control.	Pontryagin's maximum principle and dynamic programming.	Optimality conditions.	Cost-effective strategies.	Informs adaptive conservation.	[88,97]

3. Summarizing the Evidence and Interpreting the Findings

In Section 2.1, we introduced two questions. "Question 1" is clearly answered by the retained list [10–116]. Meanwhile, to answer "Question 1", we introduce two classifications pertaining to the mathematical approaches used and the topics researched. We revised the retained list and defined the following four groups according to the mathematical theories applied for mathematical modeling:

(1) Ordinary differential equations group (see Section 3.1): Here, we distinguish between three types of models based on the methodology used for modeling. First, we considered the Lotka–Volterra models for two populations (see Section 3.1.3). The works of this type are [10,11,15,22,28,30,33,47,55,68,73,76,86,88,91,96,104,106]. The second type was the Lotka–Volterra models for more than two populations (see Section 3.1.2), with the articles being [15,16,19,24,28,30,33,34,37,42,44–46,48,51,62–64,67,70,71,74,75,78,79,81,87,100,107,114,115]. The third class of ordinary differential equations systems was obtained via application of the compartmental methodology (see Section 3.1.3), and the works of this type are [14,36,38,39,41,61,83,85,89,90,93,98,105,110].

(2) Partial differential equations group (see Section 3.1): The generalization of the ordinary differential equations to include the spatial displacement of pollinators was studied in [19,26,27,31,32,50,111].

(3) Patch network (see Section 3.3): Consideration of groups of pollinators in different patches and the interaction of network concepts was conducted in [13,21,25,35,40,43,49,52,54,65,69,80,82,95,97,99,101,103,108,113]

(4) Other methodologies (see Section 3.4): Other kinds of works like letters, reviews, and emergent methodologies like fractional-order or delay models were introduced in [12,17,18,20,23,26,29,53,56–60,66,84,94,102,104,109].

Extensive details are presented below in Sections 3.1–3.4. We include a summarization of the topics covered by the studies in Section 3.5. We note that some works were considered to be in more than one group, as they examined mathematical modeling from multiple perspectives. For instance, the more typical case is that partial differential equation models are obtained as a generalization of ordinary differential equation models, and the same work can be considered to be in groups (1) and (2) (see, for example, [19]). Another example is [110], which involves multiple species and a class-structured (compartmental) framework, and it also incorporates nonlinear interaction terms that are characteristic of Lotka–Volterra systems. Therefore, it could alternatively be included in Section 3.1.2. However, it was placed in this subsection because the study's primary focus lies in disease propagation and infestation dynamics rather than population-level coexistence or competitive interactions. This aligns it more closely with classical compartmental models such as SIR, SEIR, and related formulations. Furthermore, we obtained a Cohen's kappa coefficient of 0.78.

3.1. Mathematical Models Based on Ordinary Differential Equations

3.1.1. Two Population Mathematical Models Using Lotka–Volterra-Like Methodologies

In the retained list, there are several works where the mathematical models were obtained through the interaction of two populations, such as plants and pollinators. The two basic assumptions required to obtain these mathematical models were as follows: there is an interaction between the pollinator and the plant, and the impact of various ecological and environmental variables is neglected. Then, the modeling approach focused on how the birth and death rates of both populations drove their changes or, equivalently, how the birth, death, and interactions of the populations affected population growth. Then, we deduced that

$$\begin{bmatrix} \text{rate of change of} \\ \text{plants} \\ \text{population} \end{bmatrix} = \begin{bmatrix} \text{plants} \\ \text{population} \\ \text{birth rate} \end{bmatrix} - \begin{bmatrix} \text{plants} \\ \text{population} \\ \text{dead rate} \end{bmatrix}, \tag{1}$$

$$\begin{bmatrix} \text{rate of change of} \\ \text{pollinator} \\ \text{population} \end{bmatrix} = \begin{bmatrix} \text{pollinator} \\ \text{population} \\ \text{birth rate} \end{bmatrix} - \begin{bmatrix} \text{pollinator} \\ \text{population} \\ \text{dead rate} \end{bmatrix}. \tag{2}$$

Let us denote the total population of plants and pollinators as p and a, respectively. Then, we have

$$\begin{bmatrix} \text{rate of change of} \\ \text{plants} \\ \text{population} \end{bmatrix} = \frac{dp}{dt}, \quad \begin{bmatrix} \text{rate of change of} \\ \text{plants} \\ \text{population} \end{bmatrix} = \frac{da}{dt}. \tag{3}$$

To model the birth and death rates, we must consider several assumptions. To specify the algebraic forms modeling birth and death rates, we considered the discussion provided in [10]. We note that they modeled the plant–pollinator interaction via analogy with the Lotka–Volterra or prey-predator systems, assuming that the plants are predators and the pollinators are the prey.

The deduction of plant and pollinator birth rates in [10] was obtained by considering the following two assumptions: the plants are self-incompatible, and the plant birth rate is related to flower visits by the pollinators. They considered that the birth rate is proportional to the pollinator visits, neglected some factor like the finite supply of ovules, and assumed that the number of pollinator visits was modeled by a Holling's functional such that the model was of the following form:

$$\begin{bmatrix} \text{plants} \\ \text{population} \\ \text{birth rate} \end{bmatrix} = k_1 a \begin{bmatrix} \text{pollinator} \\ \text{plant} \\ \text{visit rate} \end{bmatrix} = k_1 a \frac{\alpha p}{1 + \alpha \beta p}, \tag{4}$$

where k_1, α, and β model the number of ovules fertilized per visit, the searching rate constant multiplied by the encounter probability, and the handling time per visit, respectively. The encounter probability and the handling time depend on the energetic reward. More precisely, we have

$$\alpha = \sigma\mu, \quad \beta = \phi\mu, \tag{5}$$

where σ is the probability of an encounter, ϕ is the reciprocal speed of nectar extraction, and μ is the energetic reward. By combining Equation (5) with Equation (4), we deduce that

$$\begin{bmatrix} \text{plants} \\ \text{population} \\ \text{birth rate} \end{bmatrix} = \frac{k_1\sigma\mu a p}{1+\sigma\phi\mu^2 p}. \tag{6}$$

It is assumed that the pollinator birth depends on the density and some variables such as the competition for nest sites or protein resources such that

$$\begin{bmatrix} \text{pollinator} \\ \text{population} \\ \text{birth rate} \end{bmatrix} = a(\delta - \epsilon a), \tag{7}$$

where δ is the maximum per capita pollinator birth rate and ϵ is the density-dependent regulation constant.

The deduction of plant and pollinator dead rates introduced in [10] is as follows. In the case of plant mortality, assume that it is proportional to the plant density, i.e., we have

$$\begin{bmatrix} \text{plant} \\ \text{population} \\ \text{dead rate} \end{bmatrix} = \gamma p, \tag{8}$$

where γ is the mortality rate. Meanwhile, for pollinators, it is assumed that the mortality pollinator rate is inversely related to the rate of energy intake, which in turn is jointly proportional to the visit rate and the energetic reward:

$$\begin{bmatrix} \text{pollinator} \\ \text{population} \\ \text{dead rate} \end{bmatrix} = \lambda a - k_2 a \begin{bmatrix} \text{pollinator} \\ \text{plant} \\ \text{visit rate} \end{bmatrix} \mu = \lambda a - \frac{k_2\sigma\mu^2 a p}{1+\sigma\phi\mu^2 p}, \tag{9}$$

where k_2 is a constant of energetic transformation and λ is the maximum death rate of pollinators in the absence of plants.

By combining Equations (3), (6)–(9) with Equations (1) and (2), we obtain a system of the form

$$\frac{dx_1}{dt} = f(x_1, x_2)x_1, \quad \frac{dx_2}{dt} = g(x_1, x_2)x_2, \tag{10}$$

where

$$(x_1, x_2) = (p, a),\ f(p, a) = \frac{k_1\sigma\mu a}{1+\sigma\phi\mu^2 p} - \gamma,\ g(p, a) = \epsilon\left(\frac{\delta-\lambda}{\epsilon} - a\right) + \frac{k_2\sigma\mu^2 p}{1+\sigma\phi\mu^2 p}. \tag{11}$$

Here, we observe that $(\delta - \lambda)/\epsilon$ is the carrying capacity of the pollinator population. Other mathematical models of the general form in Equation (10) were introduced in [11,15,22,28,30,33,47,55,68,73,76,86,88,91,96,104,106]. We remark that there are eight types of models, depending on the interaction populations: plant–pollinator [10,11,15,22, 28,30,55,68,73,91,96], plant–robber [28], pollinator–secretor [33], pollinator–cheater [33], plant–novice pollinator [47], novice pollinator–expert pollinator [47], plant–plant [76,86], juvenile pollinator–adult pollinator [88,104], and honeybee–mite [106].

3.1.2. More Than Two Population Mathematical Models Using Lotka–Volterra-Like Methodologies

Analogous to the analysis developed in Section 3.1.1, we can consider that more than two species are interacting. To illustrate the concept, we consider the model introduced in [15], where the authors examined the interaction among three species—herbivores, plants, and pollinators—with populations denoted by x, y, and z, respectively. Then, by realizing a balance of birth rates, death rates, and interaction of the three populations, they deduced that the mathematical model is given by

$$\frac{dx}{dt} = bx(K - x) + \frac{g(z)k_2\mu^2\sigma xy}{1 + \phi\sigma\mu^2 y}, \tag{12}$$

$$\frac{dy}{dt} = \frac{g(z)k_1\mu\sigma xy}{1 + \phi\sigma\mu^2 y} - \gamma y - \frac{m_1 yz}{a + y}, \tag{13}$$

$$\frac{dz}{dt} = \frac{m_2 yz}{a + y} - \delta z, \tag{14}$$

where b is a density-dependent regulation constant, K measures the diversity of pollinators of plants, k_2 is a constant of energetic transformation, μ is the energetic reward, σ is the probability of an encounter, ϕ is the reciprocal speed of nectar extraction, k_1 is an efficiency constant representing the number of ovules fertilized per visit, γ is the plant mortality rate, δ is the pollinator mortality rate, a is the half-saturation constant, m_1 is the maximal ingestion rate, m_2 is the herbivore maximal growth rate, and $g(z)$ is a function depending on the herbivore population density. We observe that Equations (12)–(14) can be rewritten in the following general form:

$$\frac{dx_i}{dt} = f_i(x_1, \ldots, x_d)x_i, \quad i = 1, \ldots, d, \tag{15}$$

where x_i is the density of the ith species, d is the number of interacting species, and f_i models the birth rates, death rates, and interactions.

The mathematical models of the general form in Equation (10) with $d \geq 3$ were introduced in the following 31 works from the retained list: [15,16,19,24,28,30,33,34,37,42,44–46,48,51,62–64,67,70,71,74,75,78,79,81,87,100,107,114,115]. In these works, the interacting populations were described as follows: plant–pollinator–hervibore [15,19,24,42,67,74,107], plant–pollinator–ant [24,44–46,51], plant–pollinator–robber [28,30,34,37,48,64], pollinator–secretor–cheater [33], plant–pollinator–flower [70,71,78,87], plant–novice pollinator-expert pollinator [47], two plants and one pollinator [63], two plants and two pollinators [62], plant–pollinator–pesticide [75], plant–pollinator–predator [100], plant–pollinator– parasite [114,115], plant–pollinator–green house gases–temperature [79], plant–seed–pollinator–seed disperser [81], and three general species [16].

3.1.3. Mathematical Models Based on Compartmental Methodology

The pollination system is formed by plants and pollinators. The plant population is divided into three sub-populations: susceptible (x_v), pollinated (x_f), and infected by a fungus (x_i). Similarly, the pollinator population is divided into three states: can carry neither pollen nor fungal spores (y_n), can carry pollen (y_p), and can carry spores (y_{sp}). Moreover, we consider that the pollinated plants have a rate of return to the susceptible class. Then, by considering other assumptions regarding the interaction, Ingvarsson and Lundberg [14] introduced the following mathematical model:

$$\begin{aligned}
\frac{dx_v}{dt} &= (b+\delta)x_f - \frac{\omega\beta_1 y_p x_v}{x_v + x_f + x_i} - \frac{\omega\beta_2 y_{sp} x_v}{x_v + x_f + x_i} - \mu x_v, \\
\frac{dx_f}{dt} &= \frac{\omega\beta_1 y_p x_v}{x_v + x_f + x_i} - (\mu+\delta)x_f, \\
\frac{dx_i}{dt} &= \frac{\omega\beta_2 y_{sp} x_v}{x_v + x_f + x_i} - \mu x_i, \\
\frac{dy_n}{dt} &= kN - \nu\beta_3 x_v \frac{y_n}{N} - \nu\beta_4 x_i \frac{y_n}{N} - \gamma(y_p + y_{sp}) - k y_n, \\
\frac{dy_p}{dt} &= \nu\beta_3 x_v \frac{y_n}{N} - \nu\beta_4 x_i \frac{y_n}{N} - (\gamma+k)y_p, \\
\frac{dy_{sp}}{dt} &= \nu\beta_4 x_i \frac{y_n}{N} - (\gamma+k)y_{sp},
\end{aligned}$$

where $N = y_n + y_p + y_{sp}$ and $b, \delta, \omega, \beta_1, \beta_2, \mu, \beta_3, \beta_4, \gamma, k,$ and ν are positive parameters.

The mathematical models governed by ordinary differential equations which are based on the compartmental methodology were introduced in the following 14 works from the retained list: [14,36,38,39,41,61,83,85,89,90,93,98,105,110]. The populations modeled in the different articles are diverse. More precisely, they are a plant population with susceptible, pollinated, and infected classes and a pollinator population divided into three classes, where they can carry neither pollen nor fungal spores, can carry pollen, or can carry spores [14]; healthy bees and impaired bees [36]; an uncapped brood, capped brood, hive bees, foragers, and food [38]; plants biomass with adult insects and adult insects with larvae [39]; a pollinator with pollen, pollinator without pollen, unpollinated flowering plants, and pollinated flowering plants [41]; the interactions of adult and non-adult pollinators [61]; four types of pollinators [83]; infected bumblebees, infected honeybees, and infected flowers [85]; the dynamics of viral genotypes [89]; contamined flowers, infected bees, and virus carrying [90]; a honeybee--parasite interaction model with seasonality [93]; hive bees, unimpaired forager bees, and impaired forager bees [98]; the transmission dynamics of deformed wing virus in a honeybee colony infested with *Varroa* mites [105]; and three plants and two pollinators with juvenile, male, and female plant classes and insects [110].

3.2. Mathematical Models Based on Partial Differential Equations

The mathematical models based on partial differential equations were obtained by assuming the spatial movement of the pollinators. It is considered that the displacement of pollinators satisfies a diffusion law. Then, the partial differential equations are extensions of ordinary differential equation models. For instance, in the case of the interaction of two populations, the authors of [19] considered the ordinary differential equation for a pollinator–plant interaction system:

$$\frac{da}{dt} = a(K-a) + \frac{ap}{1+p}, \tag{16}$$

$$\frac{dp}{dt} = -\frac{p}{2} + \frac{ap}{1+p}, \tag{17}$$

where K is the carrying capacity for the pollinator population and a and p are the population densities of the pollinators and plants, respectively. Assuming that the pollinator population moves toward negative values of the gradient of the population density direction, the plants do not disperse, but their spatial distribution changes because of the interaction with the pollinator population. Then, the authors of [19] defined the new system extending the model in Equations (16) and (17) as follows:

$$\frac{\partial a}{\partial t} = D\Delta a + a(K-a) + \frac{ap}{1+p}, \tag{18}$$

$$\frac{\partial p}{\partial t} = -\frac{p}{2} + \frac{ap}{1+p}, \tag{19}$$

where the parameter $D > 0$ is the diffusivity of the pollinator population and Δ is the Laplacian operator. Similarly, the authors of [19] assumed an ordinary differential model for pollinator–plant–herbivore interactions of the following type:

$$\frac{da}{dt} = ba(K-a) + \frac{g(h)k_2\mu\sigma ap}{1+\phi\sigma\mu^2 p}, \tag{20}$$

$$\frac{dp}{dt} = -\gamma p + \frac{g(z)k_1\mu\sigma ap}{1+\phi\sigma\mu^2 p} - \frac{m_1 ph}{c_1+p}, \tag{21}$$

$$\frac{dh}{dt} = -\delta h + \frac{m_2 ph}{c_1+p}, \tag{22}$$

where K is the carry capacity for the pollinator population; a, p, and h are the population densities of pollinators, plants, and hervivores, respectively; g is a real function such that $g \in C^1[0,\infty)$, $g(0) = 1$, in which $g'(h) < 0$ and $g(h) > 0$ for all $h > 0$, modeling the reduction rate of visits of pollinators to plants due to herbivore interaction; k_1 is the number of fertilized ova in each pollinator visit; σ is the probability of visits; ϕ is a measure of the speed of nectar extraction; and μ is the energetic recompense. The generalization of Equations (20)–(22) to a partial differential system is given by

$$\frac{da}{dt} = D_1\Delta a + ba(K-a) + \frac{g(h)k_2\mu\sigma ap}{1+\phi\sigma\mu^2 p}, \tag{23}$$

$$\frac{dp}{dt} = -\gamma p + \frac{g(z)k_1\mu\sigma ap}{1+\phi\sigma\mu^2 p} - \frac{m_1 ph}{c_1+p}, \tag{24}$$

$$\frac{dh}{dt} = D_2\Delta a - \delta h + \frac{m_2 ph}{c_1+p}, \tag{25}$$

where the parameters $D_1 > 0$ and $D_2 > 0$ are the diffusivity of the pollinator and herbivore populations, respectively. Similar extensions of ordinary differential equations models were deduced via application of the compartmental methodology.

The mathematical models based on partial differential equations were considered in the following seven works: [19,26,27,31,32,50,111]. The modeled populations considered in the different articles are the following: the interaction of plant, pollinator, and herbivore populations [19,26]; harvester and scout populations Tyson [27]; multiple species of pollinators [31,32]; and plant and pollinator populations [50,111].

3.3. Network and Patch Mathematical Models

In the retained list, several works applied networks and patch concepts to model the dynamics of pollinator populations. In order to be precise, we consider [21], where the authors considered the interaction of a plant p_i and animal a_j, obtaining the following system:

$$\frac{dp_i}{dt} = \sum_{j=1}^{n}\left(c_{ij}\frac{p_i a_j}{w_j}\right)(1-d-p_i) - e_i p_i, \qquad 1 \le i \le n,$$

$$\frac{da_j}{dt} = c_j a_j(w_j - a_j) - e_j a_j, \qquad 1 \le j \le n,$$

where c_{ij} models the per capita colonization rate of a population of plants i when pollinated or dispersed by a pollinator j; c_j is the per capita colonization rate of a pollinator j; e_i and e_j denote the per capita extinction rates for a plant i and animal j, respectively; d models the fraction of patches permanently lost through habitat destruction; and w_j is the union of the patches occupied by n plant species interacting with the same j pollinator species.

In the retained list, we found that there were 20 works focused on the modeling of pollinator population dynamics using networks and patches [13,21,25,35,40,43,49,52, 54,65,69,80,82,95,97,99,101,103,108,113]. In [13], the authors applied patch concepts to study the age-structured pollinator population model considering adult and non-adult pollinators [13]. Meanwhile, in the other works, the authors used networks and patch concepts [25,35,40,43,49,52,54,65,69,80,82,84,95,97,99,101,103,108,113].

3.4. Other Methodologies

Other articles that were difficult to include in the previous classification are the following 19 articles: [12,17,18,20,23,26,29,53,56–60,66,84,94,102,104,109]. These included a letter to the editor with an opinion on the mathematical model for mutualism on a patch [12], two review articles [26,102], a study on discrete models [20], a work focused on the study of virulence [23], a study on microscopic populations by considering five types of cells [17], a study on the fractional order mathematical mould for plant–pollinator–nectar interactions [84], a study on the modelization of pollen transport [58], studies on the concept of delaying ordinary differential equations to model a plant-pollinator system [59,94], a study on adult and juvenile pollinators [104], a study on the idealization of bumble bees [56], studies on the application of stochastic differential equations [18,109], a study on the analysis of flowering [29], studies on the modellization by hybrid ordinary differential equations and partial differential equations [53,60], a study on a particular form of ordinary differential system for modeling genotypes [57], and an empirical study which developed data fitting for ordinary differential models [66].

3.5. A Summary of the Topics Studied in the Retained List

The analysis and main results of the articles of the retained list focused on the following seven topics:

(1) *Positive bounded solutions*: The variables of the mathematical models are the population or the density of the population. Then, the first question of the consistence of the mathematical model with the biological system is for analyzing if the mathematical model's solutions are positive and bounded. In this sense, the following works [15,19,20,26,28,30,41,50,54,59,65,66,71,75,78,93,102] have explicit results proving that the dynamics of the mathematical systems have positive bounded solutions.

(2) *Equilibrium and stability analysis*: In the mathematical analysis of dynamical systems, the study of linearization and asymptotic behavior is strongly related to the analysis of stability analysis. In particular, mathematical models are an important tool for characterizing the large time behavior of the system and answering other important questions, like the prevalence or extinction a species of pollinator. The works focused on the development of equilibrium and stability analysis are the following [10–12,14,15,19,20,22,26,33,34,36,39,43–48,54,55,57,60,61,63,67,68,70,71, 73–76,78,79,81,84,86,90–93,95,96,98,100,102–104,107,111,113,114].

(3) *Bifurcation*: One topic related to equilibrium and stability analysis is bifurcation analysis. Indeed, the analysis of bifurcation was introduced in [16,30,53,67,74,94,103, 106,107,114].

(4) *Mutualistic interactions*: In the case of mathematical models based on networks and patch concepts, there are several topics which have been researched, including coex-

istence [13,16,18,21,24,25,28,30,33,35,37,39,40,43–46,51–54,62,64,65,68–72,80–82,92,96, 97,99–101,103,108],
dissipation [28,33,34,48,74,78,94], and eco-evolution [59,67,68,94].

(5) *Periodicity of the solution*: An interesting question for pollinators strongly related with seasonality is what the periodicity behavior of the populations of the different variables involved in pollination models is. Indeed, the following topics have been researched: periodic orbits [30,47,65,93], non-periodic orbits [28,45,48,65,74], and oscillation [30,53,63].
(6) *Numerical solutions and comparison with empirical data*: The mathematical models are strongly nonlinear, and the analytical solution cannot be construed. Consequently, numerical solutions of the mathematical models are introduced in order to simulate and calibrate the mathematical models. In the retained list, the authors of [19,25,26,31,32,35–37,42,53,56–58,60,61,66–69,75,78,84,85,87–90,92,94,96,98–101,103–106,108,110,111,113,115,116] developed numerical simulations.
(7) *Mathematical control*: Optimal control of the pollination systems via introducing appropriate control variables was conducted in [88,97].

4. Biological and Applied Problem Typologies in the Retained Literature

The retained articles address a spectrum of biological and applied problems through mathematical modeling. These can be categorized into seven thematic domains, each reflecting distinct modeling priorities, parametrization strategies, and implications for ecological policy and management:

(1) *Biological consistency and population viability*: In this group, we consider the works addressing biological realism in population dynamics and focus on the research of positive bounded solutions. Models in this category ensure that population variables remain biologically meaningful, i.e., non-negative and bounded over time. This foundational consistency is critical for validating ecological interpretations and avoiding spurious predictions. Parametrization typically involves biologically constrained initial conditions and growth functions (e.g., logistic or saturating terms). These models support policy decisions related to conservation thresholds and population viability. Representative works include [15,19,20,26,28,30,41,50,54,59,65,66,71,75,78,93,102].
(2) *Long-term dynamics and species persistence:* In this group, the problems to study are prevalence, extinction, and asymptotic behavior. These studies examine the conditions under which pollinator populations persist or collapse, often through linearization techniques and Lyapunov-based stability criteria. Parametrization emphasizes sensitivity to reproductive rates, mortality, and interaction coefficients. The results inform long-term sustainability planning and resilience forecasting. Representative works include [10–12,14,15,19,20,22,26,33,34,36,39,43–48,54,55,57,60,61,63,67,68,70,71,73–76, 78,79,81,84,86,90–93,95,96,98,100,102–104,107,111,113,114].
(3) *Regime shifts and critical transitions*: In this group of works, the authors focus on the bifurcation analysis and address the study of threshold phenomena and qualitative change. Bifurcation studies identify parameter regimes where small changes induce qualitative shifts in system behavior, such as transitions from coexistence to extinction. These models often employ continuation methods and bifurcation diagrams to explore critical thresholds, with implications for adaptive management and early warning indicators. Representative works include [16,30,53,67,74,94,103,106,107,114].
(4) *Mutualism and network structure*: In this group, the focus is mutualistic interactions and the study of phenomena like coexistence, dissipation, and eco-evolutionary dynamics. These models incorporate spatial structure, network topology, and evolutionary feedback to explore how mutualistic systems maintain biodiversity. Parametrization in-

cludes patch-based connectivity, trait evolution, and interaction matrices. The findings support the design of pollinator corridors, agroecological zoning, and biodiversity incentives. Representative works include [13,16,18,21,24,25,28,30,33,35,37,39,40,43–46,51–54,62,64,65,68–72,80–82,92,96,97,99–101,103,108], as well as dissipation-focused studies [28,33,34,48,74,78,94], and eco-evolutionary dynamics studies [59,67,68,94].

(5) *Seasonal and oscillatory behavior*: The addressed problem is the temporal variability and seasonality, along with the study of periodicity and oscillations in model solutions. Models in this group address how seasonal forcing and intrinsic dynamics lead to periodic or chaotic population fluctuations. Parametrization incorporates time-dependent coefficients and delay terms. These insights guide seasonal pollination services, crop planning, and phenological synchronization. Representative works include [28,30,45,47,48,53,63,65,74,93].

(6) *Simulation and empirical calibration:* In this group, we consider works focused on numerical solutions and data comparison and developed for model validation and empirical integration. Due to nonlinear complexity, many models rely on numerical simulations to explore parameter spaces and fit empirical data. Parametrization strategies include optimization techniques, sensitivity analysis, and empirical calibration. These models enhance the credibility of model-based recommendations and support data-driven decision making. Representative works include [19,25,26,31,32,35–37,42,53,56–58,60,61,66–69,75,78,84,85,87–90,92,94,96,98–101,103–106,108,110,111,113,115,116].

(7) *Intervention and optimization*: There are some works on solving the problem of applied control and resource allocation, which are focused on mathematical control. These studies introduce control variables—such as habitat enhancement or pesticide reduction—to optimize ecological outcomes. Parametrization uses Pontryagin's maximum principle or dynamic programming to derive optimal strategies. The results directly inform cost-effective conservation and adaptive management protocols (see [88,97]).

5. Other Aspects of the Literature Review

5.1. Research Gaps and Future Directions for Control, Stochastic Modeling, and Network-Based PDEs

Despite the breadth of topics addressed in the retained literature, three modeling domains remain notably underdeveloped: (1) optimal control under uncertainty, (2) stochastic ecological modeling, and (3) network- or patch-based partial differential equations (PDEs) for spatially structured systems. These gaps are particularly relevant given the increasing complexity of ecological systems and the need for robust, data-informed decision making:

(1) *Optimal Control under Uncertainty:* While mathematical control was explored in [88,97], current models rely on deterministic frameworks and assume full observability of system states and parameters. These assumptions limit applicability in real-world settings, where ecological responses to interventions (e.g., pesticide reduction or habitat restoration) are uncertain and data are sparse. Neither study incorporated stochastic perturbations or feedback mechanisms, nor did they address parameter uncertainty or adaptive control strategies. This restricts the robustness and generalizability of the proposed solutions.

(2) *Stochastic Modeling*: Across the retained list, stochastic formulations are conspicuously absent. Although several studies addressed oscillatory behavior and bifurcation phenomena (e.g., [30,53,63]), they did so within deterministic systems. The lack of stochastic differential equations or probabilistic transitions limits the capacity to model demographic noise, environmental variability, and uncertainty propagation, especially in fragmented landscapes or under climate stress. This gap is critical given the increasing emphasis on resilience and risk-aware ecological planning.

(3) *Network-Based PDEs and Patch Dynamics*: Numerous studies incorporated network or patch structures in mutualistic systems (e.g., [13,21,28,33,40,65,71,72,96,100]), yet most relied on discrete or compartmental models. Continuous-space PDEs on networks or graph-based domains are rare, and when present, they often lack empirical calibration or realistic topologies. For example, the authors of [28,48,74] explored dissipation and spatial dynamics but did not integrate high-resolution landscape data or adaptive dispersal mechanisms. This limits the ecological realism and policy relevance of spatial predictions.

5.2. *Limitations of Current Findings*

Across these domains, a recurring limitation is the scarcity of longitudinal, high-resolution data for model calibration and validation. Many studies rely on synthetic simulations (e.g., [67–69,75,78]) or static parameter estimates, which constrain ecological realism and hinder generalization across systems. Furthermore, sensitivity analysis and uncertainty quantification are rarely formalized, reducing the interpretability and robustness of model outcomes. Additionally, we report at least three limitations of this research: a checklist for data extraction was not constructed, only two databases were considered, and the analysis of data was developed without using advanced methodologies.

5.3. *Methodological Pathways*

To address these gaps, future research should pursue hybrid frameworks that integrate stochasticity into control models (e.g., stochastic optimal control or robust model predictive control), and embed network-aware partial differential equations within empirically grounded landscapes. Promising techniques include the following:

(1) Graph Laplacians and metapopulation partial differential equations for dispersal modeling;
(2) Bayesian inference and ensemble simulations for uncertainty quantification;
(3) Data assimilation methods for real-time calibration.

5.4. *Potential Data Sources*

Empirical grounding can be strengthened using the following:

(1) Remote sensing data for habitat fragmentation and land use change;
(2) Citizen science platforms (e.g., iNaturalist or eBird) for species occurrence;
(3) Long-term ecological monitoring networks (e.g., Global Biodiversity Information Facility—GBIF, or Long Term Ecological Research—LTER) for population dynamics.

5.5. *Roadmap for Future Work*

A strategic agenda should include the following:

(1) Development of modular, interoperable modeling platforms that integrate control, stochasticity, and spatial structure;
(2) Co-design of models with stakeholders to ensure contextual relevance and usability;
(3) Formal incorporation of sensitivity analysis and uncertainty quantification;
(4) Establishment of typological benchmarks to compare model performance across ecological and socio-political scenarios.

Such efforts will enhance both theoretical rigor and translational impact, positioning mathematical ecology as a key contributor to adaptive management and evidence-based policy design.

5.6. Implications of the Retained Modeling Topics for Agricultural Planning, Habitat Management, and Pesticide Regulation

The seven modeling themes identified in Section 3.5 provide a rigorous mathematical foundation for informing real-world decision making in agroecological systems. Their relevance extends to agricultural planning, habitat conservation, and the formulation of pesticide policies. Below, we detail the practical implications of each topic:

(1) *Positive Bounded Solutions:* Ensuring that model solutions remain positive and bounded is essential for biological realism, particularly when variables represent population densities. This property supports the development of ecologically valid simulations that can guide agricultural interventions and pesticide thresholds, preventing unintended population collapses.
(2) *Equilibrium and Stability Analysis*: Stability analysis enables the characterization of long-term system behavior, including species persistence or extinction. In agricultural contexts, it informs crop-pollinator compatibility and resilience, while in habitat management, it supports the design of restoration strategies and ecological corridors.
(3) *Bifurcation Analysis*: Bifurcation theory reveals how small parameter changes can induce qualitative shifts in system dynamics. This is critical for anticipating nonlinear responses to environmental stressors, such as pesticide application or habitat fragmentation, and for designing adaptive management strategies that avoid tipping points.
(4) *Mutualistic Interactions*: Modeling mutualistic networks elucidates mechanisms of coexistence, dissipation, and eco-evolutionary dynamics. These insights inform the diversification of cropping systems, the conservation of keystone mutualists, and the regulation of agrochemicals that may disrupt ecological interactions.
(5) *Periodicity of Solutions*: Seasonal and periodic behaviors in pollinator populations are central to synchronizing agricultural calendars with ecological cycles. Understanding periodicity aids in optimizing planting schedules, flowering periods, and pesticide applications to align with pollinator activity.
(6) *Numerical Simulations and Empirical Validation*: Given the nonlinear nature of most models, numerical simulations are indispensable for calibration and scenario testing. These simulations support evidence-based agricultural planning and policy evaluation, enabling cost-benefit analyses of proposed interventions.
(7) *Mathematical Control*: Optimal control frameworks allow for the strategic modulation of system variables to achieve desired ecological or economic outcomes. In agriculture, this translates to resource-efficient practices that sustain pollinator populations, while in regulatory contexts, it supports dynamic policy design responsive to ecological feedback.

Collectively, these modeling approaches bridge theoretical ecology with applied decision making, offering quantitative tools for sustainable land use, biodiversity conservation, and environmental governance.

5.7. A Particular Comparative Analysis

In this subsection we develop a comparative analysis between [74] and the Thematic Synthesis in Section 3.5.

The authors of [74] offered a focused and technically rigorous contribution to the mathematical modeling of ecological systems, particularly in the context of bifurcation analysis, dissipative dynamics, and non-periodic oscillatory behavior. Chen's work employed nonlinear differential equations to explore critical transitions and qualitative shifts in population dynamics, with emphasis on parameter sensitivity and system resilience. The model demonstrates how small perturbations can lead to significant changes in ecological outcomes, contributing to the literature on regime shifts and early warning indicators.

However, when compared with the broader synthesis presented in Section 3.5, the scope of [74] appears more specialized and thematically constrained. The retained literature encompassed seven interrelated modeling domains, ranging from positive boundedness and stability analysis to mutualistic networks, seasonality, empirical calibration, and optimal control. This thematic architecture enables a more comprehensive understanding of pollinator dynamics and ecological decision making.

Notably, the synthesis in Section 3.5 integrates multiple methodological layers:

- It links biological realism (e.g., positive bounded solutions [15,19]) with long-term system behavior (e.g., stability analysis [10,44]).
- It incorporates spatial and network structure in mutualistic interactions [21,33], extending beyond the local dynamics emphasized in [74].
- It addresses empirical calibration and numerical simulation [68,75], which are not central in Chen's formulation.
- It introduces optimal control frameworks [88,97], offering policy-relevant strategies absent in [74].

In this light, the novelty of Section 3.5 lies in its integrative typology, which not only categorizes the retained studies but also reveals methodological synergies and thematic gaps. While [74] contributes valuable insights into bifurcation and dissipation phenomena, the broader synthesis provides a multidimensional roadmap for future research—bridging theoretical modeling with empirical validation and policy design.

This comparative perspective underscores the importance of typological frameworks in advancing ecological modeling, enabling researchers to situate individual studies within a structured landscape of methodological and applied relevance.

6. Conclusions

In this paper, we applied a systematic bibliographic review. Our research methodology adeptly allowed us to identify and analyze a substantial body of research on the mathematical modeling of pollinators. We retrieved and reviewed 107 works published between 1981 and 2025, leveraging databases such as the Web of Science and Mathscinet. We examined the mathematical theory and the topics analyzed. Our findings reveal a significant increase in research dedicated to the introduction or improvement of mathematical modeling to study the dynamics of pollinators. The landscape of mathematical modeling of pollinators has covered the standard topics of dynamical systems, like equilibrium and stability analysis. However, in recent years, the field has shifted dramatically, moving away to include some new topics like the fractional-order or diffusion models. In particular, modeling by using networks has promise in the future development of research. Moreover, given that the mathematical models arise from different mathematical approaches, it is essential to use an interdisciplinary approach for constructing complex models that more closely resemble pollination phenomena.

Research on the mathematical modeling of pollinators is an active area. However, there is still much to be developed in the context of addressing the challenges of pollination dynamics. We identified four issues that require further detailed exploration. First, there is the study of mathematical control theory about biological control. We identified only two works related to control theory (see [88,97]). Therefore, constructing mathematical models that incorporate the principles of optimal control is necessary. Second, the development of stochastic models is an area that needs strong attention from researchers. Third, the inclusion of network and patch concepts with mathematical models based on partial differential equations is a topic that requires attention and future development. Fourth, we reduced the present analysis to the word pollinators and two databases. Clearly, some representative works researching mathematical models for pollinators were excluded. We plan to expand our search to other specific pollinators and to other bibliographic databases, including Scopus, PubMed, BIOSIS Previews, and AGRICOLA.

Author Contributions: Conceptualization, F.H. and A.C.; methodology, F.H. and A.C.; software, E.L. and J.T.; validation, E.L. and J.T.; formal analysis, F.H., A.C. and E.L.; investigation, F.H. and A.C.; resources, F.H. and A.C.; data curation, E.L.; writing—original draft preparation, F.H. and A.C.; writing—review and editing, F.H., A.C. and E.L.; visualization, F.H. and A.C.; supervision, F.H. and A.C.; project administration, F.H. and A.C.; funding acquisition, F.H. and A.C. All authors have read and agreed to the published version of the manuscript.

Funding: This research was funded by the National Agency for Research and Development, ANID-Chile, through FONDECYT project 1230560, and the Project supported by the Competition for Research Regular Projects, year 2023, code LPR23-03, Universidad Tecnológica Metropolitana. A. C. was funded by the Universidad del Bío–Bío through projects FAPEI FP2510413 and RE2547710.

Institutional Review Board Statement: Not applicable.

Informed Consent Statement: Not applicable.

Data Availability Statement: The original contributions presented in this study are included in the article.

Conflicts of Interest: The authors declare no conflicts of interest.

Appendix A. Details on Counting the Regions

In order to count the regions, we considered the affiliations declared by the authors. For instance, in [10], the authors, Soberon and Rio, declared Mexico as their affiliation, and in [11], the author, Wells, declared USA for the affiliation. Extensive details for the retained list are presented on Table A1. We note that in the case of the same author declaring multiple affiliations, we considered only the first declared region.

Table A1. Affiliations of the authors declared on the retained list. Here, RR stands for retained reference.

RR	Affiliations	RR	Affiliations	RR	Affiliations	RR	Affiliations
[10]	Mexico (2)	[37]	USA (2), China (2)	[64]	China (1)	[91]	USA (4)
[11]	USA (1)	[38]	Australia (3)	[65]	USA (4), China (2), UK (1)	[92]	China (3)
[12]	USA (1)	[39]	Netherlands (2), France (1)	[66]	Spain (4)	[93]	USA (6)
[13]	USA (2)	[40]	Ecuador (1), Spain (2)	[67]	Mexico (2)	[94]	China (4)
[14]	Sweden (2)	[41]	USA (5), Brazil (1)	[68]	Australia (1), UK (1)	[95]	China (1), Brazil (1), Netherlands (1), France (1), Spain (4)
[15]	USA (1)	[42]	Mexico (3)	[69]	R. of Korea (3)	[96]	Germany (4)
[16]	Italy (1)	[43]	Mexico (1)	[70]	China (1)	[97]	India (3)
[17]	UK (3)	[44]	China (2)	[71]	China (1)	[98]	USA (2)
[18]	USA (2)	[45]	China (2)	[72]	USA (1)	[99]	China (2)
[19]	Mexico (2)	[46]	China (1), USA (2)	[73]	Czech Republic (2)	[100]	Japan (2)
[20]	Argentina (2)	[47]	Mexico (2)	[74]	China (3)	[101]	India (3)
[21]	Chile (5)	[48]	China (3)	[75]	India (2)	[102]	USA (4)
[22]	Israel (2)	[49]	USA (4)	[76]	China (2), USA (1)	[103]	Czech Republic (3)
[23]	UK (3), Canada (1), USA (1)	[50]	China (3)	[77]	China (3)	[104]	Bulgaria (3)
[24]	Germany (2)	[51]	France (2)	[78]	China (4)	[105]	South African (2), Netherlands (1)
[25]	Argentina (2), Germany (1)	[52]	South African (2)	[79]	India (2)	[106]	India (2)

Table A1. *Cont.*

RR	Affiliations	RR	Affiliations	RR	Affiliations	RR	Affiliations
[26]	Mexico (2)	[53]	China (1), France (1), USA (1)	[80]	China (3), USA (2)	[107]	Mexico (3)
[27]	Canada (3)	[54]	Ecuador (1), Spain (2)	[81]	USA (2), France (1), Switzerland (1)	[108]	South Africa (4)
[28]	China (1), USA (3)	[55]	Mexico (2)	[82]	USA (6), UK (1)	[109]	Netherlands (2), Canada (1)
[29]	USA (4)	[56]	USA (5), Sweden (2)	[83]	USA (6), New Zeland (6)	[110]	Japan (3)
[30]	China (3)	[57]	Canada (2), Australia (2)	[84]	Saudi Arabia (2), Mexico (1), Pakistan (1)	[111]	China (4)
[31]	USA (1)	[58]	Belgium (2), Canada (2), Netherlands (1)	[85]	USA (6)	[112]	China (2)
[32]	USA (3)	[59]	China (2), USA (1)	[86]	Czech Republic (3)	[113]	Czech Republic (1)
[33]	China (2)	[60]	Japan (1)	[87]	USA (3)	[115]	China (1), Canada (1)
[34]	China (1)	[61]	South Africa (1), France (2)	[88]	Spain (2), Hungary (6)	[114]	India (3)
[35]	Chile (3), USA (1)	[62]	Czech Republic (2)	[89]	Italy (4), Germany (7)	[116]	Czech Republic (1)
[36]	UK (5)	[63]	Mexico (2)	[90]	USA (4)		

Appendix B. Details on Journals for Retained List

We identified the journals and searched for the impact factor, H index and quartiles, which are presented in Table A2.

Table A2. List of the journals appearing in the retained reference list.

Journal	H Index	SJR	Quartile
Agronomy-Basel	114	3.7	Q1
Alexandria Engineering Journal	112	5.6	Q1
American Naturalist	236	3	Q2
Annals of Botany	215	4.1	Q1
Applied Ecology and Environmental Research	48	0.9	Q4
Applied Mathematical Modelling. Simulation and Computation for Engineering and Environmental Systems	150	4.2	Q1
Applied Mathematics and Computation	182	3.1	Q1
Applied Sciences-Basel	162	2.7	Q2
Biosystems	85	0.392	Q2
Boletín de la Sociedad Matemática Mexicana. Third Series	20	0.414	Q2
Bulletin of Mathematical Biology	101	0.702	Q1
Chaos, Solitons & Fractals	175	1.184	Q1
Chaos. An Interdisciplinary Journal of Nonlinear Science	–	–	–
Communications in Nonlinear Science and Numerical Simulation	143	0.956	Q1
Discrete and Continuous Dynamical Systems. Series A	80	1.065	Q1
Discrete and Continuous Dynamical Systems. Series B. A Journal Bridging Mathematics and Sciences	65	0.735	Q1
Discrete and Continuous Dynamical Systems. Series S	43	0.514	Q2
Ecological Modelling	189	0.896	Q1
Ecological Research	87	0.616	Q2
Ecology	345	5.5	Q1
Ecology and Evolution	109	0.858	Q1
Ecology Letters	330	9.8	Q1
European Journal of Applied Mathematics	53	0.750	Q2

Table A2. *Cont.*

Journal	H Index	SJR	Quartile
Evolution	227	3.4	Q2
Evolutionary Applications	95	1.362	q1
Evolutionary Ecology	96	0.645	Q2
Evolutionary Ecology Research	82	–	–
International Journal for Parasitology-Parasites and Wildlife	44	0.618	Q1
International Journal of Bifurcation and Chaos in Applied Sciences and Engineering	120	0.596	Q1
International Journal of Biomathematics	38	0.527	Q2
Journal of Applied Ecology	216	6.2	Q1
Journal of Biological Dynamics	46	0.597	Q2
Journal of Biological Systems	39	0.487	Q2
Journal of Ecology	219	6.1	Q1
Journal of Evolutionary Biology	148	0.921	Q1
Journal of Mathematical Biology	111	0.921	Q1
Journal of Mathematics	30	0.322	Q3
Journal of Statistical Mechanics: Theory and Experiment	95	0.373	Q3
Journal of the European Mathematical Society	68	3.043	Q1
Journal of the Royal Society Interface	177	1.025	Q1
Journal of Theoretical Biology	178	0.532	Q2
Lobachevskii Journal of Mathematics	31	0.435	Q2
Mathematical Biosciences	114	0.555	Q2
Mathematical Methods in the Applied Sciences	87	1.991	Q1
Modeling Earth Systems and Environment	66	0.654	Q1
Natural Resource Modeling	38	0.521	Q2
Nonlinear Analysis. Real World Applications. An International Multidisciplinary Journal	106	1.168	Q1
Nonlinear Studies. The International Journal	22	0.229	Q4
Oikos	210	1.438	Q1
Physica A. Statistical Mechanics and its Applications	195	0.669	Q2
Physica D: Nonlinear Phenomena	154	0.940	Q1
PLoS ONE	467	3.3	Q1
PLoS Pathogens	260	5.5	Q1
Proceedings of the National Academy of Sciences of the United States of America	896	10.8	Q1
Royal Society Open Science	92	0.795	Q1
Scientific Reports	347	4.3	Q1
Theoretical Ecology	45	0.524	Q2
Theoretical Population Biology	99	0.563	Q2

References

1. Potts, S.G.; Imperatriz-Fonseca, V.L.; Ngo, H.T. *IPBES the Assessment Report of the Intergovernmental Science-Policy Platform on Biodiversity and Ecosystem Services on Pollinators, Pollination and Food Production*; Secretariat of the Intergovernmental Science-Policy Platform on Biodiversity and Ecosystem Services: Bonn, Germany, 2017; p. 552.
2. Deeksha, M.G.; Khan, M.S.; Kumaranag, K.M. Cuphea hyssopifolia Kunth: A Potential Plant for Conserving Insect Pollinators in Shivalik Foot Hills of Himalaya. *Nat. Acad. Sci. Lett.* **2023**, *46*, 137–142. [CrossRef]
3. Devillers, J. *In Silico Bees*; Taylor & Francis, CRC Press: Boca Raton, FL, USA, 2014.
4. Winston, M.L. *The Biology of the Honey Bee*; Harvard University Press: Cambridge, MA, USA, 1991.
5. Karlin, S.; Farkash, S. Some Multiallele Partial Assortative Mating Systems for a Polygamous Species. *Theoret. Popul. Biol.* **1978**, *14*, 446–470. [CrossRef] [PubMed]

6. Belsare, P.V.; Sriram, B.; Watve, M.G. The Co-optimization of Floral Display and Nectar Reward. *J. Biosci.* **2009**, *34*, 963–967. [CrossRef]
7. Chen, J.; DeGrandi-Hoffman, G.; Ratti, V.; Kang, Y. Review on mathematical modeling of honeybee population dynamics. *Math. Biosci. Eng.* **2021**, *18*, 9606–9650. [CrossRef]
8. Lozada, E.; Guerrero-Ortiz, C.; Coronel, A.; Medina, R. Classroom Methodologies for Teaching and Learning Ordinary Differential Equations: A Systemic Literature Review and Bibliometric Analysis. *Mathematics* **2021**, *9*, 745. [CrossRef]
9. Khan, K.S.; Kunz, R.; Kleijnen, J.; Antes, G. Five Steps to Conducting a Systematic Review. *J. R. Soc. Med.* **2003**, *96*, 118–121. [CrossRef]
10. Soberon, J.; Rio, C. The Dynamics of a Plant-Pollinator Interaction. *J. Theoret. Biol.* **1981**, *91*, 363–378. [CrossRef]
11. Wells, H. Population Equilibria and Stability in Plant-Animal Pollination Systems. *J. Theoret. Biol.* **1983**, *100*, 685–699. [CrossRef]
12. Armstrong, R. A Patch Model of Mutualism. *J. Theoret. Biol.* **1987**, *125*, 243–246. [CrossRef]
13. Hastings, A.; Wolin, C.L. Within-Patch Dynamics in a Metapopulation. *Ecology* **1989**, *70*, 1261–1266. [CrossRef]
14. Ingvarsson, P.K.; Lundberg, S. The Effect of a Vector-Borne Disease on the Dynamics of Natural Plant-Populations—A Model for Ustilago-Violacea Infection of Lychnis-Viscaria. *J. Ecol.* **1993**, *81*, 263–270. [CrossRef]
15. Jang, S. Dynamics of Herbivore-Plant-Pollinator Models. *J. Math. Biol.* **2002**, *44*, 129–149. [CrossRef]
16. Dercole, F. Border Collision Bifurcations in the Evolution of Mutualistic Interactions. *Int. J. Bifurc. Chaos* **2005**, *15*, 2179–2190. [CrossRef]
17. Stewart-Cox, J.; Britton, N.; Mogie, M. Pollen Limitation or Mate Search Need not Induce an Allee Effect. *Bull. Math. Biol.* **2005**, *67*, 1049–1079. [CrossRef]
18. Smith, C.A.; Wilson, W.G. Evolutionarily Stable Dispersal with Pattern Formation in a Mutualist-Antagonist System. *Evol. Ecol. Res.* **2007**, *9*, 987–1004.
19. Sánchez-Garduño, F.; Breña-Medina, V.E. Spatio-Temporal Dynamics of a Three Interacting Species Mathematical Model Inspired in Physics. In Proceedings of the AIP Conference Proceedings, Ponta Delgada, Portugal, 18–23 May 2008; Dagdug, L., Scherer, L.G.C., Eds.; Universidad Autonoma Metropolitana: Mexico City, Mexico, 2008; Volume 978, pp. 115–134.
20. Momo, F.R.; Ure, J.E. Stability and Fluctuations in a Three Species System: A Plant with Two Very Different Pollinators. *Appl. Ecol. Environ. Res.* **2009**, *7*, 287–296. [CrossRef]
21. Valdovinos, F.S.; Ramos-Jiliberto, R.; Flores, J.D.; Espinoza, C.; Lópen, G. Structure and Dynamics of Pollination Networks: The Role of Alien Plants. *Oikos* **2009**, *118*, 1190–1200. [CrossRef]
22. Fishman, M.A.; Hadany, L. Plant-Pollinator Population Dynamics. *Theor. Popul. Biol.* **2010**, *78*, 270–277. [CrossRef]
23. Little, T.J.; Shuker, D.M.; Colegrave, N.; Day, T.; Graham, A.L. The Coevolution of Virulence: Tolerance in Perspective. *PLoS Pathog.* **2010**, *6*, e1001006. [CrossRef]
24. Oña, L.; Lachmann, M. Ant Aggression and Evolutionary Stability in Plant-Ant and Plant-Pollinator Mutualistic Interactions. *J. Evol. Biol.* **2011**, *24*, 617–629. [CrossRef]
25. Abramson, G.; Soto, C.A.T.; Oña, L. The Role of Asymmetric Interactions on the Effect of Habitat Destruction in Mutualistic Networks. *PLoS ONE* **2011**, *6*, e0021028. [CrossRef]
26. Sánchez-Garduño, F.; Breña-Medina, V. Searching for Spatial Patterns in a Pollinator-Plant-Herbivore Mathematical Model. *Bull. Math. Biol.* **2011**, *73*, 1118–1153. [CrossRef] [PubMed]
27. Tyson, R.C.; Wilson, J.B.; Lane, W.D. Beyond Diffusion: Modelling Local and Long-Distance Dispersal for Organisms Exhibiting Intensive and Extensive Search Modes. *Theor. Popul. Biol.* **2012**, *79*, 70–81. [CrossRef]
28. Wang, Y.; DeAngelis, D.; Holland, J. Uni-Directional Interaction and Plant-Pollinator-Robber Coexistence. *Bull. Math. Biol.* **2012**, *74*, 2142–2164. [CrossRef] [PubMed]
29. Gilman, R.T.; Fabina, N.S.; Abbott, K.C.; Rafferty, N.E. Evolution of Plant-Pollinator Mutualisms in Response to Climate Change. *Evol. Appl.* **2012**, *5*, 2–16. [CrossRef]
30. Wang, Y.S.; Wu, H.; Sun, S. Persistence of Pollination Mutualisms in Plant-Pollinator-Robber Systems. *Theor. Popul. Biol.* **2012**, *81*, 243–250. [CrossRef]
31. Yong, K.E. A Mathematical Model of the Interactions Between Pollinators and Their Effects on Pollination of Almonds. Ph.D. Thesis, The University of Iowa, Iowa City, IA, USA, 2012.
32. Yong, K.E.; Li, Y.; Hendrix, S. Habitat Choice of Multiple Pollinators in Almond Trees and its Potential Effect on Pollen Movement and Productivity: A Theoretical Approach Using the Shigesada-Kawasaki-Teramoto Model. *J. Theoret. Biol.* **2012**, *305*, 103–109. [CrossRef]
33. Wang, Y.; Wu, H. Invasibility of nectarless flowers in plant-pollinator systems. *Bull. Math. Biol.* **2013**, *75*, 1138–1156. [CrossRef] [PubMed]
34. Wang, Y. Dynamics of plant-pollinator-robber systems. *J. Math. Biol.* **2013**, *66*, 1155–1177. [CrossRef]
35. Valdovinos, F.S.; de Espanes, P.M.; Flores, J.D.; Ramos-Jiliberto, R. Adaptive foraging allows the maintenance of biodiversity of pollination networks. *Oikos* **2013**, *122*, 907–917. [CrossRef]

36. Bryden, J.; Gill, R.J.; Mitton, R.A.A.; Raine, N.E.; Jansen, V.A.A. Chronic Sublethal Stress Causes Bee Colony Failure. *Ecol. Lett.* **2013**, *16*, 1463–1469. [CrossRef] [PubMed]
37. Holland, J.N.; Wang, Y.S.; Sun, S.; DeAngelis, D.L. Consumer-Resource Dynamics of Indirect Interactions in a Mutualism-Parasitism Food Web Module. *Theor. Ecol.* **2013**, *6*, 475–493. [CrossRef]
38. Khoury, D.S.; Barron, A.B.; Myerscough, M.R. Modelling Food and Population Dynamics in Honey Bee Colonies. *PLoS ONE* **2013**, *8*, e59084. [CrossRef] [PubMed]
39. Encinas-Viso, F.; Revilla, T.; Etienne, R. Shifts in Pollinator Population Structure may JeopardizePollination Service. *J. Theoret. Biol.* **2014**, *352*, 24–30. [CrossRef]
40. Guerrero, G.; Langa, J.; Suárez, A. Biodiversity and Vulnerability in a 3D Mutualistic System. *Discret. Contin. Dyn. Syst.* **2014**, *34*, 4107–4126. [CrossRef]
41. Fagan, W.F.; Bewick, S.; Cantrell, S.; Cosner, C.; Varassin, I.G.; Inouye, D.W. Phenologically Explicit Models for Studying Plant-Pollinator Interactions under Climate Change. *Theor. Ecol.* **2014**, *7*, 289–297. [CrossRef]
42. Sánchez-Garduño, F.; Castellanos, V.; Quilantán, I. Dynamics of a Nonlinear Mathematical Model for Three Interacting Populations. *Bol. Soc. Mat. Mex.* **2014**, *20*, 147–170. [CrossRef]
43. Vargas-De-León, C. Global Stability for Multi-Species Lotka-Volterra Cooperative Systems: One Hyper-Connected Mutualistic-Species. *Int. J. Biomath.* **2015**, *8*, 1550039. [CrossRef]
44. Wang, Y.; Wu, H. Stability of Plant-Pollinator-Ant Co-Mutualism. *Appl. Math. Comput.* **2015**, *261*, 231–241. [CrossRef]
45. Wang, Y.; Wang, S. Persistence of Pollination Mutualisms in the Presence of Ants. *Bull. Math. Biol.* **2015**, *77*, 202–229. [CrossRef]
46. Wang, Y.; DeAngelis, D.; Holland, J. Dynamics of an Ant-Plant-Pollinator Model. *Commun. Nonlinear Sci. Numer. Simul.* **2015**, *20*, 950–964. [CrossRef]
47. Barradas, I.; Vázquez, V. Improving Pollination Through Learning. *J. Biol. Syst.* **2015**, *23*, S77–S90. [CrossRef]
48. Wang, Y.S.; Wu, H.; Wang, S.K. Invasibility of Nectar Robbers in Pollination-Mutualisms. *Appl. Math. Comput.* **2015**, *250*, 908–919. [CrossRef]
49. Jover, L.F.; Flores, C.O.; Cortez, M.H.; Weitz, J.S. Multiple Regimes of Robust Patterns between Network Structure and Biodiversity. *Sci. Rep.* **2015**, *5*, 17856. [CrossRef] [PubMed]
50. Wang, L.; Jiang, H.; Li, Y. Positive Steady State Solutions of a Pant-Pollinator Model with Diffusion. *Discret. Contin. Dyn. Syst. Ser. B* **2015**, *20*, 1805–1819. [CrossRef]
51. Georgelin, E.; Loeuille, N. Evolutionary Response of Plant Interaction Traits to Nutrient Enrichment Modifies the Assembly and Structure of Antagonistic-Mutualistic Communities. *J. Ecol.* **2016**, *104*, 193–205. [CrossRef]
52. Minoarivelo, H.O.; Hui, C. Invading a Mutualistic Network: To be or not to be Similar. *Ecol. Evol.* **2016**, *6*, 4981–4996. [CrossRef] [PubMed]
53. Liu, Z.; Magal, P.; Ruan, S. Oscillations in Age-Structured Models of Consumer-Resource Mutualisms. *Discret. Contin. Dyn. Syst. Ser. B* **2016**, *21*, 537–555. [CrossRef]
54. Guerrero, G.; Langa, J.; Suárez, A. Architecture of attractor determines dynamics on mutualistic complex networks. *Nonlinear Anal. Real World Appl.* **2017**, *34*, 17–40. [CrossRef]
55. Vázquez, V.; Barradas, I. Deceptive Pollination and Insects' Learning: A Delicate Balance. *J. Biol. Dyn.* **2017**, *11*, 299–322. [CrossRef]
56. Banks, H.T.; Banks, J.E.; Bommarco, R.; Laubmeier, A.N.; Myers, N.J.; Rundlöf, M.; Tillman, K. Modeling Bumble Bee Population Dynamics with Delay Differential Equations. *Ecol. Model.* **2017**, *351*, 14–23. [CrossRef]
57. Bouhours, J.; Mesgaran, M.B.; Cousens, R.D.; Lewis, M.A. Neutral Hybridization Can Overcome a Strong Allee Effect by Improving Pollination Quality. *Theor. Ecol.* **2017**, *10*, 319–339. [CrossRef]
58. Vallaeys, V.; Tyson, R.C.; Lane, W.D.; Deleersnijder, E.; Hanert, E. A Levy-Flight Diffusion Model to Predict Transgenic Pollen Dispersal. *J. R. Soc. Interface* **2017**, *15*, 20160889. [CrossRef]
59. Huang, J.; Liu, Z.; Ruan, S. Bifurcation and Temporal Periodic Patterns in a Plant-Pollinator Model with Diffusion and Time Delay Effects. *J. Biol. Dyn.* **2017**, *11*, 138–159. [CrossRef]
60. Yamauchi, A. Joint evolution of interspecific mutualism and regulation of variation of interaction under directional selection in trait space. *Theor. Ecol.* **2017**, *10*, 477–491. [CrossRef]
61. Dumont, Y.; Soulie, J.; Michel, F. Modeling oil Palm Pollinator Dynamics using Deterministic and Agent-Based Approaches. Applications on Fruit Set Estimates. Some Preliminary Results. *Math. Methods Appl. Sci.* **2018**, *41*, 8545–8564. [CrossRef]
62. Revilla, T.; Křivan, V. Competition, Trait-Mediated Facilitation, and the Structure of Plant-Pollinator Communities. *J. Theoret. Biol.* **2018**, *440*, 42–57. [CrossRef]
63. Vázquez, V.; Barradas, I. A Plant-Pollinator System: How Learning Versus Cost-Benefit can Induce Periodic Oscillations. *Int. J. Biomath.* **2018**, *11*, 1850024. [CrossRef]
64. Wang, Y. Global Dynamics of a Competition-Parasitism-Mutualism Model Characterizing Plant-Pollinator-Robber Interactions. *Phys. A Stat. Mech. Its Appl.* **2018**, *510*, 26–41. [CrossRef]

65. Jiang, J.J.; Huang, Z.G.; Seager, T.P.; Lin, W.; Grebogi, C.; Hastings, A.; Lai, Y.C. Predicting Tipping Points in Mutualistic Networks through Dimension Reduction. *Proc. Natl. Acad. Sci. USA* **2018**, *115*, E639–E647. [CrossRef]
66. Usó-Doménech, J.L.; Nescolarde-Selva, J.A.; Lloret-Climent, M.; González-Franco, L. Behavior of Pyrophite Shrubs in Mediterranean Terrestrial Ecosystems (i): Population and Reproductive Model. *Math. Biosci.* **2018**, *297*, 58–77. [CrossRef]
67. Castellanos, V.; Sánchez-Garduño, F. The Existence of a Limit Cycle in a Pollinator-Plant-Herbivore Mathematical Model. *Nonlinear Anal. Real World Appl.* **2019**, *48*, 212–231. [CrossRef]
68. Cropp, R.; Norbury, J. Simulating Eco-Evolutionary Processes in an Obligate Pollination Model with a Genetic Algorithm. *Bull. Math. Biol.* **2019**, *81*, 4803–4820. [CrossRef]
69. Maeng, S.; Lee, J.; Lee, D. Competition-Induced Increase of Species Abundance in Mutualistic Networks. *J. Stat. Mech. Theory Exp.* **2019**, *19*, 033502. [CrossRef]
70. Wang, Y. Pollination-Mutualisms in a Two-Patch System with Dispersal. *J. Theoret. Biol.* **2019**, *476*, 51–61. [CrossRef]
71. Wang, Y. Dynamics of a Plant-Nectar-Pollinator Model and its Approximate Equations. *Math. Biosci.* **2019**, *307*, 42–52. [CrossRef]
72. Valdovinos, F.S. Mutualistic Networks: Moving Closer to a Predictive Theory. *Ecology Lett.* **2019**, *22*, 1517–1534. [CrossRef]
73. Krivan, V.; Revilla, T.A. Plant Coexistence Mediated by Adaptive Foraging Preferences of Exploiters or Mutualists. *J. Theor. Biol.* **2019**, *480*, 112–128. [CrossRef] [PubMed]
74. Chen, M.; Wu, H.; Wang, Y. Persistence and Oscillations of Plant-Pollinator-Herbivore Systems. *Bull. Math. Biol.* **2020**, *82*, 57. [CrossRef] [PubMed]
75. Devi, S.; Mishra, R. A plant-pollinator-pesticide model. *Nonlinear Stud.* **2020**, *27*, 1003–1019.
76. Wang, Y.; Wu, H.; Wang, S. Persistence of Pollination Mutualisms under Pesticides. *Appl. Math. Model.* **2020**, *77*, 861–880. [CrossRef]
77. Wu, H.; Sun, S.; Wang, Y. Bifurcations in a Pollination-Mutualism System with Nectarless Flowers. *Discret. Contin. Dyn. Syst.-S* **2020**, *13*, 3213–3229.
78. Yuan, H.; Wang, Y.; Wu, H.; Wu, S. Persistence of Pollination-Mutualisms under the Effect of Intermediary Resource. *Nat. Resour. Model.* **2020**, *33*, e12259. [CrossRef]
79. Devi, S.; Mishra, R.P. A Mathematical Model to See the Effects of Increasing Environmental Temperature on Plant-Pollinator Interactions. *Model. Earth Syst. Environ.* **2020**, *6*, 1315–1329. [CrossRef]
80. Zhang, H.X.; Liu, X.M.; Wang, Q.; Zhang, W.D.; Gao, J.X. Co-Adaptation Enhances the Resilience of Mutualistic Networks. *J. R. Soc. Interface* **2020**, *17*, 20200236. [CrossRef]
81. Yule, K.M.; Johnson, C.A.; Bronstein, J.L.; Ferrière, R. Interactions among Interactions: The Dynamical Consequences of Antagonism between Mutualists. *J. Theor. Biol.* **2020**, *501*, 110334. [CrossRef]
82. Figueroa, L.L.; Grab, H.; Ng, W.H.; Myers, C.R.; Graystock, P.; McFrederick, Q.S.; McArt, S.H. Landscape Simplification Shapes Pathogen Prevalence in Plant-Pollinator Networks. *Ecol. Lett.* **2020**, *23*, 1212–1222. [CrossRef]
83. Peace, A.; Pattemore, D.; Broussard, M.; Fonseka, D.; Tomer, N.; Bosque-Pérez, N.A.; Crowder, D.; Shaw, A.K.; Jesson, L.; Howlett, B.G.; et al. Orchard Layout and Plant Traits Influence Fruit Yield More Strongly than Pollinator Behaviour and Density in a Dioecious Crop. *PLoS ONE* **2020**, *15*, e0231120. [CrossRef]
84. Khan, A.; Gómez-Aguilar, J.F.; Abdeljawad, T.; Khan, H. Stability and Numerical Simulation of a Fractional Order Plant-Nectar-Pollinator Model. *Alex. Eng. J.* **2020**, *59*, 49–59. [CrossRef]
85. Burnham, P.A.; Alger, S.A.; Case, B.; Boncristiani, H.; Hébert-Dufresne, L.; Brody, A.K. Flowers as Dirty Doorknobs: Deformed Wing Virus Transmitted between Apis Mellifera and Bombus Impatiens through Shared Flowers. *J. Appl. Ecol.* **2021**, *58*, 2065–2074. [CrossRef]
86. Revilla, T.A.; Marcou, T.; Krivan, V. Plant Competition under Simultaneous Adaptation by Herbivores and Pollinators. *Ecol. Model.* **2021**, *455*, 109634. [CrossRef]
87. McPeek, S.J.; Bronstein, J.L.; McPeek, M.A. The Evolution of Resource Provisioning in Pollination Mutualisms. *Am. Nat.* **2020**, *198*, 441–459. [CrossRef] [PubMed]
88. Fernández, F.J.; Garay, J.; Móri, T.F.; Csiszár, V.; Varga, Z.; López, I.; Gámez, M.; Cabello, T. Theoretical Foundation of the Control of Pollination by Hoverflies in a Greenhouse. *Agronomy* **2021**, *11*, 167. [CrossRef]
89. Paxton, R.J.; Schäfer, M.O.; Nazzi, F.; Zanni, V.; Annoscia, D.; Marroni, F.; Bigot, D.; Laws-Quinn, E.R.; Panziera, D.; Jenkins, C.; et al. Epidemiology of a Major Honey Bee Pathogen, Deformed Wing Virus: Potential Worldwide Replacement of Genotype A by Genotype B. *Int. J. Parasitol. Parasites Wildl.* **2022**, *18*, 157–171. [CrossRef] [PubMed]
90. Ng, W.H.; Myers, C.R.; McArt, S.H.; Ellner, S.P. Pathogen Transport Amplifies or Dilutes Disease Transmission Depending on the Host Dose-Response Relationship. *Ecol. Lett.* **2022**, *25*, 453–465. [CrossRef]
91. Hale, K.R.S.; Maes, D.P.; Valdovinos, F.S. Simple Mechanisms of Plant Reproductive Benefits Yield Different Dynamics in Pollination and Seed Dispersal Mutualisms. *Am. Nat.* **2022**, *2022*, 202–216. [CrossRef]
92. Wang, G.; Chen, G.; Zhang, H.-T. Resilience of Hybrid Herbivore-Plant-Pollinator Networks. *Chaos* **2023**, *33*, 093129. [CrossRef]

93. Chen, J.; Rodriguez, J.O.; DeGrandi-Hoffman, G.; Fewell, J.; Harrison, J.; Kang, Y. Impacts of Seasonality and Parasitism on Honey Bee Population Dynamics. *J. Math. Biol.* **2023**, *87*, 1–49. [CrossRef]
94. Li, L.; Zhang, Y.; Yao, J.; Wu, X. Bifurcations in a Plant-Pollinator Model with Multiple Delays. *J. Math.* **2023**, *2023*, 9950187. [CrossRef]
95. Wang, X.; Peron, T.; Dubbeldam, J.L.A.; Kéfi, S.; Moreno, Y. Interspecific Competition Shapes the Structural Stability of Mutualistic Networks. *Chaos Solitons Fractals* **2023**, *172*, 113507. [CrossRef]
96. Weyerer, F.; Weinbach, A.; Zarfl, C.; Allhoff, K.T. Eco-Evolutionary Dynamics in Two-Species Mutualistic Systems: One-Sided Population Decline Triggers Joint Interaction Disinvestment. *Evol. Ecol.* **2023**, *37*, 981–999. [CrossRef]
97. Deb, S.; Bhandary, S.; Dutta, P.S. Evading Tipping Points in Socio-Mutualistic Networks via Structure Mediated Optimal Strategy. *J. Theor. Biol.* **2023**, *567*, 111494. [CrossRef]
98. Elzinga, D.C.; Strickland, W.C. Generalized Stressors on Hive and Forager Bee Colonies. *Bull. Math. Biol.* **2023**, *85*, 112. [CrossRef]
99. Ma, Q.; Su, M. Herbivore-Induced Pollinator Limitation Increases Community Stability of Mutualism-Antagonism Continuum. *Biosystems* **2023**, *229*, 104929. [CrossRef]
100. Kawata, S.; Takimoto, G. Pollinator Predation Stabilizes Plant-Pollinator Mutualisms through the Modification of Pollinator Behavior. *Ecol. Res.* **2023**, *38*, 360–366. [CrossRef]
101. Bhandary, S.; Deb, S.; Dutta, P.S. Rising Temperature Drives Tipping Points in Mutualistic Networks. *R. Soc. Open Sci.* **2023**, *10*, 221363. [CrossRef]
102. Kiselev, A.; Nazarov, F.; Ryzhik, L.; Yao, Y. Chemotaxis and Reactions in Biology. *J. Eur. Math. Soc.* **2023**, *25*, 2641–2696. [CrossRef]
103. Marcou, T.; Revilla, T.A.; Křivan, V. Evolutionary Emergence of Plant and Pollinator Polymorphisms in Consumer-Resource Mutualisms. *J. Theoret. Biol.* **2024**, *594*, 111911. [CrossRef] [PubMed]
104. Atanasov, A.Z.; Georgiev, S.G.; Vulkov, L.G. Analysis of the Influence of Brood Deaths on Honeybee Population. *Appl. Sci.* **2024**, *14*, 11412. [CrossRef]
105. Mugabi, F.; Duffy, K.J.; van Langevelde, F. Behaviours of Honeybees Can Reduce the Probability of Deformed Wing Virus Outbreaks in Varroa Destructor-Infested Colonies. *Model. Earth Syst. Environ.* **2024**, *10*, 3745–3761. [CrossRef]
106. Karmakar, S.; Pal, N. Complex Dynamical Behaviors of a Honeybee-Mite Model in Parameter Plane. *Phys. D Nonlinear Phenom.* **2024**, *468*, 134300. [CrossRef]
107. Castellanos, V.; Dela-Rosa, M.A.; Sánchez-Garduño, F. Dynamics of a Pollinator-Plant-Herbivore Mathematical Model. *Lobachevskii J. Math.* **2024**, *45*, 5211–5231.
108. Yahaya, M.M.; Rodger, J.G.; Landi, P.; Hui, C. Emergence of Structure in Plant-Pollinator Networks: Low Floral Resource Constrains Network Specialisation. *Oikos* **2024**, *2024*, e10533. [CrossRef]
109. Saltini, M.; Barrett, S.C.H.; Deinum, E.E. Evolution from Mixed to Fixed Handedness in Mirror-Image Flowers: Insights from Adaptive Dynamics. *Evolution* **2024**, *79*, 65–79. [CrossRef] [PubMed]
110. Nakazawa, T.; Matsumoto, T.K.; Katsuhara, K.R. When Is Lethal Deceptive Pollination Maintained? A Population Dynamics Approach. *Ann. Bot.* **2024**, *134*, 665–682. [CrossRef] [PubMed]
111. Wang, J.; Yang, R.; Wang, J.; Cao, J. Threshold Dynamics Scenario of a Plants-Pollinators Cooperative System with Impulsive Effect on a Periodically Evolving Domain. *Eur. J. Appl. Math.* **2024**, *35*, 797–822. [CrossRef]
112. Zhao, Y.F.; Li, Y.K. Permanence and Extinction for the Stochastic Plant-Pollinator Model with Degenerate Diffusion. *Int. J. Biomath.* **2024**, *2024*, 2450065. [CrossRef]
113. Revilla, T. Plant–Pollinator Interaction Model with Separate Pollen and Nectar Dynamics. *J. Theor. Biol.* **2024**, *606*, 112096. [CrossRef]
114. Feng, T.; Wang, H. Complex Dynamics in Plant–Pollinator–Parasite Interactions: Facultative versus Obligate Behaviors and Novel Bifurcations. *J. Math. Biol.* **2024**, *90*, 46. [CrossRef]
115. Biswas, A.; Medda, R.; Pal, S. Dynamics of Predatory Effect on Saturated Plant–Pollinator Mutualistic Relationship. *Chaos* **2025**, *35*, 023118. [CrossRef]
116. Tang, X.; Zhou, S.; Luo, J. Spatial and Temporal Periodic Patterns in a Delayed Diffusive Plant–Pollinator Model with Memory-Based Diffusion. *Int. J. Biomath.* **2025**, *18*, 2350093. [CrossRef]

MDPI AG
Grosspeteranlage 5
4052 Basel
Switzerland
Tel.: +41 61 683 77 34

Biology Editorial Office
E-mail: biology@mdpi.com
www.mdpi.com/journal/biology

www.ingramcontent.com/pod-product-compliance
Lightning Source LLC
Chambersburg PA
CBHW061518210926
51850CB00008B/1959
* 9 7 8 3 7 2 5 8 5 7 9 7 5 *